Elementary Statistics

Elementary Statistics

FOURTH EDITION

JANET T. SPENCE
University of Texas at Austin

JOHN W. COTTON
University of California, Santa Barbara

BENTON J. UNDERWOOD
Northwestern University

CARL P. DUNCAN
Northwestern University

Prentice-Hall, Inc., Englewood Cliffs, New Jersey 07632

Library of Congress Cataloging in Publication Data

Main entry under title:

Elementary statistics.

(The Century psychology series)
Includes bibliographical references and index.
1. Psychometrics. I. Spence, Janet Taylor.
II. Series: Century psychology series (Prentice-Hall)
BF 39.E46 1983   519.5'02415   82-10177
ISBN 0-13-260141-9

THE CENTURY PSYCHOLOGY SERIES

Printed in the United States of America

10 9 8 7 6 5 4 3 2 1

Editorial/production supervision: Maureen Connelly
Cover art: Maria Termini
Cover design: Suzanne Behnke
Manufacturing buyers: Edmund W. Leone and Ron Chapman

ISBN 0-13-260141-9

Prentice-Hall International, Inc., *London*
Prentice-Hall of Australia Pty. Limited, *Sydney*
Prentice-Hall Canada Inc., *Toronto*
Prentice-Hall of India Private Limited, *New Delhi*
Prentice-Hall of Japan, Inc., *Tokyo*
Prentice-Hall of Southeast Asia Pte. Ltd., *Singapore*
Whitehall Books Limited, *Wellington, New Zealand*

Contents

Preface

Undergraduate students in psychology and related disciplines come to their first course with diverse expectations and backgrounds. Some have considerable mathematical training and quantitative aptitude and evidence a genuine interest in the subject matter. Others—perhaps the majority—are less certain about their quantitative skills and regard a course in statistics as an irksome requirement or as a necessary evil if they are to understand or to perform research in their chosen field.

This fourth edition of *Elementary Statistics*, like its predecessors, is primarily designed to accommodate the latter audience. It was written with the conviction that statistical concepts can be described informally in clear, straightforward language, without being oversimplified or assuming substantial mathematical sophistication on the part of student. It also assumes that for such an audience, understanding of statistical methods as research tools can be more effectively promoted by presenting them within the context of their application to concrete data rather than as pure abstractions. The content of the text is limited to those techniques that are widely used in the psychological literature and to the basic principles and assumptions underlying them.

Instructors of statistics are constantly challenged to change and refine their teaching methods to increase their effectiveness, and so are writers of statistics texts. The major changes in the present edition reflect our own teaching experiences as well as those of colleagues who have been kind enough to give us their suggestions. The order of presentation of a number of topics has been rearranged. For example, certain basic concepts, such as the distinction between samples and populations, have been introduced in the first chapter. The chapter on correlation appears earlier in the text, grouped with the other chapters devoted to descriptive statistics. This chapter has also been expanded to include a more extensive discussion of the topic of prediction.

Several instructional aids have also been added. A glossary of terms and symbols concludes the text of each chapter. Each chapter also includes a set of problems that test students' understanding of major concepts and provide experience in analyzing data. Answers to these problems are given in Appendix III.

Supplementing the text is a separate *Workbook for Elementary Statistics,* prepared by the text's senior author. The *Workbook* contains additional calculational problems, short answer questions for review, and supplementary explanatory material (such as a review of basic arithmetic operations and a decision tree that assists students in determining the appropriate statistical test to apply to concrete research data).

Responsibility for the present edition was assumed by the first two authors. We wish to thank Ann Deaton, Ralph DeAyala, and Josephine Ho for their meticulous checking of formulas and computations and their editorial suggestions. Our gratitude also goes to Faye Gibson and Dee Schoenfeld who skillfully and patiently typed the manuscript through its many drafts.

Finally, we are grateful to the Literary Executor of the late Sir Ronald A. Fisher, F.R.S., to Dr. Frank Yates, F.R.S., and to Longman Group Ltd. London, for permission to reprint Tables IV and VI from their book *Statistical Tables for Biological, Agricultural and Medical Research* (6th edition, 1974). We also thank a number of other authors and publishers for granting us permission to abridge or reproduce their materials. The sources of these materials are acknowledged as they occur in the text.

JTS
JWC

Elementary Statistics

Statistics: What It's About

The Collection, Classification, and Interpretation of Facts

CHAPTER 1

Suppose you are driving downtown to do some shopping. You remember your tires are low, so you stop at a service station and have them brought up to 24 pounds, and while you're at it you put 10 gallons in your gas tank. In the course of your shopping you have a lumberyard cut three 5-foot planks for some shelves you are building, order a sweater in your size, buy a 40-watt lightbulb and select a half-pound of mushrooms. On your way home you glance at your watch to see if you'll make your favorite TV program.

All these commonplace activities depend on *quantification*—the assignment of *numbers* to objects or events to describe their properties. If you had attempted to express your requirements without numbers, you would have found yourself using vague verbal descriptions, gesturing a lot, and failing to communicate your exact needs. Numbers allow you to describe phenomena with considerably more precision. If numerical quantification had never been devised, it is doubtful that cars, lightbulbs, watches, and the like would even have existed.

In scientific research we attempt to describe the properties of objects and events in the universe around us and the regularities of occurrence among them. Quantification is at the heart of the scientific enterprise, and the basic data resulting from scientific research are almost invariably a set of numbers. Having collected the data, we apply statistical methods to describe and interpret them.

A look at a hypothetical investigation will show some of the statistical challenges. The instructor of an introductory course in statistics offered by the psychology department is distressed because so many of the students who take the course are afflicted by "math anxiety." He gives common-sense examples illustrating statistical principles that everyone seems to understand. But when he translates them into statistical language, many students begin to look panic-stricken. He asks students to compute a statistic by arithmetic operations no more complicated than they learned in grade school, and a number of them stare at him helplessly. They're no good at math, they say; they are convinced that statistics is beyond them. One year the instructor decides to conduct an experiment. During the first session of the class, he asks students to volunteer for a two-week therapeutic program the college counseling service has agreed to set up to help students get over their "math anxiety."[1] Twenty students volunteer for the program, more than can be accommodated. He picks every other person on the list to take part. These 10 students form the *experimental group*. The other 10 students serve as a *control group*—that is, they receive no special treatment. At the end of the semester, he determines the final examination scores for all the volunteers, as shown in Table 1.1.

Did the treatment program have beneficial effects? The instructor could not even begin to answer the question until he brought some sort of order to these jumbled sets of scores and described them in some way. For example, what was the average exam score

---

[1] Although this experiment is hypothetical, "math anxiety" is a real phenomenon whose origins and treatment psychologists and others have become active in investigating. (See, for example, S. Tobias, *Overcoming math anxiety*, New York: Norton, 1978.) The math anxious among you should take heart from the fact that this text was written so as to be comprehensible to students without extensive training in mathematics and without a high level of quantitative aptitude. To paraphrase Franklin D. Roosevelt, students should have little to fear but fear itself.

Table 1.1 Results of a Hypothetical Experiment Showing Final Examination Scores of Treated and Untreated "Math Anxious" Students

Treated		Untreated	
Student	Score	Student	Score
A.B.	74	N.K.	76
J.M.	85	E.G.	68
H.P.	77	P.F.	52
B.F.	89	K.S.	59
T.J.	72	A.V.	71
G.C.	69	L.D.	80
L.K.	77	C.R.	57
C.D.	84	M.H.	75
W.Y.	64	R.T.	69
F.L.	79	D.W.	63

in each group? In each group, how variable were the scores—that is, were they all very similar or did they range widely?

By adding the scores together and dividing by the number of scores, the instructor determined that the average student in the experimental (treated) condition earned a score of 77 on the final exam. For students in the control (untreated) condition, the average was 67. Does this 10-point difference mean that the treatment program was effective? Or did the treated students just happen to do better? In other words, if the instructor repeated the experiment over and over with different groups of students, is it likely that the results would turn out the same way? Before he suggests that the counseling center set up a treatment program on a regular basis, he wants to be reasonably sure the therapy is effective.

Consider another example. A doctor hears a friend of hers confidently claim that the proportion of male and female babies who are born varies according to the month of the year, with January being the high point and June the low point for boys. She bets her friend that time of year has nothing to do with a baby's sex and checks last year's birth records for these two months at the local hospital. She finds that in January 54 percent of the babies were boys, and in June 52 percent were boys. Does she lose the bet?

Simply staring at the data will not bring satisfactory answers to these questions. What is needed are methods for extracting information from the results of these studies and interpreting what they mean. This is where the science of statistics comes into play. By performing arithmetic and algebraic manipulations with the data, group trends and relationships can be abstracted and described in a precise manner. With further statistical operations, it is possible to make inferences or guesses about the characteristics of still larger groups—for example, about the effects of treatment for math anxiety for introductory statistics students in general, not just for the particular group of students shown in

Table 1.1, or for male and female babies in general, not just for the babies born during a specific year at one hospital.

In subsequent chapters we will first consider a series of *descriptive statistics*, methods that permit us to discover and describe the relationships hidden in masses of number. Then we will take up topics that are related to *inferential statistics*—techniques for making guesses about the properties of what are called *populations*, based on the data from a *sample* of individuals that has actually been observed. Before proceeding further, we will elaborate the distinction between populations and samples and introduce certain other terms and concepts that will be encountered in later chapters.

POPULATIONS AND SAMPLES

We are all familiar with national surveys in which people from all over the country are questioned about their beliefs on various topics. Should women with preschool children work outside the home? Is the president doing a good job in managing the economy? Should the death penalty be prohibited? The purpose of these surveys is to try to assess the mood of the people. It would be both impractical and impossible to question everyone, but it *is* possible to obtain the views of a sample of individuals and to use their answers to infer the opinions of the entire population.

In still another example, suppose an engineer claims that a simple, inexpensive device attached to an automobile's fuel line will cut gas consumption by 10 percent. If the claim is correct, the discovery is obviously of great economic significance. A government agency equips 20 cars with the device and has each of them driven over a test course under controlled conditions. Twenty other cars that are similar except for the device are also driven over the same course, and the fuel consumption of the cars in the experimental and control groups is then compared. The government agency is interested in the effectiveness of the device for *all* cars (the entire population), not just the cars in the experiment. Since testing the population of cars is impossible, the intent of the experiment is to use the results from the sample to make guesses about what *would* happen if all cars were equipped with the device.

A population consists of all members of a group of individuals who are alike on at least one specified characteristic. Populations may consist of any kind of animate or inanimate objects or events, but for illustrative purposes we will commonly refer to people. A population can be infinite or indefinitely large in number, such as all the infants who have ever been or ever will be born, or it can be finite, such as all the infants born last year in the United States. Finite populations may be large, such as all individuals enrolled in college, or they may be small, such as all majors in Greek at a given college. Populations may be real, as in the examples immediately above, or they may be hypothetical. If, for example, we were to conduct an experiment using a new drug designed to overcome psychotic depression, we would be concerned with estimating the drug's effect on the population of *all* individuals, now or in the future, who might exhibit this condition and *could* be given the drug.

Whether real or hypothetical, the population is specified by the investigator, and the aim is to find out something about it. The populations that are the subject of scientific investigation typically are large or indefinite in number, so that only a sample can be studied. *A sample is any number of cases less than the total number of cases in the population from which is it drawn.* If we have a population of 50,000, a sample might consist of 10 cases, 15,000 cases, or 49,999. Normally, however, samples consist of only a small or a very small proportion of the population being investigated.

The purpose in observing a sample, we have said, is to use the data obtained to make inferences about the characteristics of the population as a whole. We may, of course, want to study and compare the characteristics of samples from more than one population in a single study.

There are certain dangers in generalizing from sample to population data. If we were attempting to predict the outcome of a mayoral election, for example, we would be foolish to question voters from only one occupational group or income level, since these factors are likely to be related to voter preference. What we would attempt to get is a sample of voters that is a miniature picture of the population from which it is taken. If we want to reach conclusions about a population from the data of a sample, the sample ought to be *representative* of the total population from which it is drawn.

Consider a psychological experiment in which an investigator had one sample of students solve arithmetic problems under quiet conditions and a second sample of students solve the problems while listening to a recording of a jackhammer. The investigator could come to very erroneous conclusions about the effects of noise if members of the first group were drawn from a class in calculus and members of the second were drawn from a class in English literature. This is so because students taking calculus are very likely to be better at arithmetic problems than students in an English class. In conducting experiments, investigators must attempt to obtain samples that are initially *comparable* in order to draw appropriate conclusions about the effects of conditions on populations of individuals. Comparability implies that if the samples had been tested under the *same* conditions their performance would have been similar.

The methods by which samples are selected are therefore extremely important considerations in scientific research. A number of sampling techniques have been devised in an effort to obtain representative samples from populations. We will consider three methods: random, randomized, and matched.

TYPES OF SAMPLES

Random Samples

Random sampling is a method of selecting a sample in such a way that each member *of the pool has an equal probability of being selected on any given draw and each possible* sample *of members has an equal probability of being selected.* To illustrate, suppose there is a population of 20 individuals, whom we identify by the letters A through T, from which we wish to draw a sample of 5 cases. A random sample is one in which individual A has the same chance of being selected as individual B, who in turn has the same chance of being selected as individual C, and so forth. Any given set of 5 individuals—for example, individuals A, D, E, H, and S—also has the same chance of being selected as any other set of 5 individuals. If each individual has an equal chance of being selected, it almost always happens that each possible sample also has an equal chance of being selected, but there are some peculiar circumstances under which it does not.

Our intent in obtaining a random sample, we have said, is to obtain a sample whose properties are representative of the properties of the population from which the sample was selected. There is no guarantee that the measures obtained from a random sample will accurately reflect the properties of the population as a whole, but statistical theory does allow us to make probability statements about how closely values obtained from random samples can be expected to agree with population values.

Sampling may be done with or without *replacement.* For example, in the instance of the 20 individuals described above, we might select the name of individual B on the first draw and after recording it, replace the individual's name in the pool before drawing a second time. This draw-and-replace procedure is continued until our desired sample of 5 has been obtained. With this replacement procedure, you will note, each individual has a 1 in 20 chance of being selected on *each* of the 5 draws.

When replacement is made, it is possible that the same individual's name will be selected more than once. For most research purposes, however, it is necessary or desirable for samples to consist of different individuals. One method of obtaining a sample or series of samples in which no individual is represented more than once is to sample *without replacement.* In this instance, the probability of being selected does not remain constant on every draw. In our group of 20, for example, the probability of any given individual being selected on the first draw in 1/20. If on the first draw we select individual B's name and do not replace it, each of the remaining 19 individuals has a probability of 1/19 of being selected on the second draw. On the third draw, the probability is 1/18, and so forth. However, the sample was still randomly selected from the population of 20 individuals because on each draw every individual remaining in the pool had an equal probability of being selected.

When several random samples are selected from a population, they are described as being *independent.* That is, the composition of a sample does not influence which individuals remaining in the pool will be selected for a subsequently chosen sample. Not all sampling methods produce independent samples, as we will see.

Table of Random Numbers. There are a number of methods for obtaining random samples. One time-honored procedure is to put the names of all members of the population on identical slips of paper, put all the slips in a container, stir them thoroughly, pull one out, stir the slips again, choose another, and continue this process until a sample of the desired size is obtained. But since the stirring and selecting process can go awry, perhaps the most satisfactory method is to consult a table of random numbers. Modern tables of random numbers have been generated by a computer to meet the two requirements we mentioned above. A sample page from a book of random numbers is reproduced in Table A (Appendix II). For convenience, the table is set up in columns of 5-digit numbers.

As an illustration[2] of how the table is used, suppose the Society of Comparative Psychologists has 848 members. The society's president is interested in members' opinion about whether the society should start its own scientific journal. For the sake of economy, he decides to poll a random sample of 100 individuals instead of the entire membership. The name of the first person in the Membership Directory is assigned the number 1, the second person is assigned the number 2, and so on. Next the president consults a table of random numbers. If the table has more than one page, he arbitrarily picks a page. Then he arbitrarily picks a place to start on the page, perhaps by closing his eyes and bringing the point of his pencil down. Suppose this is the fourth entry in the third column of Table A, which you can see is 29311. Since there are only 848 members of the society, he will need only the first three digits of this number. The person assigned the number 293 is the first individual in the sample. He then proceeds to consult the first three digits of the numbers that follow. Since he does not want to include anyone more than once, he writes down all numbers that do not exceed 848 until he has 100 *different* numbers: 488, 199, 60, and so forth.

As this example illustrates, the number of digits in any row that one uses for random selection depends on the size of the population. If the population consists of 50,000 cases, all five digits in each set of five would be used in selecting the sample. If the population consisted of 200,000 individuals, one would "borrow" the first digit of the five-digit number in the column to the immediate right, and so forth.

Randomized Samples

In scientific experiments, investigators test two or more groups under different conditions in an attempt to determine whether the conditions they manipulate have an effect on individuals' performance. Experimental groups could be obtained by drawing the required number of random samples from some defined population. Often, however, investigators cannot identify or do not have access to all the members of a population in which they might be interested, so they are unable to select random samples that can then be assigned to the several experimental conditions. However, often the investigator has access to a certain number of subjects and has control over what subjects are assigned to what experimental groups. For example, investigators might have a group of volunteers from their classes or a group of animals available from a biological supply house.

Such groups cannot be considered random samples from a defined population. Usually, however, members of the pool of available subjects can be assigned at random to the different conditions of the experiment. That is, an equal number of individuals can be selected for each experimental group in such a manner that each individual, and each set of individuals, have an equal chance of being assigned to a particular group. For example, the investigator might give individuals an identifying number and then consult a table

---

[2]We stress that Table A contains only a single page from a large table of random numbers and therefore only a subset of all possible five-digit numbers. Table A is therefore useful only for illustrative purposes; readers interested in actually conducting research should consult a full table of random numbers. As stated in the example in the text, when the full table involves multiple pages, the process of selecting a random sample should start with selecting a page at random and then proceeding as described.

of random numbers, assigning (in a two-group experiment) the individual with the first number in the table to the first condition, the second individual to the second condition, the third to the first condition, and so on. To distinguish it from random sampling, this procedure is called *randomization*—that is, randomized assignment of subject to group.

The randomized method of assigning subjects to experimental conditions also yields *independent* samples. Generally speaking, the fact that a particular individual or series of individuals has already been selected from the subject pool and assigned to a particular group does not influence the assignment of the other individuals remaining in the pool.

Matched Samples

Some techniques of selecting and assigning subjects to groups in experiments involve matched samples which are not independent. In this section we discuss a particular method of matching, the method of *matched pairs*, which is suitable for experiments in which there are only two experimental conditions or groups. In principle the method could be extended and used to obtain a larger number of groups, but for the sake of simplicity we will confine our discussion to the two-group case.

Our aim in assigning subjects to group in an experiment, we have said, is to produce comparable groups, groups that would have performed similarly had they been observed under identical conditions. Assuming that it is done adequately, selecting matched pairs is a particularly effective method of obtaining comparable groups. In the method of matched pairs, each individual is tested under both conditions or one individual is tested under one condition and another individual who has previously been identified as being highly similar in relevant characteristics to the first individual is tested under the other condition. The result is a series of pairs of observations.

In some experimental situations, it is reasonable to test each individual under both conditions. For example, in a study of the effects on motor coordination of a drug with short-term effects, each subject might be given a well-practiced task, once while under the influence of the drug and once while drug-free. This method is particularly desirable, since individuals are obviously the best match for themselves.

Often, however, the same individual cannot legitimately take part in both conditions. It would make no sense, for example, to teach a novice violinist by one technique and then to teach the individual by another technique to find out which is more effective, since it is obviously impossible to erase what the person has learned and to start again. In such instances, other methods of matching can be used. One could randomly select pairs of identical twins, pairs of brothers or sisters, or even pairs of unrelated individuals who are similar in characteristics that are likely to affect the performance to be measured— factors such as age, educational background, aptitudes, and so forth. Groups of subjects are then formed by randomly picking one member of each pair to be observed under one of the experimental conditions. The other member of each pair is then automatically assigned to the other condition. It is this feature that makes the samples nonindependent. Once one member of each pair is assigned to one group, composition of the other group is fixed.

PARAMETERS AND STATISTICS

If we had access to an entire population, we could proceed to describe its properties. For example, if we knew the weights of all the players in the National Football League, we could determine the league average. Or if we have access to the records of all West Point cadets, we could determine the percentage who are female. The properties of populations are called population *parameters*.

Properties of samples drawn from a population can also be described. For example, the percentage of males and females in a sample of West Point cadets could be determined, as could average weight in a sample of NFL players. Characteristics of samples are called sample *statistics*. To emphasize the distinction between a population parameter and the parallel sample statistic, population parameters are typically symbolized by Greek letters and sample statistics by letters from our regular alphabet.

The types of properties exhibited by populations and the samples drawn from them are determined by the types of numerical information obtained about members of these groups. In the section below, three major kinds of quantitative information will be considered: measurement, rank-order, and categorical data.

TYPES OF DATA

Measurement Data

How tall is each person in your statistics class? How fast could each of you memorize the definitions listed at the end of this chapter? How many children would each of you like to have? Each of these questions inquires about a property that exists along a dimension in various degrees. Each member of the group could be described by assigning a number that specifies *how much* of the property the individual exhibits. These numbers we call *measurement* or *metrical* data.

Discrete vs. Continuous Variables. Measurable properties are usually *continuous*. That is, objects or events may fall at *any* point along an unbroken scale of values running from high to low. In measuring the weight of objects, for example, we may have such values as 4.2681 grams, or 68.723584 grams, or any positive value we can think of. Similarly, time is continuous, as is the distance between geographic locations.

Discrete variables, in contrast, fall only at certain points along a scale. For example, students' scores on a 20-item true-false test would be whole numbers; they would fall only at certain points along a scale running from 0 to 20. Similarly, you might reasonably aspire to have 0, 1, 2, or 3 children, not 2.3 or 4.6 children.

In practice, the distinction between the two types of measures breaks down. When the property is continuous, we can never measure objects exactly, but only within the limits of a specified degree of accuracy. Depending on our purpose, for example, we might be satisfied to specify an object's age to the nearest century, the nearest year, the nearest hour, or the nearest millionth of a second. Whatever the degree of accuracy we attempt, the resulting numbers are still discrete. If we measured time to the nearest tenth of a second, for example, numbers such as .1, .3, or .8 second would be reported, but not .12, .32, or .86. If we measured to the nearest hundredth of a second, we might have .08 and .36, but not .084 or .362. Because most measurable properties are continuous, each number we assign to each object or event typically represents a range or *interval* of values, rather than an exact scale point. If length is reported to the nearest centimeter, for example, objects described as being 14 centimeters long are understood to fall someplace in a range extending from 13.5 to 14.5 and not all to be exactly 14.000. . . .

A Note about Rounding.   It is not uncommon to record observations with greater numerical precision than we wish to report. We might have recorded our observations using two decimal places, for example, but want to report data only to the first decimal place or in whole numbers. The traditional rule is to round to the *nearest* decimal or whole number. The value 4.38, for example, would be rounded *down* to 4 if a whole number were required or rounded *up* to 4.4 if the first decimal place were to be retained. What if the number falls at the border of two adjacent intervals? For example, is 9.5 rounded to 9 or to 10? Is 2.65 rounded to 2.6 or 2.7? A convention devised for handling such situations is to round to the nearest *even* number, so that 9.5 is rounded up to 10 and 2.65 is rounded down to 2.6. This convention is intended to avoid systematic bias when a whole group of numbers is to be rounded, since it can be anticipated that about half the numbers falling on the borderline will be assigned to the lower interval and half to the upper interval.

A second situation in which rounding is desirable is one in which the result of some statistical calculation is to be reported. For example, suppose we weigh a series of 8 objects to the nearest $\frac{1}{10}$ of a gram: 6.2 grams, 5.9 grams, 6.5 grams, and so forth. If the 8 objects weighed a total of 52.3 grams, we could divide 52.3 by 8 to find the average weight of the objects. The result of this calculation turns out to be exactly 6.5375 grams. Reporting this number would be misleading, because it implies that we have measured each object with greater precision than we actually have. A good rule to follow is to report the average, or other statistic, to *one or two digits beyond what has been measured directly.* In our example, the objects were measured to the nearest .1 gram. The average would therefore be rounded to 6.54 (one extra digit) or to 6.538 (two extra digits). In most of the illustrative data analyses in this text, we have retained two extra digits. Note that in carrying out the calculation, we avoid rounding or at least carry out our calculations to a number of extra digits until all stages of computation are complete. It is the *final answer* that is rounded to one or two digits beyond the original measurement.

Rank-Order Data

Sometimes numbers that directly indicate how much of a given property each individual possesses are impossible or inconvenient to determine. We may instead assign *ranks* that indicate the individuals' positions relative to one another. In a foot race, for example, the contestants might not be timed but instead ranked from fastest to slowest according to the order in which they crossed the finish line. Or people might be asked to indicate how prestigious 20 occupations are by ranking them from highest to lowest. Rank orders allow more-than, less-than statements to be made about pairs of objects. For example, the occupation of physician might be ranked higher than elementary school teacher, which in turn might be ranked higher than street cleaner. Rank orders, however, do not indicate how great the difference is.

Categorical Data

How many men and women are enrolled in your statistics class? How many of you are freshmen, sophomores, juniors, or seniors? How many of you have been to Europe, smoked pot, own a car, have shoes on, are blondes, brunettes, or redheads?

Each of these questions involves a set of two or more mutually exclusive *categories:* male *or* female; car owner *or* not; blonde, brunette, *or* redhead; and so on. It is meaningless to ask "how much" of the property each individual has or to rank-order the individuals according to "how much" of the property each exhibits. Rather, an individual either does or does not have the property specified by each category within a given set (that is, his or her "score" is either 0 or 1). And since the categories are mutually exclusive, each individual can be assigned to one and only one of them. More than one set of categories can, of course, be used simultaneously in classifying an individual case; for example, we can identify Samantha as a female, redheaded freshman who came to class barefoot.

It is simple to describe the results of our observations when categorical data are involved. We simply sort the cases, placing each in its proper category, then count the number in each category and present the sums. For convenience, the sums may be translated into percentages. Table 1.2, for example, gives the percentages of groups of men and of women in three different years who said "yes," "no," or "can't say" about whether they were afraid to walk alone at night. The table lets us see that in each year, more women than men reported being afraid, but that both sexes were becoming more fearful over the three-year period.

Typically, each category can easily be identified or "named" by a verbal label, such as yes, no, and can't say, or blonde, brunette, and redhead. This characteristic has led categorical data also to be called *nominal* data. Sometimes it is more convenient to give each category a numerical label, such as category 1, category 2, category 3, and so forth. In this instance the numbers are also *nominal*—that is, the numbers are used as "names" to identify categories, in the same way that football players wear identifying numbers on their jerseys. Which category is assigned which number is purely arbitrary.

Table 1.2 Men and Women Afraid to Walk Alone at Night, 1968, 1972, 1980

Year	Women			Men		
	Yes	*No*	*Can't Say*	*Yes*	*No*	*Can't Say*
1968	50	47	3	19	79	2
1972	61	36	3	22	77	1
1980	72	28	0	33	66	1

Table 1.3   Three Types of Data and the Information Obtained About the Individual Case in Each Type

Type of Data	Information obtained about individual
Measurement	Number indicating the amount of a property that exists along a continuum ("how much")
Rank order	Amount of a property expressed as a rank relative to other individuals (1st, 2nd, 3rd)
Categorical or nominal	Category to which individual belongs

This observation serves to emphasize an important point: Categories ordinarily cannot be ordered along a continuum. For example, if people are classified by current marital status as never married, widowed, separated, divorced, or currently married, the categories could not be arranged in order according to "how much" marital status each has. With categorical data, the question we ask about each *individual* is to what category the individual belongs, not how much of some property the individual has. The question we typically have about a *set of categories* is how many individuals, or what percentage of the total number of individuals, fall into each category, not "how much" of a property is represented by a particular category in comparison to other categories in the set.

The major distinctions among the three types of data are summarized in Table 1.3.

DEFINITIONS OF TERMS

Population.   All members of a group of individuals, objects, or events that have a specified property in common.   For example, all licensed barbers in the state of Wisconsin.

Sample.   A subset of cases from a population of individuals, objects, or events.   For example, 100 barbers from the population of all licensed barbers in Wisconsin.

Random sample.   A sample of a given size drawn from a population in such a way that each member of the population and each possible sample of that size have an equal probability of being selected.

Randomization.   A technique used to assign members of an available pool of individuals to conditions in experiments involving two or more conditions. Members of the pool are assigned to condition by use of a table of random numbers or some other random procedure. The resulting groups are called *randomized samples.*

Matched pairs.   A technique used in two-condition experiments in which each individual is tested twice, once under each experimental condition, or in which a series of pairs of individuals who are matched on one or more characteristics are identified prior to the experiment, one member of each pair being randomly assigned to one condition and the other member to the second condition.

Population parameter. A number representing some property of a population—for example, the average beginning salary of all assistant professors at a given university. Population parameters are typically symbolized by Greek letters.

Sample statistic. A number representing some property of a sample. Sample statistics are typically symbolized by letters from the regular alphabet.

Measurement or metrical data. Observations to which numbers have been assigned to express different magnitudes of some property that exists in various degrees. For example, the speed in number of seconds with which each of a group of 10-year-olds is able to run 100 yards.

Discrete variables. Properties existing to various degrees that in nature take only certain values along a continuum of values. For example, the number of sets won by players in a tennis match can only be whole numbers and cannot involve fractions.

Continuous variables. Properties existing to various degrees that in nature can take any possible value between an upper and lower limit. For example, the circumference of people's waists. In practice, continuous properties are measured only to a specified degree of accuracy, and the reported values thus fall only at discrete points or values. For example, length of objects measured to the nearest millimeter would take the values 1, 2, 3, etc., but not fractional values.

Rank-order data. Data in which individuals are arranged in order according to the relative amount of some property they exhibit and assigned ranks to indicate their standing, 1st, 2nd, 3rd, etc.

Categorical or nominal data. Data involving a set of properties (categories), each of which a particular individual does or does not exhibit. In sets of two or more *mutually exclusive* categories, each individual falls into one and only one category, and a group of individuals is described by counting the number in each category. For example, a group of 60 children may be described as containing 28 boys and 32 girls. Numbers used to identify categories are *nominal*—that is, they are "names" of categories used in place of verbal labels. For example, boys might be called 0's and girls called 1's.

PROBLEMS

Answers to the problems in this and later chapters are given in Appendix III.

1. For each of the following, indicate whether a *population* or a *sample* is involved and for the latter, from what population the sample was drawn.
   (a) Everyone who voted in the last U.S. presidential election.
   (b) Names of 100 individuals drawn from the residential listings of the Chicago telephone directory.
   (c) The instructor ratings obtained from the students attending their introductory psychology class on the day the rating forms are distributed.
   (d) Three-year-old boys.

2. For each of the following, indicate whether the sample or samples being described are *random*, *randomized*, or *matched*.
   (a) Grocery store patrons fill out slips of paper with their names and addresses in a drawing for free hams. The slips are placed in a barrel and 10 are blindly selected as winners.
   (b) In a study of the effects of sleeplessness on performance, a group of 10 individuals is given a simple reaction time task after members have been awake for 12, 24, and 48 hours.
   (c) Two parallel forms of a vocabulary test have been prepared. The test developer arranges to give the two forms to 100 students to determine if the tests produce comparable results. Fifty form A test booklets and 50 form B booklets are arranged in random order. The first student seated in the first row is given the booklet on the top of the pile, the next student the next booklet, and so forth.

3. Twenty people have signed up for an experiment and are to be assigned to one of two experimental conditions at random. Their names have been alphabetized and then assigned a number. Starting at the sixth number in the third column in the table of random numbers in Table A, Appendix II, select a sample of 10 to serve in the first condition.

4. For each of the following, indicate whether *categorical* (*nominal*), *measurement*, or *rank-order* data are being described.
   (a) The ordered list of the top 20 football teams in the country according to a poll of coaches.
   (b) The current marital status of each member of a sample of 200 men.
   (c) Scores on a 30-item personality test of authoritarianism.
   (d) The distance each of a group of 10-year-old girls can throw a baseball.

5. Indicate for each of the following whether the property being measured is *discrete* or *continuous*.
   (a) In a sample of married couples, the number of children who are living at home.
   (b) The time it requires individuals to solve a set of simple arithmetic problems.
   (c) The number of errors on a typing test.
   (d) The birth weight of males versus females.
   (e) The 12 noon temperatures yesterday at several locations in Indiana.

Frequency Distributions

In Chapter 1 we described measurement data: the assignment of numbers that specify *how much* of some property individuals have. We might measure, for example, the performance of young children on a task designed to determine their understanding of abstract moral principles, or the ego-strength of hospitalized schizophrenics as revealed in their responses in a standardized psychiatric interview. When a group of individuals is measured, rarely do all the measures turn out to be the same; the almost inevitable outcome is a set of numbers varying in value. In this chapter we will consider methods of ordering measurement data so that we may begin to comprehend and describe the performance of a group of differing individuals.

SIMPLE FREQUENCY DISTRIBUTIONS

One obvious method of bringing order into a collection of measures is to arrange them in order of magnitude from highest to lowest. When only a few measures are involved, this ordering is probably sufficient. But with a group of substantial size, it is no longer immediately evident how many times a particular value occurs, or whether it occurs at all. An additional step in our procedure will give us this extra information. All possible values between the highest and lowest reported measures can be listed in one column and the number of individuals receiving each value in an adjacent column, as illustrated in Table 2.1.

The numbers in the table represent scores received by a sample of 150 college women on a questionnaire assessing attitudes toward women's rights and roles. The higher the score, out of a maximum possible score of 45, the greater the belief that women ought to have the same rights and privileges as men. The first column is labeled

Table 2.1 Simple Frequency Distribution of Sex-Role Attitude Scores for 150 Female Students

Score	f	Score	f	Score	f
45	2	33	10	21	3
44	1	32	7	20	4
43	0	31	6	19	2
42	2	30	9	18	4
41	3	29	7	17	2
40	3	28	5	16	2
39	4	27	4	15	1
38	5	26	5	14	1
37	7	25	7	13	2
36	10	24	6	12	0
35	7	23	7	11	1
34	6	22	5	N =	150

Source: **The data are a representative sample taken from the 1980 group reported in R. L. Helmreich, J. T. Spence, & R. H. Gibson, Sex-Role Attitudes: 1972–1980.** *Personality and Social Psychology Bulletin* **(in press). Raw data courtesy of the authors.**

"Score." The order of highest to lowest in this column has, by custom, become the standard arrangement and should always be followed. The second column, labeled "*f*" (frequency), contains the number of cases, in this instance the number of students, receiving each score. For example, three students received a score of 41 and ten students a score of 36. At the bottom of the last frequency column you will observe the letter N, which is simply a symbol for the total number of cases in the group. In the present illustration N is 150. Such a table, in which all score values are listed in one column and the number of individuals receiving each score in the adjacent column, is called a *simple frequency distribution.*

Arranging the measures into a simple frequency distribution makes description of group performance easier. In the example in Table 2.1, the lowest score is 11 and the highest score is 45. Thus the *range* of scores—the number of score units between the highest and the lowest score—is quite wide, 34 units. It can also be seen that certain scores stand out as occurring frequently and, more important, that the scores are heavily concentrated toward the high end of the scale—not too surprising in view of the vigor of the feminist movement among college women. Collecting scores into a frequency distribution allows us to see these characteristics of group performance more easily than would just listing scores in order.

GROUPED FREQUENCY DISTRIBUTIONS

In many instances, particularly those in which the *range* of the measures (number of score units from highest to lowest) is small, a frequency distribution of the type just described yields a compact table that shows the pattern formed by a set of scores. But when the measures are scattered over a wide range, the frequency distribution is long and bulky, as in the three long pairs of columns in Table 2.1. More important, so few cases fall at each score value that the group pattern is not too clear. We can, however, simultaneously make our distribution more compact and obtain a better idea of the overall pattern of scores by bringing together the single measures into a number of groups, each containing an *equal* number of score units. Table 2.2 does this with the attitude scores from Table 2.1, grouping them into blocks of three score units each. Each block or *class interval* is

Table 2.2   Grouped Frequency Distribution Based on Data from Table 2.1
$(i = 3)$

Class Interval	f	Class Interval	f
43–45	3	25–27	16
40–42	8	22–24	18
37–39	16	19–21	9
34–36	23	16–18	8
31–33	23	13–15	4
28–30	21	10–12	1
			N = 150

identified by its "class limits"—the highest and the lowest score in the interval. Thus, the top interval in the first column of Table 2.2 includes the scores 43, 44, and 45 and is identified as class 43–45. The next class consists of the three scores 40, 41, and 42, and so forth.

We can describe these intervals by saying that each has a *width* of three, since three score units are included in each. The symbol we will be using to indicate the width of the interval in any grouped frequency distribution is i. The second column again contains frequencies (f), but these are now the number of individuals falling into each class interval rather than the number receiving each separate score. The resulting distribution is an improvement over the previous one in showing group trends. We can now see quite clearly that the scores for the students tend to cluster near the high end of the scale and trail off in numbers toward the low end.

The degree of condensation obtained by grouping data is determined by the number of score units included in each class interval, since increasing the size of i decreases the number of class intervals. In Table 2.3 the data are again represented, but i has been increased from 3 to 5 units, thus decreasing the number of intervals and making the table shorter. A comparison of these frequency distributions, all derived from the same raw data, reveals an interesting fact. As the width of the class interval, i, increases, a pattern first emerges and later disappears as more and more scores are lumped together in fewer classes. We see that as far as obtaining information about group trends is concerned, a frequency distribution can be condensed too far.

A further point to be noted about grouping data is that the identity of the individual score is lost. In Table 2.3, for example, where $i = 5$, the topmost interval contains 8 women, but we do not know exactly what score each received; we know only that each person falls somewhere within the interval. As i becomes larger, the numerical value of the individual score becomes more and more obscured.

How Many Intervals?

Before a set of data can be converted into a grouped frequency distribution, we must decide how many class intervals to employ. This decision is largely arbitrary, depending on such factors as the number of cases, the range of measures, and the purpose for which the distribution is intended. As a general principle we might say that the classes should be few enough so that the resulting distribution reveals a group pattern, but not so few as to obscure a group pattern and cause great loss in the precision with which we can identify

Table 2.3   Grouped Frequency Distribution Based on Data from Table 2.1
($i = 5$)

Class Interval	f	Class Interval	f
41–45	8	21–25	28
36–40	29	16–20	14
31–35	36	11–15	5
26–30	30		N = 150

the individual score. A rule of thumb often given is that between 10 and 20 intervals should be employed, since most data can be adequately handled within these limits. However, in choosing the number of class intervals, or indeed in deciding whether the range of measures is so small that scores should not be grouped at all, we must examine each set of data individually.

How Many Units for the Class Interval?

Once the *approximate* number of class intervals has been chosen, we must determine the width of the class interval necessary to produce the desired number of intervals. We can roughly estimate the needed i by dividing the range of the scores (the highest minus the lowest score) by the number of intervals wanted. Thus, if a set of scores ran from 90 to 143 and about 15 intervals were to be employed, the approximate width would be 3.5 score units, since

$$\frac{143 - 90}{15} = 3.5$$

Whole numbers are more convenient than fractions for widths of intervals, especially when the original measures are themselves integers. In this example we thus have a choice of 3 or 4 units for i. In most situations there will be a similar choice, since dividing the range by the number of intervals wanted usually results in a fractional value.

On the basis of practical considerations, certain widths are preferable to others. When the choice is among widths smaller than 10, odd-sized widths (3, 5, 7, 9) are often more convenient than even widths. In the example above, therefore, we would select 3 as our i instead of 4. For interval sizes of 10 and above, multiples of 5 (10, 15, 20, and so forth) are usually selected.

Nothing is sacred about any of these widths; they just turn out to have certain practical advantages. As for the larger widths, we are a "ten-minded" people and are happier when dealing with multiples of 10 or 5 than with other numbers. For smaller widths there is a different reason for using odd numbers. We frequently will be using and referring to the *midpoint*—the middle score value in any interval. For example, 11 is the midpoint of the interval 10-12, 105 is the midpoint of the interval 103-107, 47.5 is the midpoint of 46-49. Note that in the first two examples i is an odd number (3 and 5 score units, respectively) and the midpoints are whole numbers (11 and 105); in the third example i is an even number (4 units) and the midpoint, 47.5, involves a fraction. These results are not coincidental: when the scores are integers and i is an odd number, the midpoint is an integer, but when i is an even number, the midpoint of each interval is a fractional value. Since whole numbers are typically more convenient to deal with than fractions, odd-numbered widths are employed so that the midpoints will be whole numbers.

In the frequency distribution used as an illustration up to this point, the raw measures have been whole numbers. When such data are grouped, we note that there are gaps between the upper and the lower limits of adjacent classes. Thus, if we had classes 10-12 and 13-15, the first class would go up to include (have an upper limit of) 12.0 and the next would start at (have a lower limit of) 13.0. Such values as 12.1 or 12.9, if they occurred, could not be assigned to either class, since they fall in the gap between 12.0 and 13.0. These class limits with gaps in between are identified as *apparent limits.* By contrast, *real limits* take into account that the variable being measured is almost always continuous and that each discrete score value actually represents an *interval* of values. Real limits therefore extend the upper and lower apparent limits of adjacent classes equally until they meet at the point midway between them and close the gap. For example, the real limits of the class 10-12 cited above are 9.5-12.5, and those of the adjacent class 13-15 are 12.5-15.5. The point of this distinction between the two kinds of class limits will become apparent in later chapters, where we will see that real rather than apparent limits are used in certain statistical operations.

DEFINITIONS OF TERMS AND SYMBOLS

Score. A general expression for the number assigned to an individual to indicate the amount of some property that individual exhibits.

Range (R). The difference between the largest and smallest score in a distribution.

N. The number of cases in a group.

Frequency (f). For each score value in a distribution of scores, the number of cases in the group that has that score value.

Simple frequency distribution. A table summarizing the scores received by a group of individuals. The table lists all possible score values from the highest to the lowest obtained score and the number of individuals (frequency) receiving each of these scores.

Class interval. Collection of a subgroup of scores into a class of scores with a range of values. For example, the class interval 3-5 contains scores of 3, 4, and 5.

Midpoint. Middle score value in a class interval.

Apparent limits of class interval. The values of the highest and lowest scores in a class interval.

Real limits of class interval. For continuous properties, each score value represents the midpoint of a range of values. The real limits of a class interval reflect this range. For example, when lengths of objects are measured to the nearest millimeter (mm), the score of 8 represents objects ranging from 7.5 to 8.5 mm long. The real limits of the class of objects whose apparent limits are 8-10 mm are 7.5 and 10.5.

Width of interval (i). For continuous measures, i is the difference between the upper and lower real limits of the class interval. For most discrete measures, i is the number of integer score values in the score interval.

Grouped frequency distribution. A table listing scores grouped into class intervals of equal width (*i*), and the number of individuals (frequencies) falling into each class interval.

PROBLEMS

1. In an experiment on the effects of competitiveness, each of a group of 50 male college students was given an anagram task which he believed he was performing in competition with a female student. The times (in seconds) taken by the male students to complete the anagrams are listed below. Arrange these measures in a grouped frequency distribution, selecting a value for *i* that will best show the pattern of the group. Set up the distribution so that the upper limit of the highest class interval corresponds to the largest number in the group.* Add up the frequency column to make sure you have accounted for all cases.

    Be sure to use the headings "Class Interval" and "*f*" (frequency) in reporting the grouped frequency distribution you develop.

44	33	53	37	41
34	65	41	50	54
48	36	47	40	43
39	50	40	44	49
57	23	45	58	39
25	42	55	35	42
42	43	43	43	52
45	56	35	51	38
31	41	32	21	34
53	46	51	47	46

2. In a second condition of the experiment described in problem 1, members of a second group of 50 male students believed they were competing with another male student. The times to completion for these students are shown below.

    Set up two grouped frequency distributions for these data, one with *i* = 3 and one with *i* = 5. Which of these distributions do you prefer? Why?

46	29	36	26	50
35	43	22	34	34
28	34	32	40	36
41	35	59	31	27
32	45	35	33	47
36	33	35	56	35
25	41	31	37	20
48	26	40	29	38
30	37	23	24	23
34	30	42	45	39

---

*In the answers to the problems for this chapter, all frequency distributions were set up using this procedure. By using the same procedure, students can compare their solutions directly to those given in the answers. However, it should be noted that this procedure is neither necessary nor desirable in all data sets.

3. For the data in problem 1, determine (a) the value of the range, (R), (b) the real limits of the *lowest* class interval and of the *highest* class interval in the grouped frequency distribution you set up, and (c) the midpoint of each of these intervals.

4. In the following list, information is given about the N and the highest and lowest scores in each of six sets of data. For each set, decide whether it would be useful to arrange the data into a frequency distribution. If it would, indicate (a) the width of the interval (*i*) you would choose; (b) the approximate number of intervals that would result, assuming that the upper apparent limit of the highest interval has the value of the highest score in the distribution, (c) the real limits and the apparent limits of the *highest* class that would result from your chosen *i*; and (d) the midpoint of the highest interval. (*Note:* List the answers for the six sets in a table with columns labeled *i*, number of intervals, and so forth.

	Extreme Scores	N			Extreme Scores	N
(1)	50–95	7		(4)	16–25	18
(2)	11–75	94		(5)	.38–1.52	34
(3)	100–283	65		(6)	3–34	75

Graphic
Representations

CHAPTER 3

Most of us have seen graphs that present various types of statistical information: infant mortality rate in different countries, popularity of political candidates, the distribution of incomes, and so forth. Such graphs are used by popular writers because they are both more interesting and more easily interpreted than columns of figures. The graphic method is also employed extensively in technical articles, since the outstanding features of group performance are most readily grasped in this form.

GRAPHIC REPRESENTATIONS OF FREQUENCY DISTRIBUTIONS

We will discuss here two graphic methods of presenting frequency distributions: the frequency polygon and the cumulative frequency curve. Certain commonsense rules and ways of graphing data apply to both methods. These will be outlined after we look briefly at each type.

Frequency Polygon

Frequency distributions are most often presented graphically by means of a frequency polygon. Two frequency polygons, along with the grouped frequency distributions from which they were derived, are shown in Figure 3.1. We have plotted both distributions in a single graph, so that we can compare them more easily.

The grouped frequency distribution at the top right of Figure 3.1 represents the scores made by 150 college men on a sex-role attitudes questionnaire in which the maximum possible score was 45. The midpoint of each class interval is shown, rather than the limits of the interval, along with the frequency of cases falling into each interval. At the top left a similar distribution, taken from the data in Table 2.2, is shown for 150 college women. These frequency distributions contain all the necessary information about the scores of male and female students, but we can see how much easier and faster it is to obtain this information by inspecting the graphed polygons. We see, for example, that the men's scores are heavily concentrated around a score of 23, near the center of the scale. In contrast, women's scores tend to cluster at the upper end of the scale, indicating their more "liberated" attitudes.[1]

Let us now examine the technical features of a frequency polygon in detail. There are two axes: the horizontal baseline, also known as the X axis or the abscissa, and the vertical or Y axis, often referred to as the ordinate. The scores, in this example grouped into class intervals, are represented along the X axis, while the Y axis represents frequency or percentage of cases. Each axis is appropriately labeled, the numbers along the abscissa indicating the midpoints of the class intervals, those along the ordinate the frequency of cases in each interval.

As can be seen, points are plotted over the midpoints of each interval, the height of each point being determined by the appropriate frequency. The polygon is completed by

---

[1]The data are a representative sample taken from the 1980 groups reported in R. L. Helmreich, J. T. Spence, and R. H. Gibson, Sex-Role Attitudes: 1972–1980. *Personality and Social Psychology Bulletin* (in press). Raw data courtesy of the authors.

	Women			Men	
Midpoint	f		Midpoint	f	
44	3				
41	8		41	3	
38	16		38	5	
35	23		35	14	
32	23		32	17	
29	21		29	20	
26	16		26	23	
23	18		23	32	
20	9		20	21	
17	8		17	7	
14	4		14	5	
11	1		11	3	

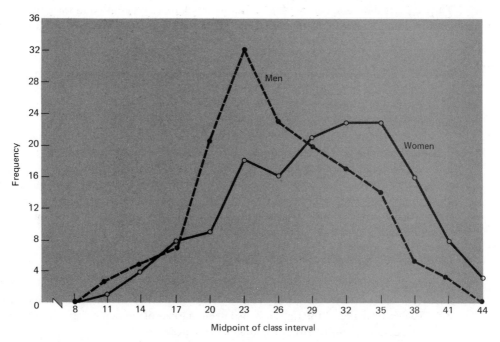

Figure 3.1 Frequency polygons showing sex-role attitude scores for male and female college students and the grouped frequency distributions from which they were drawn.

connecting adjacent points with a straight line and dropping the lines to the baseline at either end to show the upper and lower intervals at which zero frequencies begin.

Cumulative Frequency Curve

The cumulative frequency curve, though used less often than the frequency polygon, is preferred when we are interested primarily in specifying the position of an individual or individual score in a distribution rather than the general form of the group performance. The results from many tests of abilities, personality attributes, and the like are reported with cumulative frequencies or cumulative percentages and graphed in this fashion, since such scores are often used for individual diagnosis and evaluation. Before discussing this use of the cumulative curve, we will examine an example of a cumulative frequency curve and describe its construction.

Figure 3.2 shows a distribution of scores earned by 80 seventh-graders on a vocabulary test along with the cumulative frequency curve derived from it. The curve, which is also known as an *ogive*, shows the total number of cases earning a certain score or less. To plot a cumulative frequency curve we must first derive *cumulative frequencies* from the distribution. Looking at the distribution at the top of Figure 3.2, we find two individuals falling in the bottom interval, 60–62. The upper real limit of this class is 62.5. Thus 2 cases have a score of 62.5 *or less.* This figure (2) is entered in the cumulative frequency column. Four individuals fall in the interval that is second from the bottom; thus 2 + 4 or 6 cases have scores of 65.5 or less. This sum, 6, is entered next in the cumulative frequency column. For the next entry we determine that 11 individuals have scores of 68.5 or less (2 + 4 + 5 or 6 + 5). By successive addition, from bottom to top, of the entries in the frequency column, we get the cumulative frequencies. The entry in the topmost interval is equal to the total number of cases, since all individuals have a score equal to or less than the upper limit of this class.

A cumulative frequency curve represents the graphing of a cumulative distribution and is carried out in a manner similar to that for the frequency polygon. In this method, however, the points are plotted directly above the *upper real limits* of each interval rather than the midpoint. The upper real limits are used because cumulative frequency represents the number or percentage of cases that have scores equal to or below the value at the particular upper limit. The height of each point above the abscissa is, of course, determined by the appropriate cumulative frequency.

Cumulative Percentages and Percentiles

Cumulative frequency can also be translated into cumulative percentage. This translation has been made in the right-hand column of the frequency distribution shown in Figure 3.2. The vertical axis at the right has also been marked off in percentages, so that we may read the curve in terms of either frequencies or percentages.

Cumulative percentages are particularly useful for describing the relative position of a given score in a distribution. For example, inspection of the cumulative curve in Figure 3.2 shows that a child who earned a score of 80 is above average in vocabulary in comparison to others in the class—that is, approximately 75 percent of the children had a score of 80 or less; conversely, only 25 percent had a score of 80 or more.

Class Interval	f	Cum f	Cum %
90–92	2	80	100.00%
87–89	3	78	97.50
84–86	6	75	93.75
81–83	9	69	86.25
78–80	12	60	75.00
75–77	14	48	60.00
72–74	13	34	51.00
69–71	10	21	26.25
66–68	5	11	13.75
63–65	4	6	7.50
60–62	2	2	2.50
	N = 80		

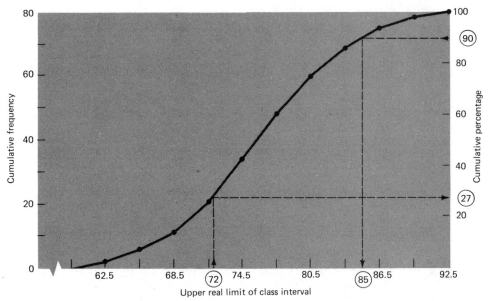

Figure 3.2 Cumulative curve of vocabulary scores for seventh-grade pupils and the frequency distribution from which it was derived.

Specifying an individual's standing in terms of these cumulative percentage figures often conveys more information than the raw score (how much would it mean to you, for example, simply to be told that on a 60-point examination, you scored 42; is that good or bad?) or even an individual's standing in terms of numerical rank (for example, 27th from the bottom in a class of 64). Cumulative percentages are therefore often used as alternatives to raw scores and are collectively identified as the *percentile scale.*

Using the cumulative percentage curve, we can go back and forth between the two scales. If we start with a raw score and find its equivalent on the percentile scale, we speak of the percentile rank of the score. *The percentile rank of any given score is a value indicating the percentage of cases falling at or below that score.* In Figure 3.2, for example, we can see that the percentile rank of the score 72 is approximately 27. Conversely, we can start with a percentage value and find its score equivalent. This score is called a percentile. *A percentile is the score at or below which a given percent of the cases fall.* In Figure 3.2, the 90th percentile is approximately 83. If the distinction between percentiles and percentile ranks seems confusing, don't worry. Just remember that one can go back and forth between raw scores and percentile scores and that percentiles represent the percentage that earn the score *or below.*

The percentile scale, derived from percentages, contains 100 units. A number of points along this scale are given special names. Often encountered are the *quartiles.* The first quartile, known as Q_1, is the special name for the 25th percentile, the second quartile (Q_2) is equal to the 50th percentile, and the third quartile (Q_3), to the 75th percentile. The 50th percentile is also called the *median* of the distribution. Occasionally certain percentiles are referred to as *deciles:* the first decile (or 10th percentile), the second decile (20th percentile), and so forth up to the tenth decile (100th percentile).

The percentile equivalents of raw scores can be determined, we have just seen, by reading them from a graph of cumulative percentages. This method has the drawback of providing only approximate values of percentiles and percentile ranks because of the necessity of estimating values from graphs. Procedures for calculating more precise values will be presented in the next chapter in our discussion of the median.

GENERAL PROCEDURES IN GRAPHING FREQUENCY DISTRIBUTIONS

With these brief descriptions of the two types of graphic representations before us, we can now discuss certain general procedures applicable to graphing frequency distributions.

Selection of Axes and Size of Units

The custom in graphing frequency distributions is to let the horizontal baseline represent scores or measures and the vertical axis frequencies or percentage of cases. This convention should always be followed, since reversal would cause confusion for those used to the customary method.

The physical distance selected to represent a score unit or frequency unit is purely arbitrary, except for convenience, since a graph can be drawn large enough to cover the wall of a room or small enough to require a microscope to be seen. Whatever the size of the graph, however, the general rule for frequency polygons has been to choose the units in such a way that the length of the vertical axis is 60–75 percent of the length of the baseline. This rule may seem picayune but there are reasons for it. Not only does the uniformity produce esthetically pleasing results, but there are practical implications as well. The latter become clearer if you examine the polygons in Figure 3.3. Each graph gives a different impression at first glance, even though each represents the same frequency distribution. These differences have been produced by manipulating the physical distances chosen to represent scores and frequency units—or, more accurately, by changing the relationship between the length of the two axes. Since the reader's impression of

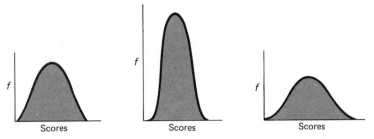

Figure 3.3   Three polygons drawn from the same frequency distribution varying in ratio of *X* to *Y* axis.

a frequency distribution can be so easily influenced by such an arbitrary factor, the adoption of an approximately standardized relationship between the two axes facilitates the interpretation of frequency polygons.

Because the highest frequency appearing in a cumulative frequency curve is greater than that in the polygon of the same data, the same rigid rule does not apply to this type of graphic representation. Most investigators, however, seem to make the baseline equal to or slightly longer than the vertical cumulative frequency axis, and it might be well for you to follow this custom whenever convenient.

Labels and Titles of Graphs

Graphs should always be completely labeled. These labels include what each axis represents, such as scores, frequencies, percentages, and the numerical values of each. The lowest values for both frequencies and scores should start at the intersection of the two axes, at the lower left of the graph. It is not necessary to indicate or number the position of every possible score along the baseline, only certain representative ones at regular intervals. Specifically, it is usually sufficient to label class limits (real or apparent) or midpoints of intervals. In frequency polygons, it is most often convenient to indicate the midpoints of intervals, and for cumulative curves the upper real limits. There is no general rule for the frequency scale; as long as enough points are identified to make interpretation easy and these are placed an equal number of units apart, labeling will be considered adequate.

Every graph should also have a concise title containing sufficient information to allow a reader to identify relevant aspects of the data, such as exactly what they measure, their source, and so forth. If several graphs are to be presented, each graph or figure should also be given its own number for further identification (for example, Figure 17). Since a graph without labels and a title is often next to meaningless, you are strongly urged to make "titling" second nature, even on statistical exercises.

FORM OF A FREQUENCY DISTRIBUTION

Frequency distributions can occur in an unlimited number of shapes and forms. Since form is often an important bit of information about a distribution, it is useful to know some of the descriptive terms used to indicate various types.

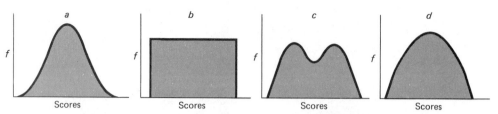

Figure 3.4   Illustrations of symmetrical frequency distributions.

Some frequency distributions, and consequently the frequency polygons derived from them, are *bilaterally symmetrical*. That is, if the polygon is folded in half, perpendicular to the baseline, the two sides are identical in shape. A variety of symmetrical distributions is shown in Figure 3.4.

Several distributions that are not symmetrical are shown in Figure 3.5. An asymmetrical distribution in which the scores "tail off" in one direction is said to be *skewed*. Such distributions can be more exactly described in terms of degree and direction of skewness. *Degree* refers to the amount that a distribution departs from symmetry. Thus, distribution *a* in Figure 3.5 is more highly skewed than *c*. *Direction* of skewness takes account of the fact that in some asymmetrical distributions the measures tend to pile up at the lower end of the scale and to trail off, in terms of frequency, toward the upper end, while in other cases the situation is reversed. Differences in the direction of skewness are indicated by saying that the distributions are positively or negatively skewed. Both as a way of explaining which is positive and which negative and as a device for remembering the difference between the two, the lower or left end of the baseline can be pictured as the negative end (values are low) and the upper or right end as positive (values are high). If the "tail" of a distribution, where scores are relatively infrequent in number, lies in the negative end of the scale, the distribution is negatively skewed. If the tail extends out toward the upper or positive end, the distribution is positively skewed. Curves *c* and *d* in Figure 3.5 are therefore negatively skewed, while *a* and *b* are positively skewed.

Certain distinctive forms of distributions that occur with some frequency are given special names. Curve *b* in Figure 3.5 is called a *J curve*, a curve that resembles either the letter J or its mirror image. This type is found primarily with certain types of socially conditioned behavior, such as a distribution of the number of major crimes committed by members of a large group of individuals. The vast majority have none to their credit, whereas a few members of the group are multiple offenders. In Figure 3.4, *b* is *a rectangular distribution*, the frequencies in every class interval being the same. Distribution *a* in

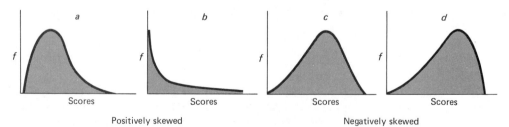

Positively skewed   Negatively skewed

Figure 3.5   Illustrations of skewed frequency distributions.

Figure 3.4 approximates *the normal curve*, a specific form of a symmetrical bell-shaped distribution. Many phenomena, including such psychological characteristics as IQ's and certain personality traits, appear to take this form. The normal curve is of such great theoretical importance in statistics that its properties will be discussed in some detail in Chapter 8.

Finally, we call attention to distribution *c* in Figure 3.4. Unlike the rest of the frequency polygons in Figures 3.4 and 3.5, this curve has two humps rather than one. This type of curve is called *bimodal*. The other curves with single peaks are called *unimodal*.

GRAPHIC REPRESENTATIONS OF OTHER TYPES OF DATA

Up to this point we have been considering methods of graphing frequency distributions. We will now discuss very briefly how to represent several other types of data commonly encountered in psychological investigations.

Categorical Membership

Table 3.1 shows the percentage of PhD's in natural sciences versus the percentage of PhD's in the social sciences, arts, and education awarded to men between 1920 and 1974 and parallel figures for PhD's awarded to women in the same period. You will recognize these percentages as categorical data, the information recorded about each individual being the field in which his or her doctorate was earned. The data obtained from each sex are shown in Figure 3.6 in the form of a *bar graph*, the most common method for representing frequency of categorical membership. The two categorical groups for each sex are shown on the baseline, the heights of the bars indicating the percentage in each category. Since neither sex nor type of degree represents any sort of continuum, which sex appears first and within sex, which field appears first is perfectly arbitrary. To emphasize the discreteness of the fields, the pairs of bars are spatially separated.

Measurement Data from Groups

Often in research we wish to compare groups on some measure—that is, on some property that exists in various degrees. For example, we might compare the average weight of the brain in different species of animals or the annual per capita consumption of beer in dif-

Table 3.1 Percentage of Men and Women Awarded the PhD in the Natural Sciences and the Social Sciences, Arts, and Education, 1920–1974

	Natural Sciences	*Social Sciences, Arts, Education*	*Total*
Men	50.4	49.6	100
Women	24.5	75.5	100

Source: The data are taken from *A Century of Doctorates*, National Sciences Foundation, 1978.

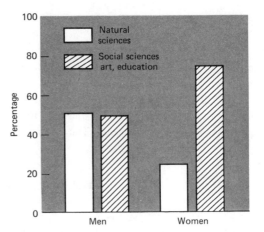

Figure 3.6 Bar graph of data in Table 3.1 showing percentage of PhD's in two different fields awarded to men and women.

ferent countries. In each of these illustrations, the various groups—species of animals or countries—represent a set of categories that do not form an ordered continuum. Unlike the section above in which our interest was in the number or percentage of cases in each categorical group, we are here concerned with comparing the groups according to how much of some measurable property is exhibited by the members within each.

Measurement data from categorical groups may also be presented in the form of a bar graph to facilitate group comparisons. An illustration of a bar graph from such a study is found in Figure 3.7. The data in the graph show the average number of hours per day spent on household tasks in the United States by married men, married women who are employed outside the home, and housewives with no paid employment.[2] Since the

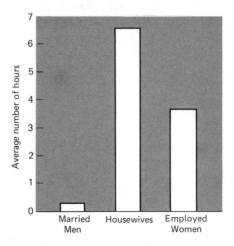

Figure 3.7 Bar graph of average number of hours per day spent on household tasks by married men, housewives, and employed married women.

_____

[2]The data are from J. P. Robinson, P. E. Converse, and A. Szalai, Everyday life in twelve countries. In A. Szalai (Ed.), *The use of time: Daily activities of urban and suburban populations in twelve countries.* The Hague: Mouton, 1974.

groups cannot be arranged along a continuum, the order in which the bars appear on the baseline is arbitrary.

In other research, we test different groups to determine how individuals' performances change with systematic increases in the value of some variable. For example, four groups of students might be asked to memorize a stanza of poetry and have their recall tested after 5, 10, 30, or 60 minutes. Or an investigator might want to determine the accuracy with which individuals can recognize words of 3, 5, or 7 letters shown to them for only $\frac{1}{100}$ of a second. In both examples, the experimental conditions represent points along a dimension: amount of time to recall or word length. In order to emphasize how the observed property (amount of recall or number of correct recognitions) varies with changes in the experimental variable, the results of such studies are usually presented in the form of a *line graph*. The experimental groups are arranged in order along the baseline and the amount of the property is represented on the vertical axis. The average amount of the property exhibited by each group is indicated by a point rather than by a bar. The adjacent points are then connected by straight lines.

An example of a line graph is shown in Figure 3.8. The graph shows the results of an experiment in which 6-year-old, 7-year-old, and 8-year-old children were given several tasks to perform. Half the children in each age group were told they had done much better than most other children, and the other half that they had not done as well as most children. Then all children were given a puzzle that was actually insoluble, and the time they spent on the puzzle before giving up was recorded. The line graph shows the average time spent on the puzzle by the previously "successful" children and previously "unsuccessful" children in each age group. Inspection of the graph shows that the three age groups were about equally persistent after success. At all three ages, unsuccessful children were less persistent. However, the detrimental effects on persistence of the lack of previous success became less marked as age increased.

In this chapter we have referred to the *median* of a frequency distribution and to the *average* of a distribution, which is more technically known as the *arithmetic mean*. The median and the mean are both examples of a measure of central tendency, a topic discussed in the next chapter.

Figure 3.8 Average time in seconds spent on insoluble task by 6, 7, and 8-year-old children who were previously successful or unsuccessful.

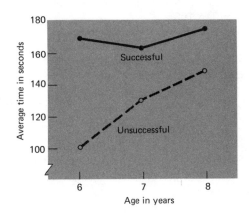

Frequency polygon. A graph based on a frequency distribution. Scores or midpoints of class intervals are shown on the horizontal (X) axis and frequencies are shown on the vertical (Y) axis. The frequency associated with each score or class interval is indicated by a point, and adjacent points are connected by straight lines.

Cumulative frequency distribution. A distribution showing for the upper real limit of each score (or of each class interval) the total number of cases earning that score value *or less.*

Cumulative frequency polygon or ogive. A graph based on a cumulative frequency distribution. Upper real limits of scores or class intervals are shown on the horizontal axis and cumulative frequencies on the vertical axis. Adjacent points, each representing a cumulative frequency, are connected by a straight line.

Symmetrical distribution. Frequency distribution whose polygon, when bisected on the baseline, forms two halves that are mirror images of each other.

Skewed distribution. Frequency distribution whose shape is asymmetrical. When frequencies are high for high score values and tail off in the direction of low score values, the distribution is described as *negatively skewed.* When frequencies tail off in the direction of higher score values, the distribution is described as *positively skewed.*

Unimodal curve. A frequency polygon in which there is only one hump (one score value with the maximum frequency).

Bimodal curve. A frequency polygon in which there are two humps or points of maximum frequency.

Percentile rank. A raw score expressed in terms of the percentage of cases in a distribution that have that score *or less.* For example, if 35 percent of the cases score 76 or less, the percentile rank of the score 76 is 35. Percentile ranks, like percentages, range from 0 to 100.

Percentile. The score at or below which a given percentage of the cases in a distribution fall. For example, if 82 percent of the cases fall at or below a score of 110, 110 is equal to the 82nd percentile.

Median. The score at or below which 50 percent of the cases fall. The median is the same as the 50th percentile.

Quartiles. Score points that divide a distribution into quarters in terms of frequency. The first quartile (Q_1) is equal to the 25th percentile, the second quartile (Q_2) is equal to the 50th percentile or median, and the third quartile (Q_3) is equal to the 75th percentile.

Deciles. Score points that divide a distribution into tenths in terms of frequency. The first decile (D_1) is equal to the 10th percentile, the second decile (D_2) to the 20th percentile, and so forth.

PROBLEMS

In the problems calling for a graph, it will be most convenient for you to draw them on graph paper. For each graph, be sure to label the axes and supply a title describing what the data represent.

1. In Chapter 2, problems 1 and 2 gave data from an experiment on males' competitiveness in which the effects on performance of the sex of partner were studied. Each problem required setting up a grouped frequency distribution with $i = 3$. (These distributions are shown in the answers for Chapter 2 on page 307.)
   (a) Draw a frequency polygon for each distribution, putting both polygons in the same figure.
   (b) Which set of data shows a more skewed distribution? In what direction is it skewed?
   (c) What do the distributions suggest about the effects of sex of partner on competitiveness? Be specific.

2. In the first session of her class, an introductory psychology instructor gives the students a true-false test of 40 common psychological misconceptions. A grouped frequency distribution for the resulting scores is shown below.

Class Interval	f	Class Interval	f
36–40	6	16–20	6
31–35	9	11–15	0
26–30	12	6–10	3
21–25	13	1–5	1
			$N = 50$

   (a) Find the cumulative frequency and the cumulative percentage for each class interval in the distribution.
   (b) Graph the cumulative frequencies against the class intervals to produce a cumulative frequency polygon. Then give an alternative labeling of the vertical Y axis, using cumulative percentages.
   (c) The *percentile ranks* of three students are shown below. Using the cumulative polygon drawn for 2(b), estimate each of their raw scores (X). (*Note:* Your estimates will not necessarily be whole numbers.) Student A: 70, student B: 32, student C: 20.

3. A test of language comprehension was given to freshmen students in a public university open to anyone with a high school diploma (open university) and to students in an elite private university with high admission standards (elite university). The cumulative percentage curves for the two groups are shown in the figure on page 36.

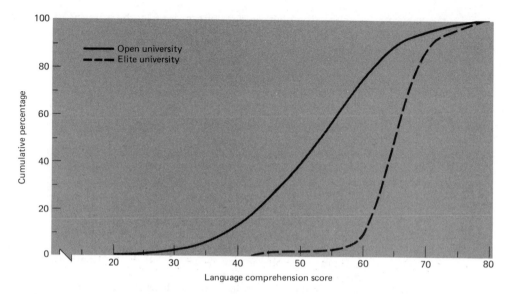

By inspecting the curves:
(a) Find the median for the open university.
(b) Find the median for the elite university.
(c) Find the 3rd quartile for the elite university.
(d) Find the 2nd decile for the open university.
(e) Mary White has a raw score of 60 and attends the open university. What is her percentile rank? What would be her percentile rank if she attended the elite university?

4. In a study of fear reduction procedures, on 10 successive days a man with a strong fear of snakes was asked to approach a glass cage containing a large but harmless snake as closely as he felt able. Shown below is the distance (in feet) from the cage at which he stopped on each day.

Day	Feet	Day	Feet
1	19	6	10
2	18	7	6
3	15	8	3
4	12	9	2
5	8	10	1

(a) Draw a line graph of these data.
(b) Describe the results of the fear reduction procedures.

5. In a particular private university, 38 percent of the students reside in the state in which the university is located, 45 percent come from other states, and 17 percent come from other countries than the United States. Draw a graph of these data.

6. From your general knowledge, sketch a frequency polygon for each set of data described below. Then describe the shape of the distribution by a label, such as symmetrical, negatively skewed, J-curve, and so on.

(a) Number of children in each of a large group of randomly selected U.S. married couples.
(b) Scores on an easy statistics exam.
(c) Number of drivers who speed up, do not change speed, momentarily slow down, or come to a full stop at a corner with a stop sign.
(d) The speed with which physically normal 10-year-old girls can run 100 yards.
(e) Attitudes toward legalized abortion on a 7-point scale (1 = strongly approve, 7 = strongly disapprove) among members of the audience attending a debate between pro- and anti-abortion groups.

Measures of Central Tendency

We have already seen how measures obtained from groups of individuals can be sorted to form frequency distributions that reveal group patterns and how to represent such distributions by graphing them. Typically, investigators are also interested in characterizing a group as a whole. They might have questions such as these: What is the typical beginning salary for graduates of engineering colleges? What is the average score on the Scholastic Aptitude Test for entering freshmen at a large state university, and how does it compare with the average score at an Ivy League university? What is the most popular size in men's hats?

Such questions ask for a single number that will best represent a whole distribution of measurements. This representative number will usually be near the center of a distribution, where the measures tend to be concentrated, rather than at either extreme, where, typically, only a few measures fall. From this fact comes the term *measure of central tendency*.

Although a number of measures of central tendency have been devised, we will be studying only three of them: the arithmetic mean ($\overline{X}$), the mode (Mo), and the median (Mdn). There is nothing magical about any of these measures of central tendency; each is simply a different method of determining a single representative number. As we shall see, the numbers resulting from each method usually do not agree exactly. The particular measure we compute in any instance depends on which one yields the number that best represents the fact we wish to convey. The specific situations in which each method is most appropriate will be discussed in a later section.

ARITHMETIC MEAN

Definitions of the Mean

The *arithmetic mean*, popularly known as the *average*, is a measure with which most of you are already familiar. The arithmetic mean, or simply "the mean" for short, is typically represented by the symbol $\overline{X}$ (pronounced "X-bar") and is the result of the well-known procedure of adding up all the measures and dividing by the number of measures. Defining the term more formally, *the mean is equal to the sum of the measures divided by their number.* A much more compact way to express this verbal definition is by using symbols, a kind of mathematical shorthand, as shown in the formula below:

$$\overline{X} = \frac{\sum X}{N} \tag{4.1}$$

where $\overline{X}$ = arithmetic mean
$\sum$ = Greek capital letter sigma, meaning the "sum of" a series of measures
X = a raw score in a series of measures
$\sum X$ = the sum of all the measures
N = number of measures

Before proceeding with a discussion of the mean ($\overline{X}$), we will take time out to explain in more detail the meaning of the Greek capital letter sigma (Σ), one of the most frequently used symbols in statistics. The presence of Σ in any formula indicates that a group of numbers is to be added or summed. Since for convenience we usually refer to any raw score in a frequency distribution as X, Σ X indicates the total obtained by adding together all the scores belonging to a distribution. Sometimes we deal with two frequency distributions—for example, scores made by business school and social science majors on a measure of political conservatism. To distinguish between the two distributions we might call one set X's (let us say, business majors) and the other Y's (the social science majors). If we wrote the symbols Σ Y, we would mean the sum of all the scores for the social science majors (the Y's). In like manner, Σf stands for the sum of all the frequencies in a distribution. Or we might arbitrarily choose the symbol A, and then Σ A is the sum of all the A scores, whatever they happened to be, and so forth.

The expression Σ X/N, then, is shorthand for the sum of all the scores in a distribution (X's) divided by N (their number). The number that results from these operations is the mean of the distribution of X's, symbolized as $\overline{X}$. This formula serves as a general expression for the mean of *any* series of quantities. In practice, the formula is often modified to indicate particular sets of quantities. For example, in the paragraph above, we distinguished between the scores of business school and social science majors by labeling those for the business students X's and those for the social science students Y's. The formula for the mean of social science students would be modified to Σ Y/N and the result symbolized as $\overline{Y}$. In Chapter 5, a formula will be encountered that reads Σx^2/N. At the moment, you do not know what either x or x^2 stand for, but even so it is important for you now to recognize two things about the formula. First, it instructs you to add together a series of squared quantities, called x^2's, and then to divide the sum by the number of x^2's. Second, the result of these operations is a *mean*, in this instance, the mean of the x^2's. Whenever you encounter a formula that instructs you to sum a series of quantities and divide by their number, the result is the *mean* of the quantities, whatever the quantities happen to be.

Another definition or characteristic of the mean ($\overline{X}$) is important for us to state, since it contributes to an understanding both of $\overline{X}$ and of the concept of variability, a topic to be considered in Chapter 5. *The mean may be described as that point in a distribution of scores at which the algebraic sum of the deviations from it (the sum of the differences of each score from $\overline{X}$) is zero.* Or, expressed differently, $\overline{X}$ is a kind of "point of balance," where the sum of the deviations of the scores above $\overline{X}$ is equal in absolute value (without regard to plus or minus sign) to the sum of the deviations below $\overline{X}$. In order to clarify these statements, we will first introduce a kind of analogy.

Imagine that the horizontal line shown in Figure 4.1 is an old-fashioned teeter-totter. Suppose further that each square drawn in Figure 4.1 represents a child, each of equal weight, seated at the place indicated. We can see that the fulcrum has been placed at a point where the board balances. But why does the board balance at this particular point? First observe that we can give each child a "value" in terms of his distance from the fulcrum (-1, -2, -3, +1, +2, +3). These distances we will call deviations from the fulcrum. The absolute sum (that is, ignoring signs) of the deviations of the three children to the right of the fulcrum is the same as the absolute sum of the deviations of the chil-

Figure 4.1 Representation of a balanced teeter-totter with "children" of equal weight placed at the distance indicated.

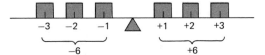

dren to its left; namely, 6 units. From painful experience, you know that the balance of the teeter-totter would be destroyed if a child were moved from one position to another, even on the same side—for example, from -1 to -3. When we disturb the values of the deviations so that their sums are not equal on both sides of the fulcrum, the balance is destroyed and the fulcrum must be moved to restore it.

Let us now relate this analogy to $\overline{X}$, which we have already said is a kind of point of balance. First, let us find $\overline{X}$ by the method of adding up the scores and dividing by N, as in the example given in Figure 4.2. Since there are 9 scores in Figure 4.2 and their sum is 450, $\overline{X}$ is 50. Next, let us find the deviation or distance each score is from $\overline{X}$. Each deviation is designated by the symbol x and is obtained by subtracting $\overline{X}$ from each score ($x = X - \overline{X}$). Obviously, scores with values greater than $\overline{X}$ will be plus deviations, those with lesser values, minus deviations. If we now add the plus deviations, those above $\overline{X}$, and the minus deviations, those below it, as was done in Figure 4.2, we find that the absolute value of the two sums is equal; retaining signs, the two sums cancel each other so that the sum of the *total* deviations is zero ($\Sigma x = 0$).

Take as another example the numbers 5, 9, and 16. $\overline{X}$ of these numbers is 10, the sum of deviations below it is -6 [i.e., $(-5) + (-1)$] and above it +6; the algebraic sum of the deviations equals 0. Thus, in any distribution of scores, $\overline{X}$ is a "point of balance," comparable to the fulcrum in the teeter-totter example, in that the algebraic sum of the deviations of the scores above and below it (Σx) always equals zero.

A simple algebraic proof can also be given to demonstrate that $\Sigma x = 0$. Since $x = X - \overline{X}$, we can rewrite the statement to be proved as $\Sigma (X - \overline{X}) = 0$. Now:

$$\Sigma(X - \overline{X}) = \Sigma X - \Sigma \overline{X}$$

Since Σ directs us to add together N quantities and $\overline{X}$ is a constant, $\Sigma \overline{X}$ can be written as $N\overline{X}$. Thus:

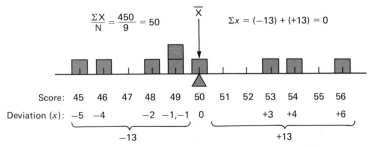

Figure 4.2 Distribution of raw scores and deviations of these scores from $\overline{X}$, showing that the algebraic sum of the deviations is zero.

$$\sum(X - \overline{X}) = \sum X - N\overline{X}$$

$$= \sum X - N\left(\frac{\sum X}{N}\right)$$

Since the N's cancel:

$$\sum(X - \overline{X}) = \sum X - \sum X$$

$$= 0$$

We have thus proved, as intended, that $\sum x$, the sum of the deviations from $\overline{X}$, is 0.

Methods of Computation

The basic formula that defines a statistical concept, such as that for $\overline{X}$, is usually relatively simple in appearance, but actually applying it to data often turns out to be long and tedious. To minimize effort, so-called computational formulas are usually derived from the basic one. These often look more difficult, as they involve more symbols, but are time-savers when any statistic is actually to be computed. In the following sections several computational formulas for $\overline{X}$ will be explained, in addition to the application of the basic one.

Computing $\overline{X}$ from Ungrouped Data. When $\overline{X}$ of ungrouped data is to be calculated, the basic definitional formula should be followed: the scores added and the result divided by the number of scores. An application of the basic formula is demonstrated in Table 4.1. This process is simple and should be followed if only a few scores are involved. If the number of measures is large, however, it may be prohibitively time-consuming. The use of frequency distributions, already found to be convenient for other purposes, will reduce our labor considerably in such instances.

Table 4.1   Calculation of $\overline{X}$ from Ungrouped Data

X
13
13
11
16
18
13
19
15

$$\sum X = \overline{118}$$

$$\overline{X} = \frac{\sum X}{N} = \frac{118}{8} = 14.75$$

Computing $\overline{X}$ from a Frequency Distribution. Suppose 70 children were given a perceptual judgment task to determine their susceptibility to optical illusion. The simple frequency distribution in Table 4.2 gives the number of errors these children made on the task. Note that most of the scores occur more than once. For example, two children made 9 errors and eight made 14. We could get ΣX by summing the 70 individual scores, duplicates and all—for example, adding two 9's, eight 14's. However, we could save time if we first multiplied each score by the number of individuals receiving it. Thus, 8×14 will result in the same answer as adding 14 eight times and give it to us quickly. If we then add all the products—each the result, you remember, of multiplying every X by its corresponding frequency, f—we will have the sum of all the Xs with far less labor than if we had summed each raw score separately. The remaining step needed to obtain $\overline{X}$ is the usual one: dividing the sum of the scores by N.

We can express these operations symbolically by amending the basic formula for $\overline{X}$ to take account of the fact that we are now dealing with a frequency distribution. The amended formula is reproduced below and its application demonstrated in Table 4.2.

$$\overline{X} = \frac{\Sigma(fX)}{N} \tag{4.2}$$

where $\quad \Sigma$ = the sum of the quantity that follows (here, all the fX's)

$\quad\quad$ X = the midpoint of a class interval

$\quad\quad$ (fX) = a midpoint multiplied by its corresponding frequency (f)—that is, by the number of cases within a class interval

$\quad\quad$ N = total number of cases, equal to the sum of the frequencies (Σf)

If necessary, the formula can also be applied to a *grouped* frequency distribution,

Table 4.2 Calculation of $\overline{X}$ from a Simple Frequency Distribution, Where i = 1

X	f	fX	X	f	fX
28	2	56	16	4	64
27	1	27	15	5	75
26	1	26	14	8	112
25	0	0	13	6	78
24	0	0	12	5	60
23	3	69	11	3	33
22	1	22	10	3	30
21	2	42	9	2	18
20	3	60	8	3	24
19	5	95	7	1	7
18	6	108	6	0	0
17	4	68	5	2	10
				$\Sigma f = 70$	$\Sigma(fX) = 1084$

$$\overline{X} = \frac{\Sigma(fX)}{N} = \frac{1084}{70} = 15.49$$

one in which the width of the class interval, *i*, is greater than one. This has been done in Table 4.3 for the data utilized in the previous table, now grouped into class intervals with a width of three. In applying the formula to grouped distributions, it should first be noted that the symbol X is the *midpoint* of a class interval. Actually, X in the formula for $\overline{X}$ of a frequency distribution is always a midpoint, even when *i* = 1 as in Table 4.2. In the latter instance we can think of a series of class intervals, each with a width of 1 (for example, class 5, 6, and 7). Thus, each raw score, X, is in a sense its own midpoint.

Going back to Table 4.3, multiplication of each midpoint by its corresponding frequency (*f*X), as done in the last column of the table, gives the (approximate) total of the scores in each interval. By adding all the entries in the *f*X column, we get Σ (*f*X). This is the (approximate) equivalent of the Σ X that would have been obtained if ungrouped scores had been added together. $\overline{X}$ is determined, of course, simply by dividing Σ (*f*X) by the total number of *cases*, N, *not* the number of class intervals.

While the foregoing method is convenient, the use of grouped frequency distributions introduces a source of error into the computation of $\overline{X}$. Underlying its use is the assumption that $\overline{X}$ of the scores in a specific interval is the same as the midpoint of the interval. If this is true, we can get the total of the scores in the interval by multiplying the number of measures (*f*) by the midpoint (X).[1] This assumption rarely is fulfilled exactly.

The net effect of such errors on $\overline{X}$ of the total distribution is not great, however, since in the intervals below $\overline{X}$ the errors tend to be in the direction of underestimation (scores usually pile up at the upper end of the interval, thus making the true $\overline{X}$ of the

Table 4.3   Calculation of $\overline{X}$ from a Grouped Frequency Distribution, Where *i* = 3

Class Interval	X	f	fX
26–28	27	4	108
23–25	24	3	72
20–22	21	6	126
17–19	18	15	270
14–16	15	17	255
11–13	12	14	168
8–10	9	8	72
5–7	6	3	18
		$\Sigma f = 70$	$\Sigma(fX) = 1089$

$$\overline{X} = \frac{\Sigma(fX)}{N} = \frac{1089}{70} = 15.56$$

Source: The data are derived from Table 4.2.

---

[1] Since $\overline{X} = \Sigma X/N$, by rearrangement of terms, $\Sigma X = \overline{X}(N)$. Thus, if $\overline{X}$ of the scores in a particular interval is equal to the value of the midpoint, we can find the total of the scores in the interval (ΣX) by multiplying the midpoint by the frequency [$\Sigma X = \overline{X}(N)$].

scores in the interval higher than the midpoint), and in intervals above $\overline{X}$ the opposite error more frequently occurs (scores tend to concentrate at the lower part of the interval). Thus, when all intervals are considered together, the errors in one direction tend to cancel the errors in the other, so that $\overline{X}$ for the total distribution is fairly close to the more accurate figure that would have been obtained had ungrouped data been used—in our example, 15.49 vs. 15.56.

Population Means versus Sample Means

A population, you recall, consists of all the individuals having some specified characteristic in common, whereas a sample consists of a subset of individuals drawn from a given population. Technically speaking, the symbol $\overline{X}$ is used to refer to the mean of a sample. To distinguish the sample statistic, $\overline{X}$, from the population parameter, a different symbol is used to indicate the mean of a population. This symbol is the Greek letter μ (mu). The population μ is found in exactly the same way as $\overline{X}$; that is, it is calculated by adding all the scores together and dividing by their number. Thus:

$$\mu = \frac{\sum X}{N} \tag{4.3}$$

What is the relationship between the value of a population μ and the values of $\overline{X}$ for samples drawn from that population? To answer this question, suppose that each member of the population of 10,000 fifth-graders in a certain school district had been given an achievement test in mathematical reasoning and the mean score turned out to be 79.6. Now suppose that a random sample of 50 children was selected from the population and the sample mean $(\overline{X})$ was obtained. It is unlikely that $\overline{X}$ would be exactly 79.6—that is, that it would take exactly the same value as μ. However, it is quite probable that $\overline{X}$ would be close to 79.6. If we were to draw a series of additional random samples from the population, it would also be unlikely that many of their $\overline{X}$'s would have exactly the same value as μ or that the $\overline{X}$'s would take values identical to one another. Some of the $\overline{X}$'s would be higher than μ and some would be lower than μ. Over the long run, however, the $\overline{X}$'s of these random samples would show no systematic tendency to be either larger or smaller than the μ of the parent population. Thus, assuming random selection, the mean of a sample can be described as an *unbiased estimate of the population mean.* This implies that if only a single random sample is available, its mean is the best single guess that can be made about the value of an unknown population mean.

MODE

The mode (Mo) may be either computed or estimated roughly from inspection of the data. Estimation by inspection is the only method we will consider here.

In ungrouped data, the mode is defined as that score which occurs most frequently. For data grouped into class intervals, Mo is the midpoint of the interval with the greatest frequency. Thus, in Table 4.1, showing ungrouped data, Mo is 13, since this score occurs three times and all others only once. In Table 4.3 Mo is 15, this score being the midpoint of the interval with the greatest frequency. The value of Mo may change, even with the same set of data, as the width of the class interval changes. Compare, for example, the Mo of Tables 4.2 and 4.3.

Sometimes a frequency distribution, like a camel, will have not one hump but two, indicating two points of maximum frequency rather than one. This type of curve, which is illustrated in Figure 3.4(c), is called *bimodal*, in contrast to the more usual *unimodal* or single-peaked curve. Both humps are identified as modes.

MEDIAN

We have already encountered the median (Mdn) in our discussion of percentiles, since the median is just another label for the fiftieth percentile. Thus, *the median may be defined as that score point at or below which 50 percent of the cases fall.* In some populations the distribution may be a peculiar one in which this definition may be satisfied by more than one point—for example, by all the values from $X = 4.0$ to $X = 5.0$. In such an instance we would have to say that the median is the *set of points* satisfying the condition stated in our definition. We will restrict ourselves here to discussing computational procedures for sample distributions yielding only one point.

Computation of the Median and Other Percentiles

Computation from Frequency Distributions. Procedures for computing the median or 50th percentile of a distribution are the same as for computing any other percentile and can be explained by referring to the grouped frequency distribution in Table 4.4. Our first step is to determine how many of the total number of cases constitute the given percentage. In finding the median of the data in Table 4.4, we must therefore take 50 percent of N, which gives us 30 cases (60 X .50). The score point we want, then, is 30th from the bottom of the distribution. Going up the *cum. f* column, we find that the 30th case falls someplace in the interval 40–44, since up to 39.5, the lower real limit of this interval, there is a total of 22 cases (too few) and up to 44.5, the upper real limit of the interval, 33 cases (too many). We need to go up through the 8th case of the 11 in the interval 40–44 to make up the needed 30 (22 + 8 = 30). The raw score value of this 8th case will give us the median.

Now we encounter a problem. The score of the 8th case cannot be determined because we do not know the placement or distribution of the 11 cases in the interval. We can, however, *estimate* its value by assuming that the cases in the interval are evenly distributed; that is, that the same number of cases fall at each score point, so that the shape of the intra-interval distribution is rectangular. Let us now examine Figure 4.3 to see the implications of this assumption. If the distribution is rectangular, then the 8th case in the interval ends 8/11 of the way up the interval. If we take 8/11 of the score units in the interval (that is, 8/11 of i), we find the "score distance" the upper end of the 8th case

Table 4.4 Grouped Frequency Distribution Used in Demonstrating Computation of Percentiles and Percentile Ranks

Class Interval	f	cum. f
65–69	1	60
60–64	2	59
55–59	5	57
50–54	7	52
45–49	12	45
40–44	11	33
35–39	8	22
30–34	7	14
25–29	4	7
20–24	3	3
	N = 60	

is from the lower real limit. This distance (8/11 of 5) turns out to be 3.6 score units. Adding 3.6 to 39.5, the lower real limit, gives us 43.1. This score, which is the 30th from the bottom of the complete distribution, is the 50th percentile or the median.

The procedures we have just discussed for obtaining the median can be described by a sort of "cookbook" recipe, which can be followed step by step and used to determine any percentile.

A. Translate the percentage of cases given in the percentile into number of cases (the *n*th case).

B. Find the score corresponding to the upper end of the *n*th case:

1. Locate by inspection of the *cum. f* column the class interval in which the *n*th case lies. This is the interval containing the percentile.

2. Subtract from the *n*th case the *cum. f* of the interval *below* the one in which the *n*th case is contained. This tells us the number of cases we need from the interval.

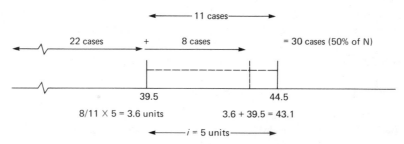

Figure 4.3 Graphic representation of obtaining a percentile.

3. Divide the value found in step 2 by the number of cases (f) in the interval containing the given percentile.

4. Multiply the quotient obtained in step 3 by i.

5. Add the number from step 4 to the lower real limit of the class in which the percentile (the nth case) is contained. This score is equal to the desired percentile.

These steps can be summarized in the following formula.

$$\text{Score} = \text{lower real limit of int.} + i \left(\frac{[n\text{th case}] - [cum.\ f\ \text{in int. below}]}{\text{no. of cases in int.}} \right) \qquad \textbf{(4.4)}$$

where nth case = the number of cases corresponding to given percent (for example, for 37th percentile when N = 40, nth case = 37 percent of 40)

int. = class interval in which the nth case is contained (determined by inspection of *cum. f* column)

int. below = class interval immediately below the one in which the nth case is contained

We can demonstrate the use of this method by applying it to another concrete example. Let us find the 80th percentile (or 8th decile) of the distribution in Table 4.4.

A. nth case = 80 percent of N or 48, where N = 60.

B. Find the score corresponding to the nth case:

$$\text{lower real limit of int.} + i \left(\frac{[n\text{th case}] - [cum.\ f\ \text{in int. below}]}{\text{no. of cases in int.}} \right)$$

thus: $49.5 + (5)([48 - 45]/7) = 51.6$.

This same method of computing percentiles can be used with simple frequency distributions with one slight modification: each score value is treated as *the midpoint of a class interval.* If the score values of a simple frequency distribution consisted of integers ranging from 123 upward, for example, 123 would be regarded as the midpoint of the interval whose real limits were 122.5–123.5, 124 the midpoint of the interval whose real limits were 123.5–124.5, and so on.

We have not discussed the strange case in which there is a zero frequency in the interval at the bottom of which a particular percentile, such as the median, falls. This means that the median or other percentile has several values rather than the single value promised in our definition of percentiles. For these unusual cases with zero in the relevant interval, we recommend defining the median or other desired percentile as the *midpoint* of the interval with zero frequency.

Computation from Ungrouped Data.   An alternate definition of the median that is useful with ungrouped data specifies that the median is the *middle score value* in a set of measures. A parallel definition may be offered for any percentile by substituting the appropriate score position. With this method, scores are first ranked in order of magnitude. If the number of cases is *odd*, the position of the median is equal to $(N + 1)/2$. We count up from the bottom until we reach the score that has the $(N + 1)/2$th position. In the set of five ordered numbers, 9, 11, 15, 16, and 18, $(N + 1)/2$ is equal to 3; the third case from the bottom is 15; this is the middle number or the median. If the number of cases is *even*, we count up to the cases that have the positions $N/2$ and $(N + 2)/2$ and then find the *midpoint* of these two scores. In the ordered set of six numbers, 3, 5, 6, 8, 12, and 13, we therefore count up to the third and fourth cases, finding the scores 6 and 8. The midpoint of these two scores is 7, the median of the distribution.

The median for ungrouped data may take a slightly different value from the one that would be found if we applied the computational method for a frequency distribution discussed earlier. This occurs because of the treatment in the latter approach of each score as an interval—for example, 9.5–10.5—rather than simply as an exact value—for example, 10.000 . . . .

Computation of Percentile Ranks

In computing percentiles, we start with a given *percentage* at or below which cases fall, for example, the lower 25 percent of the distribution, and determine the *raw score* that is equivalent to the given percentage. We can also start with a given *raw score* and calculate its *percentile rank*—that is, determine the percentage of cases that fall at or below the raw score. We often want to report individuals' scores in terms of percentile ranks instead of raw scores because the latter are seldom meaningful in and of themselves.

Suppose, for example, we wanted to know the percentile rank of the score 47, found in the distribution in Table 4.4. What has to be determined is the *number* of cases receiving a score of 47 or less, and from there, the *percentage* of cases. This percentage gives us the percentile rank.

Looking at Table 4.4, we discover that 47 is in the interval 45–49 and, from inspecting the *cum. f* column, that 33 cases fall below the lower real limit of the interval in which our score of 47 is contained. To obtain the number of cases scoring 47 or below, we have to add to the *cum. f* of 33 the number of cases lying between the lower real limit, 44.5, and our score, 47. By assuming that the 12 cases in the interval are rectangularly distributed, just as we did earlier in computing percentiles, we can determine this number quite simply. The width (i) of the interval is 5 score units and our score lies 2.5 units up from the lower real limit ($47 - 44.5 = 2.5$). We therefore estimate that $\frac{2.5}{5}$ or $\frac{1}{2}$ of the 12 cases lie between our score and the lower real limit. We now add $\frac{1}{2}$ of 12, or 6, to the *cum. f* up to the interval, which gives us 6 + 33 or 39 cases. Now the going is easy, since all we have to do is to translate number of cases into the *percentage* of N. This we do by dividing our number by N and multiplying the result by 100, which gives us $\frac{39}{60} \times 100$. Hence the answer to our problem is that the score of 47 has a percentile rank of 65.

These procedures for obtaining percentile ranks from frequency distributions can be described by a step-by-step "recipe."

A. Determine the number of cases falling below the given score:
  1. Subtract from the score the lower real limit of the interval in which it is contained.
  2. Divide the result of step 1 by the width of the interval (i).
  3. Multiply this quotient by the number of cases (f) in the interval.
  4. Add the product obtained in step 3 to the *cum. f* of the interval *below* the one in which the score falls.

B. Translate the number of cases below the score into percentages of cases:
  5. Divide the result of step 4 by N and multiply the product by 100. This gives us the percentile rank of the score.

These steps can be summarized in the following formulas, which should be easy to apply after a few practice problems have made the terms in them familiar.

(1) No. of cases = *cum. f* in int. below

$$+ \left(\frac{\text{given score} - \text{lower real limit of int.}}{i} \right) (f \text{ in int.}) \qquad (4.5)$$

(2) Percentile rank = $100 \left(\dfrac{\text{no. of cases}}{N} \right)$

where   int. = class interval that contains the given score whose percentile rank is being found

      int. below = class interval immediately below the one which contains the given score

To illustrate the use of these formulas, we will find the percentile rank of the score 28 in Table 4.4.

  1. Number of cases = $3 + ([28 - 24.5]/5)(4) = 5.8$
  2. Percentile rank = $100(5.8/60) = 9.67$

COMPARISON OF THE MEAN, MEDIAN, AND MODE

In any unimodal symmetrical distribution the values of $\overline{X}$, Mdn, and Mo are the same. This is true because the same point on the baseline occurs most frequently (Mo), divides the number of cases into the upper and lower 50 percent (Mdn), and is the point of balance ($\overline{X}$). As a curve departs from symmetry and becomes skewed, however, the values of the three measures of central tendency show differences. The relative positions of the three measures, however, will usually be the same for all single-peaked nonsymmetrical distributions, as shown in Figure 4.4. Mo in any skewed distribution occurs, of course, at the highest point of the curve. $\overline{X}$ falls someplace toward the "tail" of the distribution. This can be understood if you remember that $\overline{X}$ is a "point of balance" and is therefore

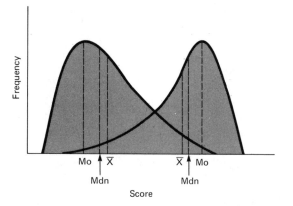

Figure 4.4 Relationship between X̄, Mdn, and Mo in skewed distributions.

very sensitive to the extreme deviations contained in the tail. Mdn in a skewed distribution lies between the $\overline{X}$ and Mo. Thus, the order of the three measures, when we start at the hump and proceed toward the tail, is (1) Mo, (2) Mdn, and (3) $\overline{X}$.

Because there is this relationship among the values of the three measures, we ordinarily can tell whether a distribution is skewed, and if so, the direction, merely by being told the values of any two measures of central tendency. If we read that $\overline{X}$ of a distribution was 26 and that Mdn was also 26, we would infer that the distribution was probably symmetrical since in this case the measures coincide. But if we were told that $\overline{X}$ was 109 and Mo from the same distribution 102, we would infer that the distribution was positively skewed. That is, since $\overline{X}$ always lies closest to the "tail" in a skewed distribution and in this case is greater than Mo, the tail is probably in the upper, positive end of the scale. The value of Mdn in this example could be roughly estimated—greater than 102 and less than 109—since Mdn always falls between $\overline{X}$ and Mo.

When to Use the $\overline{X}$, Mdn, or Mo

In deciding which measure of central tendency to compute, we must consider a number of factors, including how much time is available, the characteristics of the data involved, and the purpose for which the measure is intended. Some of these considerations are discussed below.

Stability of Measures. The three measures of central tendency differ with respect to their consistency or stability from sample to sample. That is, if we obtained a number of random samples from a population, the values of $\overline{X}$, as well as the values of Mdn and Mo, would vary from sample to sample. For example, we might test several samples of middle-aged businessmen to determine their cholesterol level. We would find that the mean was not the same for each group, nor was the Mdn or Mo. But we would discover that $\overline{X}$ showed less variability from sample to sample than the other two and that Mo was the least stable. That is, if we took a large number of samples from a large group and computed $\overline{X}$, Mdn, and Mo for each, we would find that the values of the $\overline{X}$'s differed less among themselves than the Mdn's and Mo's, whereas the Mo's varied the most. Since we frequently test small groups or samples in order to estimate the characteristics of the larger group or population to which the samples belong, stability is a desired virtue for a statistical measure. With respect to consistency, then, it should be remembered $\overline{X}$ is most satisfactory, Mo is least satisfactory.

Subsequent Manipulations of the Data. Of the three measures of central tendency, only $\overline{X}$ is a measure that is determined by the algebraic value of every score. This algebraic property makes $\overline{X}$ particularly useful in many further statistical operations that yield additional kinds of information about data. In contrast, once Mo and Mdn have been computed, little more can be done with them, especially Mo. Since information beyond a measure of central tendency usually is to be obtained from data, computation of $\overline{X}$ becomes almost obligatory.

Time Factors. A trivial but often practical consideration is the time taken to compute each measure. If a measure of central tendency is needed in a hurry, Mo can be rapidly obtained by inspection and would therefore be preferred over the other measures.

Characteristics of the data. (1) *Skewed distributions.* When a distribution is markedly nonsymmetrical, it is possible to give a distorted impression of the data when reporting central tendency. Take, for example, a frequency distribution of incomes in the United States. Such a distribution is extremely skewed in a positive direction, incomes trailing off, in terms of frequency, from the modest modal point to a few fabulous incomes of several million dollars each. The modal and median incomes will be much lower than $\overline{X}$, which is drawn far out toward the tail. It might be good capitalistic propaganda to give $\overline{X}$ as the "average income in the U.S.," but it would give an uninformed reader, who is likely to assume that this figure represents the most frequent income, or the income of the ordinary person, a very inflated view of our personal wealth. A Marxist, on the other hand, might select Mo to represent us, since this measure would have the smallest value. Neither way is completely honest, of course, and what is usually done in the case of extremely skewed distributions is to report all three measures. From these the reader can infer the direction and amount of skewness and interpret the data properly.

(2) *Some special cases.* In some instances, unusual characteristics of the data, other than skewness, dictate the measure of central tendency to use. A peculiarity that sometimes occurs is that one extreme of a distribution is not available for testing. For example, if we gave an intelligence test to a sample of schoolchildren in order to estimate the central tendency of IQ's of *all* children of similar age, the severely mentally retarded would not be represented in our sample, as these children are not in school.

In such a situation, computation of $\overline{X}$ would yield a distorted estimate of $\overline{X}$ for children in general, since this measure is highly sensitive to extreme deviations and these are missing at the lower end. Mo or Mdn would be more appropriate, since each would give a closer estimate of the value of central tendency that would have been obtained if the lower extremes had been available.

Occasionally, representative cases are available for testing, but exact scores cannot be obtained from cases falling in one or both extremes of the distribution. In experiments involving animals, for example, running time from one end of a path to the other may be used as a response measure. Often a few animals refuse to run and after sitting at the starting point for, say, five minutes are removed from the apparatus. Their exact running times are therefore unknown, since they might have run had they been left a second longer or they might have stayed there forever. Since $\overline{X}$ requires the exact value of every measure, it should not be computed. As long as exact values are available for more than half the animals, Mdn could be used, however, since an approximate score (five minutes plus) for the extreme cases is sufficient.

Many further examples of very special situations in which one measure is appropriate or inappropriate could be given. Instead, you will merely be reminded that every set of data should be examined to determine whether any specific problems exist and which measure of central tendency would be most appropriate.

Specific Purpose for Which Measure Is Intended. Sometimes a measure of central tendency is intended for some special use rather than as a purely scientific description of data. Such a purpose will often determine the measure to be employed, especially when a distribution is skewed so that the values of the $\overline{X}$, Mdn, and Mo disagree markedly. Occasionally, the "typical case" is wanted, thus calling for Mo. A furniture manufacturer, hoping to make money on volume of sales, might want to know the size of the typical living room, the modal dimensions, to design furniture scaled to suit the largest single consumer market. At other times, the "middle case" is desired, the median individual. Test grades are sometimes given on this basis, Mdn being the dividing line between B's and C's.

In summary, $\overline{X}$ is the most generally preferable measure of central tendency, particularly in a nearly symmetrical distribution, since it has the greatest stability and lends itself most to further statistical manipulations. Mdn is the middle case and is usually considered most appropriate when the distribution shows peculiarities: marked skewness, missing cases, and the like. Finally, Mo is utilized in situations in which a quick, rough estimate of central tendency is sufficient or the typical case is wanted. Although these are good general rules, remember that they are not exhaustive and that each set of data should be examined to see exactly which measure or measures are best in a specific situation or for a particular need.

DEFINITIONS OF TERMS AND SYMBOLS

Before offering definitions of the major terms and symbols that have been introduced in this chapter, we call attention to Appendix I. All numbered formulas that occur in this chapter and in all subsequent chapters are reproduced in Appendix I. This formula list should be useful not only for purposes of review, but for solving problems presented in later chapters that require the use of formulas introduced earlier.

Measure of central tendency. A single number that is used to characterize an entire distribution of measures. Typically this number is at or near the center of the distribution. The mean, median, and mode are all measures of central tendency.

Mode (Mo).   The score that occurs most frequently in a simple frequency distribution or the midpoint of the class interval with the greatest frequency in a grouped frequency distribution.

Median (Mdn).   The 50th percentile. See Chapter 3.

Arithmetic mean.   The arithmetic mean (usually referred to simply as the mean) is obtained by summing all the measures in a distribution and dividing the result by the total number of measures. The arithmetic mean is also known as the average.

M $\overline{X}$.   A general expression for the mean of a sample of measures (X's) drawn from some population.

ϕ μ **(mu).**   The mean of a population of measures.

Σ.   The capital Greek letter sigma, meaning "the sum of." For example, ΣX is the sum of all the X's.

Deviation score (x).   The difference between a score and the mean of the distribution to which the score belongs, obtained by subtracting the mean from the score. The algebraic sum of the deviation scores obtained from all scores in a distribution equals zero ($\Sigma x = 0$).

Percentile and percentile rank.   See Chapter 3.

PROBLEMS

1. The subjects in an experiment were given two prose passages to study and after each of them were asked a series of questions about the content. The errors made by the eight subjects on each passage (labeled X and Y) are shown below. Describe what each of the following quantities represent and then find the value of each quantity for the data that are given.   (a) ΣX; (b) $\overline{X}$; (c) ΣY; (d) $\overline{Y}$; (e) $(\Sigma X)^2$; (f) ΣX^2; (g) Σx; (h) ΣXY.

Subject	X	Y	Subject	X	Y
A	4	5	E	9	8
B	1	2	F	2	2
C	7	6	G	6	5
D	4	3	H	3	1

2. A grouped frequency distribution, with $N = 30$, is shown below. Find the following statistics: (a) Mode; (b) Mean; (c) Median; (d) Percentile rank of $X = 19$; (e) 10th percentile.

Class Interval	f	Class Interval	f
21–23	3	9–11	6
18–20	2	6–8	4
15–17	5	3–5	0
12–14	7	0–2	3

3. In a random sample of 100 cases, $\overline{X}$ is calculated to be 63.4. What is your best single estimate of μ, the mean of the population from which the sample was drawn? That is, would you guess μ had the same value as $\overline{X}$, some value greater than $\overline{X}$, or some value less than $\overline{X}$? Why?

4. For ungrouped data, it is useful to define the median as the middle score value in an ordered set of numbers. Using this definition, find the median for each of the following: (a) 20, 12, 14, 10, 25, 18, 9; (b) 7, 32, 16, 25, 20, 12, 14, 24; (c) 125, 319, 238, 161, 148, 170.

5. The values of two measures of central tendency are given below for several distributions. For each distribution, estimate whether it is approximately symmetrical, negatively skewed, or positively skewed. (Assume each distribution is unimodal.) Sketch a figure showing what you expect each distribution looks like and where you expect *all three* measures of central tendency (mean, median, mode) would fall. (a) Mo = 50; $\overline{X}$ = 58; (b) Mdn = 71.4; $\overline{X}$ = 71.4; (c) Mdn = 104; $\overline{X}$ = 96.

6. Give a concrete example of data in which the (a) mean, (b) median, or (c) mode would be the most preferable measure of central tendency to report.

Variability

Consider the three distributions in Figure 5.1 They are all unimodal symmetrical distributions, are based on the same number of cases, and have the same mean—yet they are quite different. What do these differences imply and how might they be described? Rather than answering these questions in the abstract, let us suggest some examples of what the distributions might represent.

Assume that three friends, Al, Ben, and Carl, spent many hours together on a golf driving range. They decided to keep a record of the distance they each hit 500 balls. Markers on the range allowed them to estimate their drives to the nearest 20 yards. Grouping the 500 drives into class intervals gave the three distributions shown in Figure 5.1. Looking at the distributions, we see that Al is quite inconsistent; he sometimes gets off a drive that a pro would envy, but sometimes he barely rolls the ball off the tee. Carl, on the other hand, is more predictable; he rarely dubs a drive badly, but he never gets the tremendous distance Al sometimes does. Ben is in between; most of his shots are very similar, but occasionally he goes wild, with results that are either very good or disastrous.

Let us assume instead that the distributions represent the scholastic aptitude scores of freshmen at three different universities. At Apollo University some of the students look like budding geniuses, while others will probably have a tough time making it through the first semester. At Caligula University there are no such extremes; as a group the freshman are much more alike in aptitude than those at Apollo. Bacchus University is in between; occasional freshmen are their professors' delight or despair, but most of them have a respectable, quite unexceptional degree of aptitude.

What terms might be used to describe the differences among the trio of distributions in Figure 5.1? We might say that the members of distribution C are more *homogeneous* than those in the other two distributions. Conversely, members of distribution A are more *heterogeneous* than those in the other distributions. We might also say that dis-

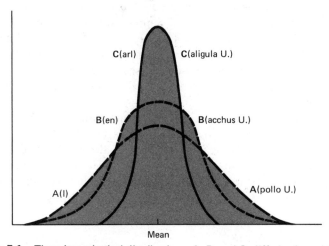

Figure 5.1 Three hypothetical distributions, A, B, and C, differing in variability.

tribution A has a greater spread than the others—a greater scatter or dispersion or, most commonly, greater *variability*. Distribution C exhibits the least variability among the measures, and B an intermediate amount. In the first illustration (the three golfers) the variability is *within* an individual, so that the curves may be said to represent differences in *intra-individual variability*. That is, repeated measures of the same activity were taken on the same person and each curve indicates how much that person varied from occasion to occasion on the same task. In the case of the second set of distributions (scholastic aptitude), the values represent *interindividual* variability—differences in aptitude among individuals at a particular college.

These examples illustrate that variability is an important characteristic of distributions. If we are to describe distributions adequately, we must include an expression that reflects the amount of variability. We shall discuss several statistics devised for this purpose.

RANGE (R)

As we learned in Chapter 2, *the range (R) is the "distance" from the lowest to the highest score in a distribution*. Or, highest score minus lowest score is R. If the lowest aptitude score in Bacchus U. is 400 and the highest 600, R is 600 - 400 = 200. If the shortest drive in Al's 500 shots is 40 yards and his longest 290, his golf range is 250 yards.

As a measure of variability R has much the same status as Mo does as a measure of central tendency. It is useful primarily as a gross descriptive statistic, although, as we shall see later, it can form the basis for a quick and easy estimate of another statistic that requires considerable computation.

The most obvious difficulty with R as a measure of variability is that its value is wholly dependent upon the two extreme scores. This causes several problems. First, one or both of the extreme scores may be capricious, so that R will be equally capricious. This becomes quite evident when we are working with a small sample of measures, as we often do in psychological research. Suppose we had ten measures as follows: 11, 14, 14, 16, 19, 20, 21, 24, 26, 42. R is 42 - 11, or 31. Now if we eliminate the tenth case, R becomes 26 - 11, or 15. The elimination of one score halved the range. Of course, any measure of variability must represent changes in variability when these occur, but an ideal measure would not fluctuate so much as R does when a single score is eliminated, as in the case above.

Even if the two extreme scores are representative, there is another difficulty with R, which we may illustrate by comparing distributions A and C in Figure 5.1. Both distributions have the same R, but the bulk of the scores in distribution C cluster more closely together than those in distribution A. The R, besides being unstable, is insensitive to the shape of the distribution of the scores between the two extremes. For these reasons, R is thought of as only a rough index of variability.

SEMI-INTERQUARTILE RANGE

A measure of variability that overcomes some of the problems of the range is the *semi-interquartile range*. This measure is obtained by first calculating Q_1 and Q_3, which you recall are also the 25th and 75th percentiles. Subtracting Q_1 from Q_3 gives us the inter-quartile range, or the number of score units between which the middle 50 percent of the cases fall. The interquartile range is then divided by 2 to yield the semi-interquartile range. Thus:

$$\text{Semi-interquartile range} = \frac{Q_3 - Q_1}{2} \tag{5.1}$$

The semi-interquartile range is not a measure to which the exact values of all scores contribute, and for this reason its usefulness is limited. However, when the distribution of scores is extremely skewed, the semi-interquartile range is often employed as the measure of variability in conjunction with the median as the measure of central tendency.

STANDARD DEVIATION AND VARIANCE

Because they are algebraic, the measures that are almost universally used as indicators of variability are the standard deviation and the variance. These related measures require several computational steps and represent statistical abstractions whose meaning is not easy to grasp. We will therefore work into them gradually.

In Chapter 4 we discussed the deviation of raw scores from $\overline{X}$. For each raw score we can calculate a deviation score by subtracting the mean from the score, $(X - \overline{X})$. Each of these deviations, you recall, is symbolized by x, so that $X - \overline{X} = x$. If all of the scores are closely clustered around the mean (if the distribution is quite homogeneous), then the raw scores deviate very little from the mean and the x's therefore are low in value. In a heterogeneous distribution in which scores are widely dispersed, on the other hand, the x's are more variable in value, since a number of scores lie far from the mean. Deviation scores, then, reflect the *variability* among a collection of raw scores. How might we capitalize on this fact to get a single number that will allow us to describe the amount of variability in a particular distribution?

We learned in Chapter 4 that the mean was a representative number used to characterize a distribution of measures of varying values. It might occur to you that we could find the mean of the x's to serve as a representative number to describe a distribution of x's. That is, add up all the deviations and divide this sum by the number of cases $(\Sigma x/N)$. This is an excellent beginning, but the procedure suffers from a lethal embarrassment. The sum of the deviations from the mean, we discovered in Chapter 4, is *always* equal to zero—that is, the sums of the deviations above and below the mean are equal except for sign. Thus the mean of the deviations, $\Sigma x/N$, will also equal zero—hardly an informative figure. We could, however, disregard the signs of the deviations and treat the x's as absolute numbers. For example, for $\overline{X} = 10$, the x's of the raw scores 13 and 7 would each be treated as 3 rather than $+3$ and -3. We could then go ahead and compute the mean of these absolute x's. The resulting statistic has come to be known as the *average deviation*.

Suppose that for one distribution the average deviation is 4 and for a second distribution it is 8—let us say the distributions of exam scores for two sections of a course in elementary statistics. We know from comparing the two averages that the raw scores in the first group cluster more closely around their $\overline{X}$ than the raw scores in the second group around their $\overline{X}$. The performance of students in the first section of statistics was much more homogeneous than the performance of students in the second section.

The average deviation provides a very satisfactory description of the amount of variability in a distribution of numbers. However, its further usefulness is limited because it is based on absolute deviations in which signs are ignored. For this reason, the average deviation is rarely used as a measure of variability. However, a few steps beyond the average deviation lead to a pair of statistics that are the preferred measures of variability, the *variance* and the *standard deviation*.

In obtaining the variance and the standard deviation, the problem of having the plus and minus x's cancel each other is avoided by squaring each x, so that all the resulting numbers are *positive*. The squared deviations or x^2's are summed and then divided by N, thus giving the mean of the squared deviations. This mean is given a special name, the *variance*. In this section, we will discuss the variance of a sample of scores, which we will identify as S^2. To repeat, the *sample variance* or S^2 is *the mean of the squared deviations of the scores from* $\overline{X}$. Thus:

$$S^2 = \frac{\sum x^2}{N} = \frac{\sum (X - \overline{X})^2}{N} \tag{5.2}$$

The square root of the variance (S^2) is called the standard deviation, which is symbolized as S. The *standard deviation* (S) is thus defined as *the root of the mean squared deviations from* $\overline{X}$. Expressing this definition symbolically, we have:

$$S = \sqrt{\frac{\sum x^2}{N}} = \sqrt{\frac{\sum (X - \overline{X})^2}{N}} \tag{5.3}$$

The variance has a great deal of use in advanced statistical techniques and will be discussed in more detail in Chapter 12. Our primary emphasis until that chapter will be on the standard deviation.

It may be useful at this point to run through a complete computational procedure for obtaining the variance and the standard deviation. Table 5.1 shows, on the left, the scores of eight 4-year-old girls on a test of manual dexterity and, on the right, the scores of eight 4-year-old boys on the same test. The basic computational steps, which are shown in the table for each distribution, can be summarized as follows:

1. Find the $\sum X$ and divide by N to obtain $\overline{X}$.
2. Subtract $\overline{X}$ from each raw score to obtain x. In Table 5.1 this has been done for each distribution in the columns labeled $x = X - \overline{X}$. As a check on the accuracy of these calculations, the sum of the x's has also been found; as it should, each $\sum x$ equals 0.[1]

---

[1]In Table 5.1, neither $\overline{X}$ was rounded. It may often be necessary or convenient to use a rounded figure for $\overline{X}$. For example, an $\overline{X}$ calculated to be 13.62358 might be reported as 13.62. If this rounded figure were used to obtain x's, slight errors would occur in each x and $\sum x$ would not be exactly zero. However, it would be very close to zero.

Table 5.1 Basic Steps in Calculating a Standard Deviation

Girls			Boys		
X	$x = X - \overline{X}$	x^2	X	$x = X - \overline{X}$	x^2
15	4.5	20.25	18	7.5	56.25
14	3.5	12.25	16	5.5	30.25
12	1.5	2.25	14	3.5	12.25
10	− .5	.25	13	2.5	6.25
10	− .5	.25	9	− 1.5	2.25
9	− 1.5	2.25	6	− 4.5	20.25
8	− 2.5	6.25	5	− 5.5	30.25
6	− 4.5	20.25	3	− 7.5	56.26
$\sum X = 84$	$\sum x = 0.0$	$\sum x^2 = 64.00$	$\sum X = 84$	$\sum x = 0.0$	$\sum x^2 = 214.00$

$$\overline{X}_G = \frac{\sum X}{N} = \frac{84}{8} = 10.5 \qquad\qquad \overline{X}_B = \frac{84}{8} = 10.5$$

$$S_G = \sqrt{\frac{\sum x^2}{N}} = \sqrt{\frac{64.00}{8}} = \sqrt{8.00} \qquad\qquad S_B = \sqrt{\frac{214.00}{8}} = \sqrt{26.75}$$

$$= 2.83 \qquad\qquad\qquad\qquad\qquad\qquad\qquad = 5.17$$

3. Square each x and sum the x^2's.
4. Divide $\sum x^2$ by N to obtain the mean x^2, or variance, symbolized as S^2.
5. Obtain the square root of S^2 to obtain the standard deviation, or S.

Before leaving Table 5.1, let us go over the results of our calculations. You will observe that although the mean of the boys and the girls turn out to be identical in this hypothetical example, inspection of the two distributions suggest that the boys' scores are more variable than those of the girls. The range of the boys' scores, for example, is 15 score units, the top score being 18 and the bottom score being 3. In contrast, the range of the girls' scores is 9 score units, the top score being 15 and the bottom score being 6. This difference in variability is reflected in the standard deviation. For boys, S is 5.17 score units, whereas for girls, S is 2.83 score units.

Corrected Estimates of Population Variance and Standard Deviation

Recall from Chapter 4 that the symbol $\overline{X}$ is used to identify the mean of a sample of X's and the symbol μ is used to identify the mean of a population. You will also recall that $\overline{X}$'s of random samples are unbiased estimates of μ of the population from which the samples were taken. That is, in a series of random samples, $\overline{X}$'s are typically variable in value but are not systematically larger or smaller than μ. Any given $\overline{X}$ can thus be described as an unbiased estimate of μ.

We now consider the symbols used to designate the population variance and standard deviation and the relationships between these population parameters and the corresponding sample statistics. The population standard deviation is symbolized by σ, which is the small Greek letter sigma. Similarly, the population variance is symbolized by σ^2 or sigma squared.

Unlike $\overline{X}$ as an estimate of μ, the sample S^2 and S are *biased estimates* of σ^2 and σ. More particularly, the S^2's and S's from a series of random samples tend to be *smaller* than the σ^2 and σ of the population to which they belong. This implies that if we were to use S^2 or S as our guess about σ^2 or σ, we would be likely to underestimate the population values.

In explaining why the sample standard deviation and variance tend to be smaller than the population values, we will find it simpler to discuss the matter in terms of the variance. Let us consider first the components of the formula for the variance, namely Σx^2 and N. As the N of a sample increases, not only the denominator of the formula but also the numerator, Σx^2, increases. That is, as we add a case to N, we also have its x^2 to add to Σx^2. Of course, if these two components increased proportionately, the variance would remain constant. This is pretty close to what happens but not quite. As we select more and more cases, it becomes increasingly likely that extremely deviant scores will be included within our sample. For example, we would be more likely to find a man over seven feet tall—or a man well under five feet—in a group of 500 men than in a group of twenty. These extreme cases add more to Σx^2 than they do to N, simply because their deviations from the mean are bigger than those we have gotten for cases closer to the middle of the distribution. So we may expect some increase in the variance with increases in N. However, when N is relatively large to begin with, say 30 or 40 cases, we expect less change in the variance with increases in N than when it is small.

Since a sample is, by definition, smaller than the population from which it was obtained, we can now see why a sample variance tends to be smaller than the variance of its population. To obtain an unbiased estimate of the population variance from sample data, we include a *correction factor* in the basic formula for the variance by dividing the Σx^2 obtained from our sample by N – 1 rather than by N. It is customary to use the symbol s^2 to indicate an unbiased estimate of a population variance from the data of a sample.[2] Thus:

$$s^2 = \frac{\Sigma x^2}{N - 1} \tag{5.4}$$

where, just as before, N is the number of cases in the sample and Σx^2 is the sum of the squared deviations of the scores from the sample mean.

Since we lower the value of the denominator by subtracting one from the sample N, you can see that s^2, our estimate of the unknown population variance, will always be

---

[2] Statisticians are inconsistent in the symbols they use to designate the sample standard deviation and variance and the corrected estimates of the population standard deviation and variance based on sample data. In accord with the majority of advanced statistical texts, and of a committee appointed to make recommendations about statistical symbols (M. Halpern, H. O. Hartley, and G. P. Hoel. Recommended standards for statistical symbols and notations. *American Statistician*, June 1965, 12-14), we have elected to use the symbols s and s^2 to refer to the *corrected estimates* of the population σ and σ^2, in which N – 1 is used in the denominator of the definitional formulas. We have chosen the symbols S and S^2 to indicate the uncorrected *sample* values. Students who read other statistics texts should be warned to note the particular system of symbols the authors have adopted.

slightly larger than the variance of the sample. Further, the difference between s^2 and S^2 will be greater in samples with small N's than with large ones. This is as it should be, since the degree to which a sample variance underestimates the population variance is inversely related to sample size, so that a greater correction is required when small samples are used to estimate population variances than when large ones are used.

In order to obtain an estimate of the population σ from sample data, we simply find the square root of s^2. Thus, the **preferred formula** for estimating the population σ from a sample is:

$$s = \sqrt{\frac{\sum x^2}{N - 1}}$$

(5.5)

Formula 5.4 for s^2 yields a completely unbiased estimate of the population σ^2. For a rather technical mathematical reason, Formula 5.5 for s does not yield a completely unbiased estimate of the population σ. The degree of bias that remains, however, is small.

The procedure of using sample data to obtain a single estimate (or point) of the value of a population parameter, in this instance σ^2 and σ, is technically known as *point estimation*. We also call your attention to another technical concept called *degrees of freedom*, symbolized as *df*. The expression N – 1 in the denominators of the equations for s^2 and s is equivalent to the number of degrees of freedom (*df*) associated with the sample. The denominator of the equation for estimating the population variance can thus be written as *df* instead of N – 1, so that $s^2 = \sum x^2/df$. Since degrees of freedom are frequently used in the process of making inferences about population parameters, we digress at this point to explain the concept further.

Degrees of Freedom (*df*)

The number of values that are free to vary in a distribution, once certain requirements are placed upon the series of quantities, is known as degrees of freedom. To illustrate, assume a distribution of five numbers, 17, 18, 19, 22, and 24. The $\overline{X}$ of these numbers is 20. Assume that this value of 20 is definitely true. How many of the original set of five numbers can be changed and still meet the requirement that $\overline{X} = 20$? This question can be answered most simply by first noting that if $\overline{X}$ is to be equal to 20, the sum of the five numbers must equal 100. Suppose we change the first four numbers to 10, 15, 16, and 21. Their sum is 62. If $\sum X$ is to be 100 and $\overline{X}$ is to be 20, as before, then the fifth number must be 38. Or suppose that we have three numbers whose $\overline{X}$ is 10. Pick any two numbers, let us say 14 and 31, and the third is fixed. In this instance it must be – 15 so that $\sum X = 14 + 31 - 15 = 30$ and $\overline{X} = 10$. As these examples illustrate, after the mean is known we are free to vary the values of all but one score in a distribution and still obtain the specified mean. The degrees of freedom (*df*) in a single distribution of numbers after the mean is known is therefore N – 1.

A Brief Review of the Measures

By this time you may be confused by the various kinds of standard deviations we have discussed and the different symbols we have used to refer to them. As you become experienced with problems involving samples and populations, the distinctions among them should become quite clear. We can at least try to hasten the process by stating once more the three kinds of measures we have presented and the formulas for each:

A. Standard deviation of a *population* of measures (σ):

$$\sigma = \sqrt{\frac{\sum x^2}{N}} = \sqrt{\frac{\sum(X - \mu)^2}{N}} \tag{5.6}$$

where μ is the population mean and N is the number of cases in the population.

B. Standard deviation of a *sample* of measures from some population (S):

$$S = \sqrt{\frac{\sum x^2}{N}} = \sqrt{\frac{\sum(X - \overline{X})^2}{N}} \tag{5.3}$$

where $\overline{X}$ is the sample mean and N is the number of cases in the sample.

C. *Estimate* of an unknown population σ from the data of a sample (s):

$$s = \sqrt{\frac{\sum x^2}{N-1}} = \sqrt{\frac{\sum(X - \overline{X})^2}{N-1}} \tag{5.5}$$

where -1 in the denominator is a correction of the sample S for underestimation of the population σ.

We return now to a consideration of the standard deviation as a purely descriptive index of the amount of variability among a group of individuals whom we have actually measured. For convenience, we will assume in our discussion that we are dealing only with *samples* from some population.

FURTHER COMPUTATIONAL TECHNIQUES

Computation of S from Raw Scores

The use of deviation scores (x) is fairly tedious and, because decimals usually result when $\overline{X}$ is subtracted from X to obtain x, inaccuracies due to rounding and calculational errors are likely to creep in. Instead of using the deviation method of computing the standard deviation, it is usually more convenient to use what is commonly called a raw-score formula. This formula is as follows:

$$S = \sqrt{\frac{\sum X^2}{N} - \left(\frac{\sum X}{N}\right)^2} = \sqrt{\frac{\sum X^2}{N} - \overline{X}^2} \tag{5.7}$$

Origin of Raw-Score Formula. It will be worthwhile to see just how this formula came about. Knowing this, you will have at least one formula that need not be taken on faith and you will get a little understanding of the derivation of formulas. We start, of course, with the basic formula:

$$S = \sqrt{\frac{\sum x^2}{N}}$$

The first step is to square both sides of the equation, obtaining the variance $S^2 = \sum x^2/N$. We know that $x = X - \overline{X}$, so we substitute this to get:

$$S^2 = \frac{\sum (X - \overline{X})^2}{N}$$

Now, to expand the term, by squaring $X - \overline{X}$:

$$S^2 = \frac{\sum (X^2 - 2\overline{X}X + \overline{X}^2)}{N}$$

Placing the summation sign and N with each term gives:

$$S^2 = \frac{\sum X^2}{N} - \frac{2\overline{X}\sum X}{N} + \frac{\sum \overline{X}^2}{N}$$

Looking at the second term on the right-hand side, we see that a portion of it is $\sum X/N$, which we know equals $\overline{X}$. So, substituting $\overline{X}$ for $\sum X/N$, we get:

$$S^2 = \frac{\sum X^2}{N} - 2(\overline{X})(\overline{X}) + \frac{\sum \overline{X}^2}{N}$$

which reduces to:

$$S^2 = \frac{\sum X^2}{N} - 2\overline{X}^2 + \frac{\sum \overline{X}^2}{N}$$

Now, look at the numerator in the last term, $\sum \overline{X}^2/N$. We know that $\sum$ means summation from the first to the last score in a distribution, the sum of N scores. Thus $\sum \overline{X}^2$ says that we should sum $\overline{X}^2$ together N times, so the last term may be rewritten $N\overline{X}^2/N$. If these N's are canceled, we have:

$$S^2 = \frac{\sum X^2}{N} - 2\overline{X}^2 + \overline{X}^2$$

and then, $-2\overline{X}^2 + \overline{X}^2 = -\overline{X}^2$, so the variance is given by:

$$S^2 = \frac{\sum X^2}{N} - \overline{X}^2$$

Taking the square root of both sides produces the formula for raw-score calculation of the standard deviation:

$$S = \sqrt{\frac{\sum X^2}{N} - \overline{X}^2}$$

The final step is to give concrete proof that this raw-score formula is equivalent to the basic formula, $\sqrt{\sum x^2 / N}$. In Table 5.2 we have used the raw-score method to compute S for the girls' data shown in Table 5.1. The S in Table 5.2 is exactly the same value (2.83) we found by the deviation method in Table 5.1.

Computation of s from Raw Scores

It is also possible to use a raw-score formula to calculate s, a statistic that, you will recall, is an estimate of the population sigma based on sample data. This formula is as follows:

$$s = \sqrt{\frac{\sum X^2 - (\sum X)^2 / N}{N - 1}} \tag{5.8}$$

The application of this formula may be illustrated by the data of Table 5.2, in which $\sum X$ and $\sum X^2$ have already been determined. Entering these values into the equation, we have:

$$s = \sqrt{\frac{946 - (84)^2 / 8}{8 - 1}} = \sqrt{\frac{946 - 882}{7}} = \sqrt{9.14} = 3.02$$

Table 5.2 Calculation of the Standard Deviation by the Raw-Score Method

X	X²
15	225
14	196
12	144
10	100
10	100
9	81
8	64
6	36
$\sum X = 84$	$\sum X^2 = 946$

$$\overline{X} = \frac{84}{8} = 10.5$$

$$S = \sqrt{\frac{\sum X^2}{N} - \overline{X}^2} = \sqrt{\frac{946}{8} - (10.5)^2}$$

$$= \sqrt{118.25 - 110.25} = \sqrt{8} = 2.83$$

Since the formula for s attempts to correct for the fact that the S of a sample tends to underestimate the population sigma, the estimate of the population value we just obtained is, as expected, somewhat higher than the standard deviation of the sample found in Table 5.2 (3.02 versus 2.83).

Raw-Score Formulas for Grouped Data

The raw-score formulas may also be modified to be used with grouped data. For S, the raw-score formula is as follows:

$$S = \sqrt{\frac{\sum fX^2}{N} - \left(\frac{\sum fX}{N}\right)^2} \qquad (5.9)$$

where X is the midpoint of each class interval.
For s, the formula is:

$$s = \sqrt{\frac{\sum fX^2 - (\sum fX)^2/N}{N - 1}} \qquad (5.10)$$

An illustration of the application of these formulas is worked out in detail in Table 5.3 In this table we have placed the f column before the X column, which shows the midpoints of the class intervals, so that multiplication is always for adjacent columns. After

Table 5.3 Calculation of S and s by Raw-Score Method with Grouped Data

Class Interval	f	X	fX	fX²
23–25	1	24	24	576
20–22	3	21	63	1323
17–19	5	18	90	1620
14–16	7	15	105	1575
11–13	8	12	96	1152
8–10	4	19	36	324
5–7	2	6	12	72
	$\sum f = 30$		$\sum fX = 426$	$\sum fX^2 = 6642$

$$\bar{X} = \frac{\sum fX}{N} = \frac{426}{30} = 14.20$$

$$S = \sqrt{\frac{\sum fX^2}{N} - \bar{X}^2} \qquad\qquad s = \sqrt{\frac{\sum fX^2 - (\sum fX)^2/N}{N - 1}}$$

$$= \sqrt{\frac{6642}{30} - (14.20)^2} \qquad\qquad = \sqrt{\frac{6642 - (426)^2/30}{30 - 1}}$$

$$= \sqrt{221.4 - 201.64} \qquad\qquad = \sqrt{\frac{6642 - 6049.2}{29}}$$

$$= \sqrt{19.76} = 4.45 \qquad\qquad = \sqrt{20.44} = 4.52$$

multiplying each X by its corresponding f, we sum the fX's here and enter the result in the formulas. Our next task is to determine ΣfX^2. To do this, we add the fX^2 column. The entries in this column are obtained by multiplying each X by its corresponding fX, since $(X)(fX)$ is equivalent to $(f)(X)(X)$ or fX^2. We sum these quantities to get ΣfX^2, which is then substituted directly into the formulas. A comparison of the values that have been calculated for S and s in Table 5.3 indicates that just as it should be, s, the corrected estimate of σ, is larger than the uncorrected S, 4.52 versus 4.45.

z SCORES

We now know what the standard deviation is: the square root of the mean of the x^2's. The standard deviation is thus a rather special kind of "average," telling us the "average" number of *score units* by which individuals deviate from the mean. Once we have calculated the standard deviation of the distribution, is there any way we can picture or visualize it? Can we find any additional use for the standard deviation besides describing a distribution's variability? Again, we will work up to the answers slowly.

The distribution of final exam scores in a large biology class turns out to have a mean of 65 and S of 10. Thus, the "average" amount by which individuals deviate from $\overline{X}$ is 10 units. Lou has a score of 85. Is that good or bad? We see that Lou's score is 20 points above the mean ($x = 85 - 65 = 20$); Lou deviates from the mean twice as much as the "average" individual. With a score of 85, Lou did quite well indeed. How about Pat with a score of 67? Pat's performance was middling, slightly above the mean but deviating from it by only 2 points ($x = 67 - 65 = 2$). This deviation is far less than an "average" amount—to be precise, only $\frac{2}{10}$ of an average amount. Finally, what about Bill, with a score of 35? We can only conclude that Bill bombed biology. With a score 30 points below the mean, three times worse than the "average" deviation, Bill's performance was obviously abysmal.

What we have done in each of these examples is to use the standard deviation as a kind of measuring unit, translating each raw score into a "standard deviation score." Lou, with a raw score of 85, had a standard deviation score exactly 2 units above the mean (a deviation of 20 points from the mean is twice the S of 10). Bill, with a raw score of 35 and x of 30 ($x = 65 - 35$), scored 3 standard deviation units below the mean. Pat's score of 67 fell $\frac{2}{10}$ of a standard deviation unit above the mean of 65. We could, of course, take *any* raw score from the distribution and express its position in standard deviation units. What we have created, then, is an alternative to the raw-score scale, using the standard deviation as the unit of measurement instead of the raw score.

This is not our first discussion of an alternative to the raw-score scale. In previous chapters we discussed using percentiles for this purpose and methods for translating raw scores (points on the raw-score scale) to percentiles (points on the percentile scale) and back again. We can also formally describe a method that permits us to go back and forth between the raw-score scale and the standard deviation scale. Before doing so, let us see how this scale can be pictured and how to lay one out.

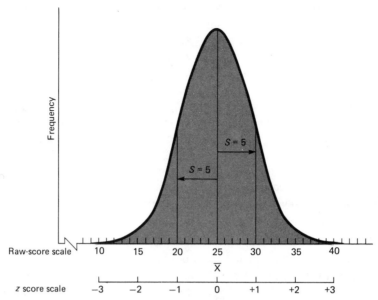

Figure 5.2 Raw-score scale and standard deviation or z score scale for a sample distribution in which $\overline{X}$ = 25 and S = 5.

Figure 5.2 shows a frequency polygon whose mean has been calculated to be 25 and whose S is 5. As usual, the baseline of the polygon shows raw scores; that is, the baseline represents a *scale*, divided into steps that have a width of one raw-score unit. Below the raw-score scale we have placed a scale whose steps have a width of *one standard deviation unit*. Exactly how wide is this unit? It is, of course, as wide as S has been calculated to be for a given distribution. In Figure 5.2, in which S is 5, one standard deviation unit is therefore 5 raw-score units wide. (This is comparable to saying that a foot is 12 inches wide.)

The standard deviation scale is also known as the z scale and the scores on it as z scores. In marking off the z scale, we start at the *mean* of the distribution and work in both directions. In the distribution in Figure 5.2, which has a mean of 25, and, by definition, a mean z value of zero, we therefore start at the point that is equivalent to 25 on the raw-score scale. The S is 5 units. In laying off the upper portion of the scale, we therefore go up from $\overline{X}$ to a point that is equivalent to 5 units on the raw-score scale. The distance between the mean and this point is one standard deviation unit, and the point has a value, in z units, of +1. Going up the scale the equivalent of 5 more raw-score units brings us to a z score of +2, and another 5 units to +3. The raw-score equivalents of these three z scores, we can see, are 30 ($\overline{X} + S = 25 + 5$), 35 ($\overline{X} + 2S = 25 + 10$), and 40 ($\overline{X} + 3S = 25 + 15$).

Notice in Figure 5.2 that almost all the cases that lie above $\overline{X}$ fall between the $\overline{X}$ and a z score of +3. Assuming that the total N is large, this is typically true of all symmetrical bell-shaped distributions, such as the idealized curve shown in the figure. Thus, at least when N is some finite number, approximately three standard deviations will cover all or almost all the distance between $\overline{X}$ and the highest score in the distribution.

We use the same procedure to mark off a succession of standard deviation widths below $\overline{X}$. Going 5 raw-score units below $\overline{X}$ brings us to a z score of -1, 10 units to a z score of -2, and 15 units to -3. Since the distribution is symmetrical, we are not surprised to find that three standard deviations also cover the distance between $\overline{X}$ and the lowest raw score in this sample of measures.

We are now aware that the standard deviation is a unit of width that is some multiple of the raw-score width. For any given distribution, the standard deviation is a calculable value of fixed size, but it may vary from one distribution to another. Now, we are accustomed to think of our measuring scales for linear dimensions such as width to be fixed and unalterable. A yard is a yard and a meter is a meter, and neither varies in size. In the case of the standard deviation as a width measure, however, we must change this conception. To repeat, the standard deviation is a given value for a specific distribution, but may vary from distribution to distribution.

Finding z from X

We are now ready to discuss the procedures we use to translate back and forth between the scales, finding the z-score equivalent of raw scores and vice versa. Take first a simple and familiar example: If we want to translate feet into yards, we divide the number of feet by 3. If we want to change ounces to pounds, we divide the number of ounces by 16. In the same fashion, if we want to change a raw score into a z score, we find the distance that score is from $\overline{X}$ and divide it by the standard deviation. In Figure 5.2, for example, where $\overline{X} = 25$ and $S = 5$, what is the z-score equivalent of a raw score of 28? This score lies 3 raw-score units above $\overline{X}$ ($28 - 25 = +3$); dividing this distance by the S of 5 gives us a z score of $+.60$. If we put these steps into equation form, we have:

$$z = \frac{X - \overline{X}}{S} \tag{5.11}$$

But, since $X - \overline{X}$ is the definition of x, we can also write the formula as:

$$z = \frac{x}{S}$$

Let's work out a few examples from a distribution with a mean of 42 and S of 8. For a raw score of 54:

$$z = \frac{X - \overline{X}}{S} = \frac{54 - 42}{8} = \frac{12}{8} = +1.5$$

If the raw score is below $\overline{X}$, the z score takes a minus value. For example, in the same distribution, what is the z score of a raw score of 36?

$$z = \frac{36 - 42}{8} = \frac{-6}{8} = -.75$$

Finding X from z

We have just seen how to translate a raw score into a z score. We can reverse this procedure, so that if we know a z score we can obtain a raw score. We could use exactly the same formula as before and solve for an unknown raw score. However, it is more convenient to rearrange the formula so that the raw score X becomes the unknown on the left-hand side of the equation. This formula is:

$$X = \overline{X} + z(S) \tag{5.12}$$

For example, with $\overline{X} = 42$ and $S = 8$, what is the raw-score equivalent of a z score of -1.5? Applying Formula 5.12, we find:

$$X = 42 + (-1.5)(8) = 42 - 12$$
$$= 30$$

We do not change raw-score distances into z scores and back again merely to get practice in arithmetic. These z scores have very definite uses. For example, the need sometimes arises to compare the same persons with themselves in two different distributions of scores. If a group of people take a test of clerical aptitude and a test of mechanical aptitude, we might want to know the relative standing of an individual on both tests. If both tests have the same $\overline{X}$ and S, and the distributions are similar in shape, the raw scores can be compared directly. This would be a rare case. More likely, the two distributions will have different $\overline{X}$'s and S's. Thus, a particular raw score on one test will probably mean something quite different from the same raw score on another.

One satisfactory way of comparing an individual across distributions is by comparing ranks. This, in effect, is what a percentile rank is. If a person has a percentile rank of 78 on one test and of 62 on the other, we know the person has done better on the first test than on the second. We may use z scores for this same purpose of comparing two scores in distributions in which the $\overline{X}$'s and S's are different. Suppose that a group is given the mechanical and clerical aptitude tests as suggested above, with the following result:

Test	$\overline{X}$	S
Mechanical	100	10
Clerical	60	6

A man gets a raw score of 69 on the clerical test and 75 on the mechanical test. His z score for the former is:

$$z = \frac{x}{S} = \frac{9}{6} = +1.5$$

and for the latter:

$$z = \frac{-25}{10} = -2.5$$

He is much worse on the mechanical aptitude test than on the clerical test, even though his raw score is higher for the mechanical test. That is, his z score on the mechanical test is nearly at the bottom of the distribution, whereas it is considerably above $\overline{X}$ on the clerical test. The pattern of aptitude shown by this individual suggests that he might do well in a job requiring clerical skills but ought to leave anything mechanical to other people.

Properties of z Scores.   It is useful to know certain properties of a set of z scores. First, the *mean of a set of z scores is always zero*. This relationship is expressed as

$$\bar{z} = \frac{\sum z}{N} = 0 \tag{5.13}$$

This relationship can be verified by considering the z score formula: $z = (X - \overline{X})/S$. When X takes the same value as $\overline{X}$, this equation can be rewritten: $z = (\overline{X} - \overline{X})/S$. Since the numerator is 0, it follows that the z of a score that falls at $\overline{X}$ is 0. This in turn implies that the mean of the z scores is zero.

The second property concerns the variance and standard deviation of a set of z scores. The value of the variance and the standard deviation of a set of z scores is always one. Thus:

$$S_z{}^2 = S_z = 1 \tag{5.14}$$

This value can be demonstrated mathematically by first rewriting the general formula for S^2, $\Sigma(X - \overline{X})^2/N$, to indicate that we are now dealing with z scores: $\Sigma(z - \bar{z})^2/N$. Since $\bar{z} = 0$, this reduces to $\Sigma z^2/N$. It can be demonstrated by a proof we will not show here that Σz^2 is equal to N. Thus:

$$S_z{}^2 = \frac{\sum z^2}{N} = \frac{N}{N} = 1$$

and

$$S_z = \sqrt{1} = 1$$

To restate these properties, the mean of a set of z scores is always equal to 0, and the standard deviation and variance are always equal to 1. These relationships are illustrated in Figure 5.2, in which it is shown graphically that $\overline{X}$ has a z score equivalent of 0 and the scores one sigma unit from $\overline{X}$ have z score equivalents of ±1.00.

Standard Scores.   A z score is also called a *standard score* because it indicates relative standing in a distribution of measures without regard to the particular units of measurement in the original set of observations. A z score of +.50, for example, indicates a score that is somewhat above average, whether the distribution represents the heights of a group of individuals measured in inches, their weights in pounds, or their performance on an 80-item personality test.

Other types of standard scores based on z scores are often used to report the results of widely used tests, such as clinical tests of personality or nationally used tests of academic achievement. Typically, the distribution of raw scores from which the z scores are derived involve extremely large samples so that, for all practical purposes, the mean and standard deviation of the raw scores can be considered to be population values.

One type of standard score derived from z is known as the T score. T scores are obtained by multiplying each z score by 10 and then adding 50 to each of the resulting scores:

$$T = 10(z) + 50$$

This procedure removes decimals and negative numbers and results in a distribution with a mean of 50 and a standard deviation of 10. That is, when $z = 0$, $T = 10(0) + 50 = 50$, and when $z = \pm1.00$, $T = 10(\pm1) + 50 = 40$ or 60.

A second system of standard scores that is often used multiplies each z by 100 and adds 500 to the result: $100(z) + 500$. The outcome is a distribution of scores with a mean of 500 and a standard deviation of 100.

As may be inferred from these examples, it is possible to transform any set of z scores into a set of standard scores in which the mean (μ) and the standard deviation (σ) each take any value that one specifies beforehand. The general transformational formula is thus:

$$\text{Standard score} = (\text{specified } \sigma)\, (z) + (\text{specified } \mu) \tag{5.15}$$

In all such conversions, the shape of the original distribution of scores is retained and the interpretation of the scores remains the same as for z scores. For example, a z score of -2.00 indicates a performance that is much below average. The equivalent score T is 30; that is, with a standard deviation of 10 and a mean of 50, 30 is two standard deviation units below the mean. If the standard deviation had instead been set at 50 and the mean at 200, the equivalent standard score would be 100, still a score that is two standard deviation units below the mean.

A CHECK ON CALCULATIONAL ERRORS: STANDARD DEVIATION AND THE RANGE

In discussing z scores, we noted that the sizes of the range (R) and the standard deviation are related. This suggests that we can check on computational errors by comparing the calculated S with the range of scores in the distribution. We have already seen that in a sample with a large N and a symmetrical bell-shaped distribution, such as the one in Figure 5.2, all or almost all the scores are encompassed by six standard deviation units, three on either side of the mean. For large samples that approximate this curve shape, therefore, you can check your calculated S to see if it is approximately one-sixth of R. After you have calculated several S's you will begin to get the "feel" of them in relationship to the distribution of scores from which they have been calculated. You will quickly find out that if R of your distribution is 24 and your calculated S is 25, something is wrong with your calculations, since S is larger than R. Or, if you have R of 24 and you obtain S of 1, something also is wrong.

But we must stress now, and stress emphatically, that we do not calculate S by taking one-sixth of R; rather, *after* we have calculated S we check to see if its value is approximately one-sixth of R. This provides only a very crude check, unless the number of measures is very large and the distribution quite symmetrical and shaped like Figure 5.2. However, so many sets of data are distributed like those of Figure 5.2 that the check has considerable usefulness.

DEFINITIONS OF TERMS AND SYMBOLS

Semi-interquartile range. A measure of variability obtained by finding $(Q_3 - Q_1)/2$, half the score distance from the first quartile to the third quartile.

Deviation (x). The difference between a given score and the mean of the distribution to which it belongs: $(X - \overline{X})$.

Average deviation. A measure of variability obtained by finding the mean (average) of the *absolute* deviations of the scores in a distribution from $\overline{X}$: $\Sigma |x|/N$.

Variance. A measure of variability obtained by finding the mean of the *squared* deviations of the scores from $\overline{X}$: $\Sigma (x)^2/N$.

Standard deviation. The square root of the variance.

σ and σ^2. Symbols for the standard deviation and variance of a population.

S and S^2. Symbols for the standard deviation and variance of a *sample* of measures. S and S^2 tend to underestimate the standard deviation and variance of the parent population and are referred to as uncorrected estimates of the population values.

s and s^2. Corrected estimates of the population standard deviation and variance based on sample data, obtained by the formulas $\sqrt{\Sigma x^2/(N-1)}$ and $\Sigma x^2/(N-1)$. The formula for s^2 yields an unbiased estimate of σ^2—that is, on the average, s^2 will equal σ^2. A slight bias remains in s.

Degrees of freedom (df). Number of scores in a distribution whose values are free to vary when the total set of scores meets some constraint, such as the value of the mean. After the mean is known, the number of degrees of freedom in a single distribution of raw scores is N – 1.

z score.   A score expressed in standard deviation units. A z score is obtained by finding the deviation of the score from the mean of the distribution and dividing the result by the standard deviation. For a sample, $z = (X - \bar{X})/S$ and for a population, $z = (X - \mu)/\sigma$. The mean of a distribution of z scores is 0, and the variance and standard deviation are each 1.

Standard scores.   Standard scores are z scores or scores based on z scores in which the transformed distribution has a specified mean and standard deviation and the score corresponding to each z is obtained by the formula: (specified standard deviation) (z) + specified mean. Standard scores are so named because they indicate relative standing in a distribution without regard to the unit for measurement in the original raw scores.

PROBLEMS

1.   A series of pairs of distributions is presented below. For each pair, indicate whether distribution 1 or 2 is *more* variable.
   (a) Dist. 1: 5, 6, 7, 7, 8, 9; Dist. 2: 7, 9, 10, 11, 12, 14.
   (b) Dist. 1: 16, 21, 25, 30, 35, 44; Dist. 2: 144, 146, 149, 153, 154, 160.
   (c) Dist. 1: 23, 25, 27, 31, 32, 34; Dist. 2: 5, 26, 28, 29, 30, 33.
   (d)   Dist. 1   Dist. 2

X	f	X	f
62	1	28	1
61	2	27	1
60	2	26	4
59	5	25	7
58	9	24	11
57	7	23	7
56	4	22	6
55	5	21	2
54	3	20	0
53	2	19	1
	$\Sigma f = 40$		$\Sigma f = 40$

2.   For each of the distributions above, determine the Range (R).

3.   For the pairs of distributions in problem 1(c) and (d), compare the values of R. What do these comparisons tell you about the limitations of R as a measure of variability?

4.   Determine the semi-interquartile range for distribution 1 in problem 1(d).

5.   What do the following trios of symbols represent: σ^2, S^2, and s^2; and σ, S, and s?

6.   Find the variance (S^2) and standard deviation (S) for the distribution of scores below, using deviation (x) scores and Formulas 5.2 and 5.3: 1, 2, 4, 5, 5, 7, 7, 8, 10, 11.

7.   For the distribution in problem 6, find S, using the raw-score method in Formula 5.7. How does this compare with your answer in problem 6?

8. For the distribution in problem 6, find s^2 and s, using Formulas 5.4 and 5.5.

9. For the same distribution, find s by the raw-score method in Formula 5.8. If your answer does not agree with your answer in problem 8, one or both of your calculations are in error.

10. Students were given data from several frequency distributions, each of which had a large N and was close to being symmetrical and bell-shaped, and asked to compute S for each distribution. The answers of one student, along with the range (highest minus lowest score), are given below. Based on your knowledge of the relationship between R and S in symmetrical, bell-shaped distributions, determine for each distribution whether the student's calculated S is probably accurate, too small, or too large.
    a. $R = 99 - 51; S = 8.5$.
    b. $R = 34 - 13; S = 7.2$.
    c. $R = 105 - 46; S = 9.3$.
    d. $R = 49 - 3; S = 3.8$.
    e. $R = 149 - 106; S = -3.6$.

11. A physical anthropologist administers a battery of tests to 200 male high school seniors. The means and standard deviations of these students for a test of reaction time (RT) and a test of lung capacity (LC) are shown below. Each distribution was approximately symmetrical and bell-shaped (N = 200).

$$\overline{X}_{RT} = 34 \qquad \overline{X}_{LC} = 193$$
$$S_{RT} = 4 \qquad S_{LC} = 20$$

    a. Sketch a figure for each distribution, marking off on the baseline the raw-score values of $\overline{X}$ and the raw scores that are one, two, and three standard deviations above and below $\overline{X}$. Below this raw-score scale, add a z-score scale showing the corresponding z values for the raw-score values you have plotted.
    b. The raw scores for three of the men are shown below. For each of these scores, find the z-score equivalent.

	RT	LC
Alex	39	210
Max	31	183
Perry	42	155

    c. On the graphs drawn for (a), show where each student fell in the distribution and compare the relative standing of each student on the two tests.
    d. Three other males received the following z scores. What are their raw scores? Sam: $z_{RT} = +2.50$; Jim: $z_{RT} = -1.75$; Boris: $z_{LC} = +.25$.

12. An arithmetic test has been administered to a very large national sample of fourth-grade students. The mean for the group has been found to be 65 and the standard deviation to be 10.
    a. If raw scores were changed into T scores, what is the T-score equivalent of 80?
    b. If raw scores were converted into standard scores with a mean of 500 and a standard deviation of 100, what is the standard score equivalent of a raw score of 41?
    c. If standard scores were obtained with a mean of 200 and a standard deviation of 50, what is the raw-score equivalent of a standard score of 280?

Correlation

CHAPTER 6

Are the political attitudes of college men related to their fathers' attitudes? That is, do students who are liberal or conservative in comparison to their peers tend to have fathers whose attitudes are liberal or conservative for members of their generation? Do more highly educated women tend to have more or fewer children than less highly educated women? Two professors independently read the essay exams of a class of 10 students and rank them from best to worst. How strongly do they agree in their judgments? Each of these questions asks about the relationship or *correlation* between two variables—*the degree to which the variables are associated or covary.* Each involves distributions of paired scores—for example, each father's score on a political attitudes questionnaire and his son's score on the same test.

Correlations between two variables vary in *direction.* Suppose we ask each member of a group of married women how many children and how many years of education she has. If women with more education have more children than women with less education, there is a *positive* correlation between the two variables, education and family size. If, on the other hand, women with more education have fewer children, there is a *negative* correlation between the variables. The third possibility is that women with more education tend to have neither more children nor fewer children than those with less education; in this case, there is a *zero* correlation between the variables.

Hypothetical examples of these three classes of correlation, positive, negative, and zero, are shown graphically in the three parts of Figure 6.1. The graphs in the figure are called *scatter plots* because they show how the points scatter over the range of possible scores. Every point represents two values, an individual's score on one variable, identified as X, and the same or a paired individual's score on a second variable, identified as Y. Both variables are plotted on the same graph because we are interested in the relationship between them. As established in Chapter 3, values of the X variable are always displayed along the horizontal axis and values of the Y variable along the vertical axis.

Figure 6.1(a) shows the relationship between scores on a measure of self-esteem and a measure of assertiveness. This graph shows that amount of self-esteem is related to amount of assertiveness; individuals with high scores on one measure tend to have high scores on the other, and those with low scores on one tend to have low scores on the other. This, of course, is an example of a *positive* correlation. The next graph, Figure 6.1(b), shows the relationship between high school students' grade point average and average amount of time per week spent watching television. In this scatter plot there is a tendency for students with *high* X scores to have *low* Y scores and vice versa. We therefore say that there is a *negative* correlation between X and Y; those who spend a lot of time watching TV tend to have a low grade point average and those with a good average tend to spend little time watching TV.

Figure 6.1(c) shows a scatter plot of two variables with virtually no relationship between them. Each X score represents the time taken to complete an introductory psychology test, and the Y score represents the same person's grade on that test. Examination scores do not vary consistently in relation to the length of time spent by the students taking the examination. We describe this lack of relationship as reflecting an essentially *zero* correlation between X and Y.

We have just seen that the data of Figure 6.1(a) and 6.1(b) exhibit different *direc-*

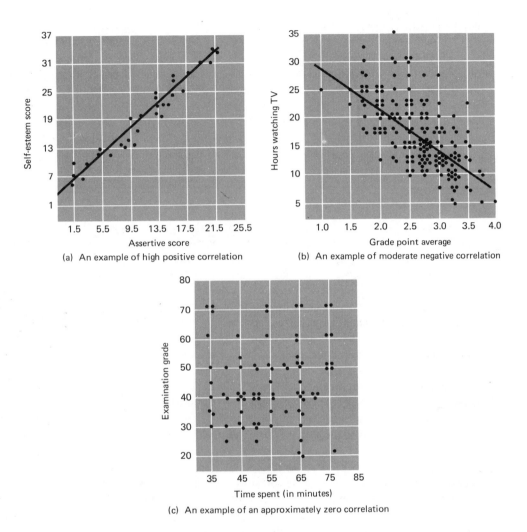

(a) An example of high positive correlation

(b) An example of moderate negative correlation

(c) An example of an approximately zero correlation

Figure 6.1 Scatter plots showing three types of correlation between X and Y.

tions of correlation—positive and negative, respectively. When we look at the graphs carefully, we note another difference. In Figure 6.1(a) we have drawn a straight line in such a way that it falls close to as many points as possible. Most of the points do not lie far from the straight line, which indicates a very strong positive relationship between the two variables. If the relationship were perfect—that is, could not be improved upon—all the points would fall exactly along a straight line.[1] Similarly, all the points in a perfect nega-tive correlation would fall along a straight line, except that high scores on the X measure would be paired with low scores on Y, and vice versa.

If we look at Figure 6.1(b), however, we observe that the straight line drawn through the points describes them fairly well, but that a number of points lie at quite a distance from that line. The two variables, we see, are not as strongly correlated as in Fig-ure 6.1(a). In Figure 6.1(c), no straight line would be particularly successful. The rela-tionship between the variables is essentially zero. Thus, the second way in which scatter plots may differ is with respect to the amount or *degree* of correlation. Correlations may go from a perfect relationship, in which all the points lie on a straight line, to a zero rela-tionship, in which the points are so scattered that no X shows a greater tendency to be paired with one Y more than any other, and vice versa.

Correlation and Causation

The demonstration that two variables are correlated demonstrates *only* that they covary, that changes in the values of one tend to be associated with changes in the other. We can-not infer from the correlation alone that one of these variables has caused or brought about the other. There are, of course, instances in which additional evidence suggests that variations in X do have some causal effect on Y. For example, the time since individ-uals last ate (X) is correlated with how hungry they report they are (Y). We know that hunger pangs are brought about by lack of food, so it is plausible to assume that X causes Y but not the reverse: changing your report about how hungry you feel does not affect when you last ate. But, as this example illustrates, the causal path may not be direct or immediately obvious. A host of physiological events are taking place over time to affect the sensation of hunger. We do not know from the correlation, then, exactly how depri-vation of food (X) brings about changes in reported hunger (Y).

While correlations may reflect some kind of causal link between X and Y, they more frequently occur because X and Y are both influenced by a common or a similar set of factors. The number of pairs of shoes a child has may be negatively correlated with the number of his brothers and sisters, but it is obvious that neither variable directly brings about the other. From our general knowledge we know that both are affected by a group of factors that centrally involve family income per family member.

We repeat our earlier statement. The mere demonstration that variables are corre-lated does not, in and of itself, tell us anything about causality; it shows us that variables tend to covary but not why. The "why" must be sought in additional kinds of data or in additional kinds of analyses of networks of correlations.

Unfortunately, there is often a temptation to infer a cause-effect relationship from no more than the correlation itself. A ludicrous example of the confusion resulting from an attempt to show a causal relation between two variables by correlational research was

---

[1]We actually are discussing here correlations that are *linear* in nature. That is, a straight line de-scribes the points in the scatter plot better than a curved line. [See pages 85–86 for a discussion of curvilinear correlation and Figure 6.3(b) for an example.] With a perfect *linear* correlation, all points fall along a straight line.

once given by Willoughby.[2] In attacking another scholar's argument that a high positive correlation between vocabulary and college grades meant that improvement in vocabulary would produce an improvement in grades, Willoughby stated that by the same reasoning a high positive correlation between the height of boys and the length of their trousers would mean that lengthening trousers would produce taller boys. In neither of these cases have we evidence proving that the manipulation of one variable directly controls the other.

Although correlation coefficients do not show what factors are responsible for the relationship between two variables, correlational techniques are nevertheless valuable for describing such relationships. The primary purpose of this chapter is to describe two of the most frequently used correlational techniques and to show how to make predictions of one variable from another when the correlation between the two is known. The latter technique enables us to do such things as predicting students' success in college from their entrance test scores or the adult heights of children from their heights at age 6. Predictions of some accuracy are possible in these instances, because the pairs of variables are known to be correlated.

The Pearson Product-Moment Correlation Coefficient (*r*)

Earlier in this chapter we developed in a general way the meaning of the terms positive and negative correlation and of relationships that are high, relatively low, or zero in magnitude. We also saw how, by inspecting a scatter plot, we often can specify the direction of the correlation between X and Y and make some kind of statement about its magnitude. Descriptions based on observing scatter plots, however, are not very precise. To overcome this deficiency we introduce a new concept, the *correlation coefficient.* This coefficient will have a specific numerical value for any given set of paired data. Positive values will correspond to what we have called positive correlation, negative values to negative correlation. Furthermore, high values of the coefficient, regardless of whether they are positive or negative, will correspond to what we have called high correlation, and low values to low correlation. What would be called zero correlation when seen on a scatter plot may not in fact have a correlation coefficient of exactly zero, although the value will surely be near zero.

In practice, statisticians use several different correlation coefficients, depending on the type of data involved. In this chapter we will discuss two coefficients that describe the linear (straight line) correlation between variables. The first correlation coefficient we will discuss is used with measurement data—that is, numerical scores indicating the amount of the characteristic each individual exhibits. The technical name of this coefficient is the *Pearson product-moment correlation coefficient*, named in honor of Karl Pearson, one of the great pioneers of statistics, and symbolized as *r*. Because product-moment correlation coefficient is a highly technical term whose meaning is not essential to the research worker, we will not explain its origin. The three expressions—Pearson cor-

---

[2]R. R. Willoughby, Cum hoc ergo propter hoc. *School and Society*, *51*, 1940, 485.

relation coefficient, product-moment correlation coefficient, and *r*—will be used interchangeably in this chapter.

Basic Formula for *r*

The Pearson correlation coefficient, *r*, is defined as the *mean of the products of the z scores for the X and Y pairs.* The mean of any series of quantities, you recall, is equal to the sum of the quantities divided by their number. Thus, the definition of *r* is stated algebraically as:

$$r = \frac{\sum z_X z_Y}{N} \tag{6.1}$$

$$\text{where} \quad z_X = \frac{X - \overline{X}}{S_X}$$

$$z_Y = \frac{Y - \overline{Y}}{S_Y}$$

N = the number of pairs

In determining *r*, we first calculate the *z* score for each X and for its paired Y. Next we multiply each z_X score by its z_Y score, giving us a ($z_X z_Y$) product for each pair of scores. Then we add up all these cross-products; that is, find $\sum z_X z_Y$, and divide by N. Thus, as we stated above, *r* is the mean of the *z*-score cross-products.

Now that we have a definition of *r*, it is appropriate to ask if the *r* values we obtain with this definition will have the same size and direction that our early statements about correlation suggest. We have said, for example, that a positive correlation will occur if relatively high X's and Y's tend to be paired and if relatively low X's and Y's tend to be paired. Can we say that *r* will be positive when these conditions are met? To answer this question, we will study how the value of *r* changes for different arrangements of *z* scores.

Three scatter plots are shown in Figure 6.2, the first showing a correlation that is close to zero, the second showing a moderate positive correlation, the third a high negative correlation. The scores that are plotted represent *z* scores, rather than raw scores, and each plot is marked off into quadrants. The *z* scores falling in the quadrant at the upper right of each plot are above the mean on both the X and Y variables and are therefore both positive in sign. The *z* score cross-products of the pairs of scores falling in this quadrant are thus positive as well. The *z* scores falling in the quadrant at the lower left are both negative and, since the multiplication of two negative numbers results in a positive number, their cross-products are all positive. In the two remaining quadrants, one of the *z* scores is positive and the other is negative, so that all the cross-products are negative in sign.

In Figure 6.2(a), which represents a correlation near zero, we see that the points are quite evenly distributed among the four quadrants. When the cross-products are added together, the negative cross-products will cancel out the positive products, and the sum will be very close to zero. The mean of the cross-products, *r*, will also be approximately

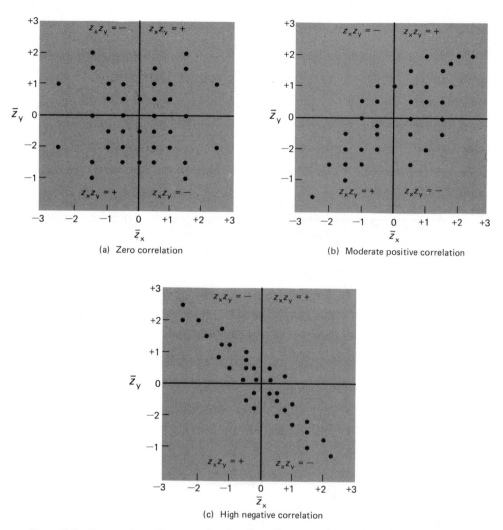

Figure 6.2 Scatter plots of *z*-scores showing signs of *z*-score cross-products in each quadrant.

zero. In Figure 6.2(b), most of the points fall into the two positive quadrants but quite a few fall into the two negative quadrants. The sum of the cross-products will therefore be positive in sign but, because of the frequency of negative cross-products, rather low in magnitude. Correspondingly, the mean of the cross-products, r, will also be positive but low in value. In Figure 6.2(c), almost all the points fall into the two negative quadrants. The sum and the mean of the cross-product will therefore be negative in sign and higher in absolute value than in the case of Figure 6.2(b).

A perfect positive correlation takes the value of +1.00 and a perfect negative correlation the value of -1.00. In order to understand why this is so, recall that in the scatter diagram of a perfect linear correlation, all the points lie along a straight line. When the correlation is perfect positive, this implies that the position of each paired X and Y in their respective distributions is the same: the individual who scores highest on X is highest on Y, the individual who scores at the mean of the X distribution also scores at the mean of Y, and so on. This in turn indicates that for each pair, $z_X = z_Y$, so that $z_X z_Y$ can be expressed as z^2. When there is a perfect positive correlation, the formula for r therefore can be rewritten as $\Sigma z^2/N$. This expression is also the formula for the *variance* of a set of z scores. The variance of a set of z scores, we demonstrated in Chapter 5, is always equal to 1.00. The value of r for a perfect positive correlation therefore is $\Sigma z^2/N = 1.00$. Parallel reasoning leads to the demonstration that a perfect negative correlation takes the value of -1.00. The value of r, then, ranges from $+1.00$ for a perfect positive correlation through zero for no correlation to -1.00 for a perfect negative correlation. This restriction of the range of possible coefficient values is extremely fortunate, because *it permits a direct comparison of* r *values obtained from widely different sets of data without a correction for the size of the original X and Y values.* In this respect r is similar to z.

Two examples of the computation of r with the basic formula are presented in Table 6.1. In case 1 most of the pairs of z's for X and Y are either both positive or both

Table 6.1   An Example of the Way r Depends upon the Pairing of z Scores

Case 1: Most $(z_X z_Y)$ products positive			Case 2: Most $(z_X z_Y)$ products negative		
z_X	z_Y	$(z_X z_Y)$	z_X	z_Y	$(z_X z_Y)$
1.4	1.3	1.82	-1.4	1.3	-1.82
1.9	1.3	2.47	-1.9	1.3	-2.47
1.0	1.1	1.10	-1.0	1.1	-1.10
.0	.5	.00	.0	.5	.00
$-.7$	.4	$-.28$	.7	.4	.28
$-.8$	$-.4$	.32	.8	$-.4$	$-.32$
$-.2$	$-.5$	.10	.2	$-.5$	$-.10$
$-.9$	-1.1	.99	.9	-1.1	$-.99$
$-.8$	-1.3	1.04	.8	-1.3	-1.04
$-.9$	-1.3	1.17	.9	-1.3	-1.17

$$\Sigma (z_X z_Y) + \text{'s} = +9.01$$

$$\Sigma (z_X z_Y) - \text{'s} = -.28$$

$$\Sigma (z_X z_Y) = 8.73$$

$$r = \frac{\Sigma z_X z_Y}{N}$$

$$r = \frac{8.73}{10}$$

$$r = .873$$

$$\Sigma (z_X z_Y) + \text{'s} = +.28$$

$$\Sigma (z_X z_Y) - \text{'s} = -9.01$$

$$\Sigma (z_X z_Y) = -8.73$$

$$r = \frac{\Sigma z_X z_Y}{N}$$

$$r = \frac{-8.73}{10}$$

$$r = -.873$$

negative, producing positive values of $(z_X z_Y)$ and a positive $\sum (z_X z_Y)$, which in turn produces a positive r of .873.[3] In case 2 the z_X values are the same as those of case 1, except that the z_X values that were positive have been changed to negative and those that were negative have been changed to positive. In other words, persons whose X scores were above $\overline{X}$ in case 1 are now below $\overline{X}$ in case 2, and persons whose X scores were below $\overline{X}$ in case 1 are now above $\overline{X}$ in case 2. Consequently, most of the $z_X z_Y$ products in case 2 are negative, producing a negative r value of $-.873$. The size of r is the same as before, but its direction has been changed from positive to negative. This illustrates that the direction of correlation depends upon whether positive z_X values are paired with positive z_Y values and negative z_X values are paired with negative z_Y values or the reverse.

The Computational Formula for *r*

Despite the ease with which our basic formula for r can be remembered, an equation that permits raw scores (X and Y) to be used in place of z scores is more convenient for computing r. Of the several computational formulas available, a very useful version is shown below:

$$r = \frac{\sum XY - N\overline{X}\,\overline{Y}}{\sqrt{\sum X^2 - N\overline{X}^2}\,\sqrt{\sum Y^2 - N\overline{Y}^2}} \tag{6.2}$$

All the components in the formula are familiar except for the expression $\sum XY$. This value is found by multiplying each X by the Y value paired with it and summing these XY products.

Table 6.2 shows how to use the computation formula to find r. The table shows the scores of 10 students in a statistics class on their first and final examinations. The r for these two distributions is .363. We may conclude that there was a moderate tendency for students who did well in comparison to their classmates on the first exam to continue to do well on the final, and for students who initially did not do well to continue their poor performance on the final.

What *r* Means

The Pearson product-moment correlation coefficient for a particular set of data measures a specific type of relationship: the *linear correlation* between two variables. By this, we mean that r measures the degree to which a *straight line* relating X and Y can summarize the trend in a scatter plot. Figure 6.3 presents two scatter plots, part (a) exhibiting a linear correlation and part (b) a nonlinear (or curvilinear) relationship between X and Y.

What would happen if we computed r, a measure of linear correlation, for data that have a curvilinear relation? The answer is that we would be measuring how well those

---

[3]In this book no sign is attached to an r value unless it is negative. Consequently, an r value of .873 is positive and an r value of $-.873$ is negative.

Table 6.2 Demonstration of the Use of the Computational Formula for *r*

Student	First exam X	X^2	Final exam Y	Y^2	XY
1	31	961	31	961	961
2	23	529	29	841	667
3	41	1681	34	1156	1394
4	32	1024	35	1225	1120
5	29	841	25	625	725
6	33	1089	35	1225	1155
7	28	784	33	1089	924
8	31	961	42	1764	1302
9	31	961	31	961	961
10	33	1089	34	1156	1122
	$\sum X = 312$	$\sum X^2 = 9920$	$\sum Y = 329$	$\sum Y^2 = 11003$	$\sum XY = 10331$

$$\overline{X} = \frac{\sum X}{N} = \frac{312}{10} = 31.2 \qquad \overline{X}^2 = 973.4$$

$$\overline{Y} = \frac{\sum Y}{N} = \frac{329}{10} = 32.9 \qquad \overline{Y}^2 = 1082.4$$

$$r = \frac{\sum XY - N\overline{X}\,\overline{Y}}{\sqrt{\sum X^2 - N\overline{X}^2}\,\sqrt{\sum Y^2 - N\overline{Y}^2}}$$

$$= \frac{10331 - 10(31.2)(32.9)}{\sqrt{9920 - 10(973.4)}\,\sqrt{11003 - 10(1082.4)}}$$

$$= \frac{10331 - 10264.8}{\sqrt{186}\,\sqrt{179}} = \frac{66.2}{13.64(13.38)} = \frac{66.2}{182.50}$$

$$= .363$$

data approximate a linear relationship. When the curvilinearity is small, a straight line may have some usefulness for summarizing the relationship between X and Y. When there is marked departure from a straight line, however, *r* may be low or even drop to approximately zero. For example, the data of Figure 6.3(b) are best described by the dotted curved line of that figure. The relationship between the variables is actually quite high, but it is so strongly curvilinear that a straight line does not fit the data at all well and Pearson *r* would be small.

Before computing a Pearson *r*, it is therefore wise to inspect the scatter plot to determine the nature of the relationship. When the data show a pronounced curvilinearity, methods for determining the amount of the curvilinear relationship should be applied. These methods, which may be found in more advanced statistical texts, will not be discussed here.

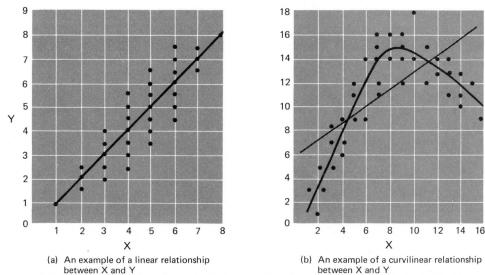

(a) An example of a linear relationship
between X and Y

(b) An example of a curvilinear relationship
between X and Y

Figure 6.3 Scatter plots of two hypothetical sets of data showing different kinds of relationships between X and Y.

USE OF *r* IN PREDICTION

We select at random from a college directory the names of 40 male students. How tall is each of them? If forced to guess, our most sensible estimate is that each of their heights is the same as the mean height of all male students which, let us say, is 68 inches. Men's heights are variable, so it is highly unlikely that all 40 men would be 68 inches; some would be taller and some would be shorter. But our guess of the mean will result, overall, in smaller errors of overestimation or underestimation than any other value. Now suppose we had an additional bit of information about each of these men: the number of pairs of shoes they own. As far as is known, the number of shoes a college student owns has nothing to do with his height, so we would still guess that each of the men is 68 inches tall. But suppose instead that we had information about each of their weights. We are told that Sam Smith, for example, weighs 120 pounds. Commonsense observation suggests that height is related to weight—that is, that there is some degree of correlation between these two variables. We also know that 120 pounds is well below average for male students. We would therefore adjust our guess downward, predicting that Sam is of less than average height. Similarly, if Tom Jones weighs 220 pounds, we would adjust our prediction about his height upward from the mean.

These examples serve as an informal introduction to the use of the correlation coefficient, *r*, in predicting the value of an unknown variable from the value of a known variable. In practice, correlations are often used for this purpose. For example, a college registrar may determine the correlation between college grades (Y) and scores made by high school seniors on a measure of academic aptitude (X). With this correlation, the registrar can predict college grades from aptitude scores, using these predictions to determine which applicants for admission would be most likely to succeed in college and which would be most likely to fail. With this information, the registrar is able to admit only those students with a reasonable chance of success. This is exactly what is done in many colleges. Or an analyst may use an economic indicator, such as number of new housing starts last month, to forecast the state of the economy at some future point, based on knowledge of the correlation of the indicator and later economic conditions.

To predict scores on one variable from scores on another (Y from X or X from Y), we first obtain pairs of scores in which one X and one Y appear in each pair. Let us suppose we want to predict Y, given X. We then determine the best Y value to predict for each X value if the actual Y value were not known to us. The predicted Y's can be compared with the obtained Y values to indicate the success of the predictions. The ultimate usefulness of this technique will be in predicting unknown Y values when only X values are available.

Now that we know our goals in the prediction of one variable from another, we can turn our attention to the method of making such predictions. We will work into this method gradually by first discussing its general logic.

Logic of the Regression Line

In Figure 6.4, we show a scatter plot that represents the relationship between college students' scores on a measure of achievement motivation and their grade point average (GPA). Inspection of the scatter plot suggests that achievement scores (X) and GPA's (Y) are moderately correlated. Also shown in the figure is the mean GPA for the total sample ($\overline{Y}$) and the mean GPA for each set of students who have the same achievement score (X score). These latter means are shown in each column. Note that as the X score increases, the mean Y score in the columns also increases. This is exactly what we would expect, given the positive correlation between achievement score and GPA. We have also drawn a straight line through the scatter plot that comes closest to the data points in the columns. This straight line is called a prediction or *regression* line. Note that the column means also fall quite close to this regression line.

We could use the column means to predict Y scores. For example, in Figure 6.4, the mean GPA of all those who earn a score of 9 on the achievement (X) measure is approximately 2.00; we might predict that anyone whose X = 9 will have Y = 2.00. However, it is reasonable to assume that if we had obtained achievement scores and GPAs for the entire population of students, the column means—the mean GPA for each subgroup of students with the same achievement score—would have fallen exactly on a straight line, and that the deviations of our obtained values from this line are largely due to sampling error. Thus, points on the regression line can essentially be regarded as a set of means and used for prediction instead of the calculated column means.

To illustrate, suppose we know only that a particular student had a score of 17 on the achievement measure. What would we predict her GPA to be? Locating the score 17 on the baseline of Figure 6.4, we go up to the regression line and then over to the Y axis. Our prediction is a GPA of 3.00. We can see that everyone who scores 17 on the achieve-

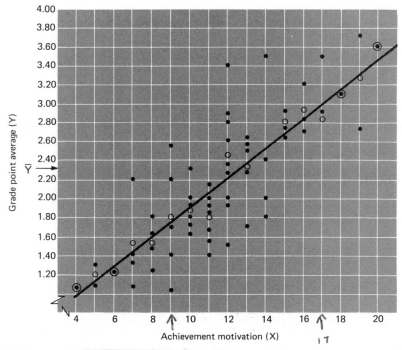

Figure 6.4 Scatter plot relating achievement motivation score (X) to grade point average (Y). Open circles indicate mean Y score for individuals with each X score.

ment measure does not earn this GPA; some have better grades and some have worse. But using the regression line, which gives us the best single estimate of the GPA of all students whose X score is 17, gives more accurate predictions than would predicting the overall $\overline{Y}$ of 2.30.

Figure 6.4 represents a positive correlation of moderate value. An example of a perfect negative correlation ($r = -1.00$) is shown at the left of Figure 6.5. As we learned earlier, a perfect correlation is one in which *all* the points in the scatter plot fall along a straight line. Unlike the situation in Figure 6.4, in which the correlation was less than 1.00, everyone in Figure 6.5 who earned a given X score earned the same Y score. If we were given an individual's X score and wanted to predict the individual's Y score, we could therefore do so with complete accuracy.

Finally, the plot at the right of Figure 6.5 represents an example of a correlation that is close to zero. Note that the regression line—the line that comes closest to the column values—is horizontal to the baseline and takes the same value of the overall $\overline{Y}$. Therefore, whatever a person's X score, we always predict Y as equal to $\overline{Y}$.

These examples illustrate the basic logic of using the regression line to predict Y when only X is known. They also illustrate how the accuracy of prediction varies with the magnitude of the correlation between X and Y. When $r = 0$, there is no increase in accuracy of predicting Y from X over and above that obtained by predicting $\overline{Y}$ for all

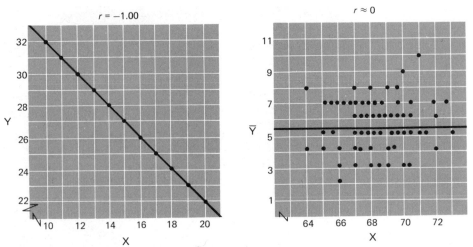

Figure 6.5 Scatter plots and regression lines for predicting Y from X for data in which $r = -1.00$ and $r \approx 0$.

individuals. As *r* departs from zero, *in either direction*, predictions become increasingly accurate. When *r* reaches ±1.00, prediction is perfect; no errors occur.

We have so far discussed only the situation in which we are given X and want to predict Y, or as it is more technically described, as determining the regression of Y from X. Using the same general logic, we can also determine the regression line of X from Y, thus allowing us to predict X from a given Y. The regression line of X from Y is basically the straight line that best fits the data points in the *rows* of the scatter plot.

Relationship of the Two Regression Lines. The statements above imply that there are *two* regression lines, one allowing prediction of Y from X and the other of X from Y. The relationships between the two regression lines for different magnitudes of *r* are illustrated in Figure 6.6. You will note that both regression lines always pass through the point representing the mean of the X and Y variables ($\overline{X}$ and $\overline{Y}$). Panel A shows the regression lines for $r = 0$. The two lines are at right angles to each other, the regression line of Y from X being parallel to the baseline and taking the value of the overall mean of the Y scores and the regression line of X from Y being parallel to the Y axis and taking the value of the overall mean of the X scores. In short, when $r = 0$, knowing X does not improve our prediction of Y; conversely, knowing Y does not improve our prediction of X. As the correlation increases from zero to a positive value, the regression lines pivot around the point of intersection and approach each other, as shown in panel B. When $r = +1.00$, the two regression lines are identical, going from lower left to upper right, as shown in panel C. As the correlation goes from zero to a negative value, the regression lines pivot in the opposite direction and become identical when $r = -1.00$, as shown in panels D and E. We remind you that accuracy of prediction also improves as *r* increases in magnitude and becomes perfect when $r = \pm1.00$.

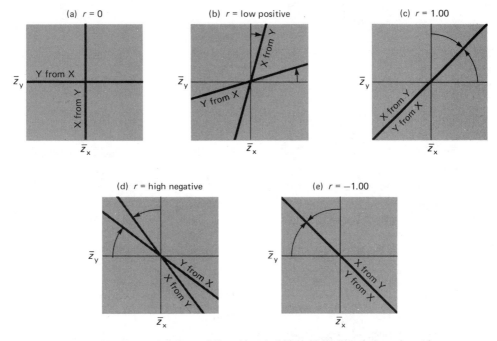

Figure 6.6 Regression lines of X on Y and of Y on X for illustrative values of *r*.

Determining the Linear Regression Line

Determining regression lines by eye is not a satisfactory procedure, since many lines could be drawn that would seem to be acceptable. Similarly, determination of predicted values of X from Y or of Y from X by inspecting regression lines in a graph might yield different estimates, depending on the observer. To make sure that everyone will make the same predictions, we must have a procedure for finding the "best" regression lines and the equation for those lines.

Equation for a Straight Line.   Before presenting the equation for a linear regression line, we need to know a little about straight lines in general.

We start with an example. A race car driver is driving at a steady pace of 100 miles per hour (mph) and then begins to accelerate evenly at a rate per second of 10 mph. Acceleration continues, let us say, until he reaches a speed of 200 mph. Let Y refer to the speed of the car at any given moment and X refer to time in seconds during the interval in which the car is accelerating. When we plot speed (Y) against time (X), the result is a straight line, as shown at the left in Figure 6.7. How can we describe this line? The first important characteristic of a straight line is its *slope constant*, which tells us how fast the line climbs or declines. In our example in which speed (Y) is plotted against time (X) in seconds, the slope constant is 10 (that is, an increase of 10 mph per second). The second characteristic we need to know is what is called the *Y intercept*. The Y intercept is

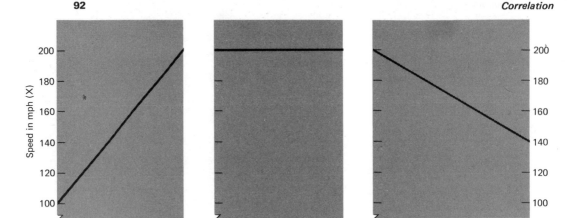

Figure 6.7   Speed of race car related to time illustrating three straight lines with different slope constants (*a*) and Y-intercepts (*b*).

simply the value of Y when X = 0. In our example, the initial speed of the car was 100 mph. At the start of the acceleration period, the Y intercept is therefore 100.

Knowing the slope constant and the Y intercept, we can write an equation that describes any specific straight line and allows us to plot a figure showing its values, such as Figure 6.7. The general form of this linear equation is:

$$Y = aX + b$$

where   *a* = value of the slope constant

b = value of the Y intercept

When X = 0, the equation reduces to Y = *b*, which is just as it should, since *b* is the value of the Y intercept when X = 0. If *b* = 0 (for example, if our race car driver made a dead start), then the equation reduces to Y = *a*X. This shows that when *a* is positive, Y increases *a* times as fast as X. Similarly, when *a* is negative (for example, if our driver were decelerating at a steady rate from his initial speed), there is a corresponding rate of decline. So *a* does act like the slope constant that it is. We can illustrate how to use the equation for a specific straight line by our race car example. Since *a* = 10 and *b* = 100, we have:

$$Y = 10X + 100$$

With this formula, we can predict Y (speed in mph) for any X (time). For example, with X = 3.5 seconds:

$$Y = 10(3.5) + 100 = 35 + 100$$

$$= 135 \text{ mph}$$

As an illustration of the use of the linear equation when *a* is negative (when the line is going from high to low rather than from low to high), suppose that after reaching and maintaining for a while a speed of 200 mph, our driver begins to decelerate each second at the rate of 5 mph. This is shown graphically at the right of Figure 6.7. The equation for this period of deceleration is:

$$Y = -5(X) + 200$$

Setting X = 0 when deceleration begins, after 6 seconds (X = 6) his speed will be:

$$Y = -(5)(6) + 200 = 170 \text{ mph}$$

Finally, we will call your attention to the line that is generated when *a* = 0: it is flat—horizontal to the baseline. This is illustrated in the middle panel of Figure 6.7, in which the driver is maintaining a constant speed of 200 mph so that *b* = 200 and *a* = 0.

The Equation of a Regression Line. We now have the background information that permits us to introduce the equation of a regression line. The particular method statisticians have adopted for finding the best-fitting regression line for predicting Y from X (or X from Y) is known as the *least-squares method.* This method attempts to specify the line that keeps at a minimum the squares of the deviations of all the points in the scatter diagram from the line (and hence keeps the sum of these squares at a minimum). The equation for the straight line yielded by this least-squares method is known as the *regression equation.*

We remind you first of the general equation for a straight line:

$$Y = aX + b$$

Statisticians have shown that *a*, or the slope constant, in the regression line of Y on X is given by:

$$a = r\left(\frac{s_Y}{s_X}\right) \tag{6.3}$$

where s_X and s_Y are the corrected estimates of the population σ values derived from the samples of X scores and Y scores, respectively, and *r* is the Pearson correlation between the pairs of X and Y scores.

The Y intercept, *b*, is given by the formula:

$$b = \overline{Y} - a\overline{X}$$

Substituting the right side of Formula 6.3 for *a*, we have

$$b = \overline{Y} - r\left(\frac{s_Y}{s_X}\right)\overline{X} \tag{6.4}$$

Substituting the quantities on the right of the equal sign in Formulas 6.3 and 6.4 into the equation for a straight line, we thus have:

$$Y_{pred} = r \left(\frac{s_Y}{s_X}\right)(X) + \overline{Y} - r\left(\frac{s_Y}{s_X}\right)(\overline{X})$$

where Y_{pred} is the predicted value of Y, given the value of X.

This formula is awkward to use but can be rearranged to yield the following more convenient equation:

$$Y_{pred} = r \left(\frac{s_Y}{s_X}\right)(X) - r\left(\frac{s_Y}{s_X}\right)\overline{X} + \overline{Y} \qquad (6.5)$$

To restate what Formula 6.5 represents, it is the equation for the regression line of Y from X which permits us to predict the value of Y for any given X value, providing we know the values of r, s_X, s_Y, $\overline{X}$, and $\overline{Y}$.

Application of the same statistical logic yields a parallel equation for the regression line of X from Y, which allows us to predict X for any given value of Y. Thus:

$$X_{pred} = r \left(\frac{s_X}{s_Y}\right)Y - r\left(\frac{s_X}{s_Y}\right)\overline{Y} + \overline{X} \qquad (6.6)$$

Note carefully that the expression s_X/s_Y in Formula 6.6 is the *inverse* of the corresponding expression in Formula 6.5, which is s_Y/s_X.

The use of these equations may be illustrated by the results of a study in which 20 eighth-graders were given, several days apart, two parallel forms of a mathematics achievement test. The scores on each test (X and Y) are listed for each student in Table 6.3, along with the values of the statistics that enter into the regression equations. You will note that $r = .80$, indicating that students' scores on the parallel X and Y tests were highly similar. Also listed in the table are the predicted Y score for each X score and the predicted X scores for each Y score. We will now demonstrate how these predicted scores were determined.

We can develop the regression equation for predicting scores on test Y from scores on test X by entering the values shown in Table 6.3 into Formula 6.5:

$$Y_{pred} = .80 \left(\frac{5.10}{5.47}\right) X - .80\left(\frac{5.10}{5.47}\right) 16.85 + 16.90$$

After the appropriate multiplications and divisions have been performed, this equation reads:

$$Y_{pred} = .75X - 12.56 + 16.90$$

Table 6.3 An Application of the Regression Equations for X from Y and Y from X in Predicting Scores on Two Forms X and Y of an Aptitude Test

Student	X (actual)	X_{pred}	Y (actual)	Y_{pred}
1	12	11.81	11	13.34
2	20	17.83	18	19.34
3	16	15.25	15	16.34
4	6	10.95	10	8.84
5	19	16.97	17	18.59
6	15	13.53	13	15.59
7	29	21.27	22	26.09
8	10	8.37	7	11.84
9	19	18.69	19	18.59
10	17	22.13	23	17.09
11	13	16.11	16	14.09
12	18	20.41	21	17.84
13	16	16.97	17	16.34
14	11	12.67	12	12.59
15	22	26.43	28	20.84
16	15	16.97	17	15.59
17	27	22.99	24	24.59
18	17	17.83	18	17.09
19	14	14.39	14	14.84
20	21	16.11	16	20.09

$$\overline{X} = 16.85 \qquad \overline{Y} = 16.90$$
$$s_X = 5.47 \qquad s_Y = 5.10$$
$$r = .80$$

Finally, subtraction of 12.56 from 16.90 leads us to the final form for this regression equation:

$$Y_{pred} = .75X + 4.34$$

This last equation was used to determine the predicted Y scores shown in Table 6.3. (In practice, of course, we would predict Y from X only if Y were unknown). Comparison of the predicted Y scores with the actual Y scores indicates that the regression equation for these data leads to rather successful predictions, as we would expect from the substantial correlation between X and Y. For example, when X = 12, $Y_{pred} = 13.34$ and the actual Y is 11. When X = 20, $Y_{pred} = 19.34$ and the actual Y is 18.

We can similarly develop the regression equation for predicting X scores from Y scores by substituting the values from Table 6.3 into Formula 6.6:

$$X_{pred} = .80 \left(\frac{5.47}{5.10}\right) Y - .80 \left(\frac{5.47}{5.10}\right) 16.90 + 16.85$$

$$= .86Y - 14.50 + 16.85$$

$$= .86Y + 2.35$$

Application of the last equation yielded the predicted X scores shown in Table 6.3. Inspection of the scores in X_{pred} and actual X columns also indicates that overall, the regression equation is quite successful in predicting X scores.

However, not all predictions of X from Y or Y from X in Table 6.3 are equally successful. You can observe from inspecting the table that in some instances predicted scores are markedly discrepant from actual scores, despite the high correlation between the two sets of scores. This observation brings us to the topic of errors of prediction.

Errors of Prediction

As we have noted, when r is ±1.00, use of the regression equation to predict X from Y or Y from X will yield perfectly accurate predictions, since all the data points in the scatter plot fall on the single regression line. Another way of stating this fact is to say that variations in X completely explain variations in Y and, conversely, that variations in Y completely explain variations in X. On the other hand, when $r = 0$, knowing X does not improve the accuracy of our predictions of Y (and vice versa); variations in one variable explain *none* of the variations in the other.

As these statements imply, the amount of variability in one factor that can be accounted for or explained by variability in the other increases as r goes from 0 to ±1.00. A statistic called the *coefficient of determination* allows us to state the amount of explained variability. More precisely, the coefficient of determination tells us the proportion of the total variability among one set of scores that can be explained by variability among the other set of scores. The coefficient of determination, which is the same for X from Y as for Y from X, is equal to the square of the correlation that has been computed between X and Y. Thus:

$$\text{Coefficient of determination} = \frac{\text{explained variation}}{\text{total variation}} = r^2 \qquad (6.7)$$

Conversely, the proportion of *unexplained* variability among a set of measures is given by $1 - r^2$. This is called the *coefficient of nondetermination*. Thus:

$$\text{Coefficient of nondetermination} = \frac{\text{unexplained variation}}{\text{total variation}} = 1 - r^2 \qquad (6.8)$$

When $r = 0$, $r^2 = 0$; this is as it should be, since variation in X explains none of the variation in Y, and vice versa, when there is no correlation between X and Y. Similarly, when $r = \pm1.00$, $r^2 = 1.00$, which indicates that the variability among scores is completely explained. Taking a few representative values between these extremes, when $r = \pm.30$, $r^2 = .09$; only 9 percent of the variability among one set of measures can be accounted for by variations among the other. When $r = \pm.50$, $r^2 = .25$ and when $r = \pm.80$, $r^2 = .64$.

As these examples illustrate, the proportion of explained variability increases as r increases but lags behind the value of r until r reaches ±1.00.

Standard Error of Estimate. In earlier chapters we have learned that when a group of individuals is measured, the resulting scores are seldom all identical, and that the mean is a single number useful in characterizing the whole distribution of numbers. The standard deviation adds information about the variability in the distribution, representing as it does a kind of average of the amount by which scores deviate from the mean.

The points on a regression line, we have said, are comparable to means. If we were to predict Y from X using the regression line and found, for example, that for X = 103, $Y_{pred} = 66$, we would essentially be saying that for all individuals having an X score of 103, 66 is the most representative Y score. Assuming that r is not ±1.00, we know that all individuals earning an X score of 103—or any other X score—do not earn exactly the same Y score. For any given X score, there is some degree of variability in the distribution of Y scores. The same statements can be made, of course, about the distribution of X scores for any given Y.

It would be informative to have a measure of variability of Y scores around the predicted Y scores (or of X scores around the predicted X scores) that is comparable to the standard deviation. Such a measure would be useful when we predict one score from another, since it gives us an idea of the range of values within which the unknown score might reasonably be expected to fall.

The *standard error of estimate* (s_{est}) is such a measure of variability and represents the standard deviation of the scores in a distribution around the regression line at any specific value of the predictor score.

When predictions are made from X to Y, the standard error of estimate of Y is given by the following formula:

$$s_{est\ Y} = s_Y \sqrt{1 - r^2} \tag{6.9}$$

When predictions are made from Y to X, the standard error of estimate of X is given by the parallel formula:

$$s_{est\ X} = s_X \sqrt{1 - r^2} \tag{6.10}$$

To illustrate the application of the standard error of estimate, suppose that in a certain college in which the same final examination is given each year in the introductory psychology course and in the elementary statistics course, the following data are obtained for students who have taken both courses:

	Intro. Psych. (X)	Elem. Stat. (Y)
	$\overline{X} = 108$	$\overline{Y} = 68$
	$s_X = 9$	$s_Y = 8$

$$r = .35$$

The correlation indicates that those who did well on the final exam in one course also tended to do well on the other final. One year the statistics instructor decides at the beginning of the semester to predict the final exam scores for all her students, using the appropriate regression equation. These predictions cannot be expected to be completely accurate, of course. In fact, the coefficient of determination, r^2, shows that introductory psychology scores account for only 12.25 percent of the variability in statistics scores.

The standard error of estimate of Y applied to these data turns out to be:

$$s_{\text{est Y}} = s_Y \sqrt{1 - r^2} = 8 \sqrt{1 - (.35)^2}$$

$$= 8 \sqrt{.8775} = 8(.9367)$$

$$= 7.49$$

The estimated standard deviation of all the statistics exam (Y) scores for any single X score is thus 7.49 score units.

What should the standard error of estimate be when $r = 0$? In the case of predicting Y from X, we expect that except for sampling error, the mean Y for any given X is equal to $\overline{Y}$ and that the variability of Y scores around the mean Y should be the same for individuals with any given X as for the group as a whole. Thus the standard deviation of Y scores for any given X should be the same as the standard deviation for the Y distribution as a whole (s_Y). This in turn implies that $s_{\text{est Y}} = s_Y$. Inspection of Formula 6.9 shows that this is indeed the case:

$$s_{\text{est Y}} = s_Y \sqrt{1 - 0^2} = s_Y$$

In parallel fashion, when $r = 0$, $s_{\text{est X}}$ is equal to s_X, the standard deviation of the total distribution of X scores.

On the other hand, when $r = \pm 1.00$, there is *no* variability among the Y scores for any X (and vice versa); that is, for any given X or Y score, all scores on the other variable are the same. This implies that each standard error of estimate should be 0. Inspection of Formulas 6.9 and 6.10 shows that this is so:

$$s_{\text{est X}} = s_X \sqrt{1 - (\pm 1)^2} = 0$$

and

$$s_{\text{est } Y} = s_Y \sqrt{1 - (\pm 1)^2} = 0$$

THE RANK-ORDER CORRELATION COEFFICIENT (r_S)

Occasionally sets of data either are reported by their rank orders only, or it seems desirable to assign ranks to them and work with the ranks rather than the raw scores. Thus, a collection of rocks might be ranked in order from hardest to softest; or we might convert the performance of baseball teams, first reported as percentages of games won, to ranks by giving the team with the highest percentage rank number 1, the next team rank number 2, and so on. In these situations the *r* between the ranks themselves could be determined by means of our computational formula. However, a new formula will give exactly the same results with even less computational work. This formula, called the formula for the *rank-order correlation coefficient*, symbolized by r_S, is shown below. The symbol for the rank-order coefficient, you will note, differs from the one we used for the product-moment coefficient (*r*) by the addition of the subscript *S*. The letter *S* has been chosen for this purpose in honor of Charles Spearman, who popularized the method.

$$r_S = 1 - \frac{6 \sum d^2}{N(N^2 - 1)} \qquad (6.11)$$

where N is the number of pairs of ranks and *d* is the difference between a pair of *ranks* (never the difference between a pair of *scores*). You should notice that the 1 and the 6 in the formula are always used when r_S is to be found. They are constant regardless of the values of N or of *d*.

We may illustrate the application of this formula for the rank-order correlation coefficient by showing the computations required to find the correlation between two movie critics' rankings of 10 movies that were released last year. Table 6.4 presents these computations. The ranks of one critic are listed as X and those of the other as Y. The difference, *d*, for each movie is equal to X - Y. The value of r_S proves to be .636. This coefficient indicates a substantial agreement between the two critics.

The Case of Tied Ranks

The rules for calculation of r_S must include what to do if there are ties for some ranks. What if, for example, we were correlating the final standings of the teams in the Big Ten Football Conference in the past two seasons and found that Indiana and Northwestern tied for sixth last year? We need to know how to treat their ranks before finding the cor-

Table 6.4   Computation of r_S for the Rankings of 10 Movies by Two Movie Critics

Movie	(Critic 1) X	(Critic 2) Y	$d = X - Y$	d^2
A	1	6	−5	25
B	2	2	0	0
C	3	3	0	0
D	4	4	0	0
E	5	7	−2	4
F	6	1	5	25
G	7	5	2	4
H	8	9	−1	1
I	9	8	1	1
J	10	10	0	0
				$\sum d^2 = 60$

$$r_S = 1 - \frac{6 \sum d^2}{N(N^2 - 1)}$$

$$= 1 - \frac{6(60)}{10(99)} = 1 - \frac{360}{990} = .636$$

relation.   A simple method is to assign each of the two tied teams the rank of 6.5, the average of the sixth and seventh ranks, since Indiana and Northwestern together must account for the ranks 6 and 7—the ranks *between* 5 and 8. Then r_S is computed as before.

We illustrate the assignment of values for the case of tied ranks by the following example, listing the intelligence quotients (IQ's) and ranks of 10 persons whom we designate by the letters A through J:

Persons

	A	B	C	D	E	F	G	H	I	J
IQ	130	128	128	122	115	110	100	100	100	95
rank	1	2.5	2.5	4	5	6	8	8	8	10

First, notice that B and C tied for second place in their IQ's. Since B and C together account for the second and third persons in order of rank, 2 and 3 were averaged, giving 2.5 for B and 2.5 for C. Then D was assigned the fourth rank, since three persons were superior to him. Finally, G, H, and I were all assigned rank 8, because they were all tied and together represent the three ranks of 7, 8, and 9, which have an average of 8; and J, the last person, received a rank of 10 because nine persons were superior to him. (See footnote, page 254, for another approach to tied ranks.)

Interpreting the Rank-Order Correlation
Coefficient

When r_S is found for data originally measured in ranks, it has the same value that r would have (except where ties have occurred) and may therefore be interpreted as a measure of the amount of linear correlation between ranks. If the original data were score values that were then converted to ranks, the rank-order correlation, r_S, would *not* be equal to the r between the original scores. However, r_S and r will have similar values so that it is possible to infer the *approximate* size of the r between the original variables from the size of r_S.

DEFINITIONS OF TERMS AND SYMBOLS

Scatter plot.   A graph in which points are plotted, each point representing the values of a pair of X and Y scores.

Linear correlation.   The line that best describes the points in a scatter plot of X and Y scores is a straight line.

Nonlinear correlation.   The line that best describes the points in a scatter plot of X and Y scores is a curved line.

Pearson product-moment correlation coefficient (r).   Pearson r, which is defined as the mean of the z-score cross-products of paired X and Y scores ($\Sigma z_X z_Y /N$), indicates the magnitude and direction of the linear relationship between the two variables. The value of r ranges from +1.00 to −1.00.

Perfect correlation.   In a perfect linear correlation, all the points in the scatter plot lie along a straight line. When the correlation is perfect positive, $r = +1.00$; when it is perfect negative, $r = -1.00$.

Regression or prediction line.   The regression line used to predict Y from X is the straight line that comes closest vertically to the columns of Y scores in a scatter plot, and the regression line used to predict X from Y is the straight line that comes closest horizontally to the rows of X scores. When $r = \pm 1.00$, the two regression lines are identical; but as r departs from ± 1.00, they diverge. When $r = 0$, they are at right angles to each other, the line for Y from X being parallel to the X axis (baseline of the scatter plot) and taking the value of $\overline{Y}$, and the line for X from Y being parallel to the Y axis and taking the value of $\overline{X}$.

Coefficient of determination (r^2).   The proportion of the total variability in one set of scores (X or Y) that is explained by variations in the other set of scores.

Coefficient of nondetermination $(1 - r^2)$. The proportion of the total variability in one set of scores that is unexplained by variations in the other.

Standard errors of estimate $(s_{est\,X}$ and $s_{est\,Y})$. The standard deviation of the scores in a distribution of X scores around the regression line of X from Y $(s_{est\,X})$ or in a distribution of Y scores around the regression line of Y from X $(s_{est\,Y})$.

Spearman correlation coefficient (r_S). The correlation between pairs of ranks. The value of r_S ranges from +1.00 to –1.00.

PROBLEMS

1. For each of the following examples, state whether or not you would expect a relationship (correlation) between the two variables and, if yes, whether you expect the relationship to be linear or nonlinear (curvilinear). In the case of an expected linear relationship, state whether the Pearson correlation (r) would be positive or negative.
    (a) Lung capacity and number of cigarettes smoked per day.
    (b) Weight and state of general health in adult women.
    (c) Years of education and annual income in 40-year-old men.
    (d) Intelligence and neuroticism.

2. Ten college students were given a set of four seemingly simple puzzles to solve and for each, allowed 3 minutes in which to find a solution. Unknown to the students, the puzzles were actually unsolvable. They were then given a group of anagrams, and the number they correctly unscrambled in 2 minutes was determined. Scores were also available for each subject on a personality measure of depression, the investigator reasoning that after a failure experience, depressed individuals might do less well than nondepressed individuals. The two sets of raw scores are shown below, along with the sum and sum of squares for each distribution. Shown at the right are the z-score equivalents and the z-score cross-products.

	Raw Scores		z Scores		
Subject	X (No. correct)	Y (Depression)	z_X	z_Y	$z_X z_Y$
1	5	14	0	– .26	0
2	3	11	– .71	–1.03	.73
3	9	22	1.41	1.79	2.52
4	1	12	–1.41	– .77	1.09
5	4	16	– .35	.26	– .09
6	2	10	–1.06	–1.28	1.36
7	6	20	.35	1.28	.45
8	10	19	1.77	1.03	1.82
9	3	12	– .71	– .77	.55
10	7	14	.71	– .26	– .18

$$\sum X = 50 \qquad \sum Y = 150$$
$$\sum X^2 = 330 \qquad \sum Y^2 = 2402$$

    (a) From inspection of the pairs of z scores, what direction and approximate magnitude (high, medium, low) do you expect Pearson r to have?

(b) Make a scatter plot of the data, using z scores. Then draw a straight line that seems to come closest to all the points. Does the correlation appear to be linear?

(c) Using the z-score formula for r (Formula 6.1), find the correlation between X and Y.

(d) Find r again, using Formula 6.2 involving raw scores. [Your answer should agree with your answer to (c), above.]

3. Convert each of the distributions shown in problem 2 into ranks and compute Spearman r between the two sets of ranks. If there is a discrepancy between this r_S and the r you previously computed, explain.

4. A school psychologist has noticed that children who are self-confident seem to have more playmates than children who are more timid and unsure of themselves. To determine whether formal evidence will confirm this observation, she asks 60 fourth-graders to write down the names of children with whom they play after school. She also has observers watch the children in several settings and obtains an overall rating of each child's self-confidence. The mean and s of each distribution of measures and the correlation between the two sets of measures are shown below.

Self-confidence	Playmates
$\overline{X} = 16.8$	$\overline{Y} = 5.2$
$s_X = 2.0$	$s_Y = 1.4$
$N = 60$	$r = .50$

(a) A teacher who learned of the results of the study suggested that instituting special play groups in which timid children would have opportunities to make friends would be an excellent way to raise their self-confidence. Comment on this proposal.

(b) Find the regression equation for predicting Y from X—that is, number of playmates from self-confidence ratings.

(c) For three of the children, their X scores were 14, 17, and 19. For each of these three X scores, use the regression equation determined in (b) to predict their Y scores.

(d) The actual Y scores for these three children were, respectively, 2, 4, and 7. How well do the predictions obtained in (c) compare to the scores these three children actually obtained?

(e) How much of the variability in one set of scores is explained by variations in the other set of scores?

5. Shown below are the scores of 20 male college professors on a measure of satisfaction with their work (X) and a measure of satisfaction with their life in general (Y). Also shown are the means, s values, and r.

Work (X)	Life (Y)	Work (X)	Life (Y)	Work (X)	Life (Y)	Work (X)	Life (Y)
44	57	49	57	49	61	43	55
53	64	54	62	39	48	57	68
47	60	40	53	60	69	47	56
50	59	52	66	46	58	51	62
36	49	48	55	48	61	45	55

$\overline{X} = 47.90 \quad s_X = 5.90 \qquad\qquad \overline{Y} = 58.75 \quad s_Y = 5.64$

$N = 20 \qquad\qquad\qquad\qquad r = .94$

(a) Find the regression equation for predicting work satisfaction (X) from life satisfaction (Y).

(b) Predict work satisfaction (X) scores from the Y scores of 50 and 70.

(c) Graph the 20 pairs of X and Y scores in a scatter plot. Then plot the regression
line determined in (a). *Note:* This can be done by first plotting the point representing Y = 50 and the X_{pred} calculated in (b) for this Y value and then plotting
the point for Y = 70 and its X_{pred}. The straight line obtained by connecting
these two points is the regression line.

(d) Compute $s_{est\ X}$. What does this statistic represent? Relate it to what you have
plotted in the graph.

Probability

In observing samples of individuals, we have noted, our scientific aim is to discover something about the properties of populations as a whole. Even if samples are selected randomly or in some other unbiased manner, the statistics obtained from a number of independent samples seldom agree exactly. For example, in the case of categorical data, the same percentage of individuals does not fall into any given category in all samples. Nor, in the case of measurement data, are all sample means identical.

Since the samples differ, no single sample can automatically be assumed to reflect exactly the properties of the population. Most samples are reasonably similar in their properties to the population, but on occasion they may be highly deviant.

Consider the following example. If a perfectly balanced coin is tossed a very large number of times—technically, an infinite number of times—a head (H) will come up half the time and a tail (T) the other half. That H and T occur equally often is in fact our *definition* of a perfectly balanced or *fair* coin. Now, suppose that instead of obtaining the entire *population* of tosses of a fair coin (possible only in theory anyway, since the population is infinitely large), we obtain a *sample* of only 10 tosses, tallying the total number of H and T for the series of 10. We repeat this procedure over and over again, gathering data on a substantial number of samples of 10 each. What do we expect the number of H and T to be in these samples? Our best *single* guess is 5 H and 5 T. But we wouldn't be at all surprised to find unequal splits, such as 6 H and 4 T or 3 H and 7 T, occurring quite frequently. Very occasionally, samples as deviant as all H or all T may occur. Samples, then, do not always mirror faithfully the population from which they were drawn.

But statistical theory does permit us, given certain facts about the population, to state the exact frequency with which events can be expected to occur. These same theoretical principles also permit us to describe the likelihood or *probability* of occurrence of some event or series of events. For example, if you toss an unbiased, balanced coin a single time, what is the probability of occurrence of H? Conversely, what is the probability of T? You know that these two events are *equally likely*, so you might describe H and T as having a 50-50 chance of occurring. Setting the sum of the probabilities of all possible events at 1.00, we therefore can describe the probability of H as .50 and of T as .50. Suppose you are about to toss the coin 5 times. What is the probability of getting 3 H and 2 T? You will discover in later discussion that, over the long run, 3 H and 2 T can be expected to occur 10 times out of 32 or approximately 31 percent of the time. The theory of probability allows us to state that the probability of getting 3 H and 2 T in a *single* series of 5 tosses therefore is .31.

Now consider another example. We have a coin of unknown origin and ask whether it is a balanced, unbiased coin or a biased one. That is, we want to know whether an infinite number of tosses of the coin (the entire population of tosses) would produce half H and half T, or some other ratio. Not having infinite patience or time, we obtain instead a sample of only 50 tosses, finding 22 H and 28 T. Is the coin unbiased? No conclusive answer can be given, but we can make a reasonable guess by determining the probability that a sample with this H-T split or an even more extreme split would occur *if the coin were in fact unbiased*. If the probability is low—that is, the likelihood is very small that a sample this different from an even split would occur if the coin were unbiased—then we reject our assumption about the coin's fairness and conclude instead that it is biased. Conversely, if the probability is high of getting a result this extreme or more so with a fair coin, we conclude that this particular coin is likely to be unbiased. What we have done in this example is to use probability theory to *test a hypothesis* about a population characteristic and to reach some conclusion about it. Probability theory thus permits us to make reasonable guesses or inferences about populations, based on the data from samples.

In this chapter we will review some of the elementary principles of probability to lay the foundation for discussion in later chapters of hypothesis testing and the statistics of inference. It is a time-honored tradition in discussing probability to use simple, understandable examples such as coins, dice, cards, or balls in a bowl, and we have already begun our discussion in this way. Although these illustrations may be of interest only to gamblers, they are useful explanatory devices. With occasional exceptions, we will therefore continue to employ them throughout the chapter.

SETS AND PROBABILITIES

We start with an example: Take five balls and number them 1, 2, 3, 4, and 5. These numbered balls constitute what is called a *set* and the individual balls the *elements* of the set. In this example the elements of the set are *mutually exclusive*. That is, each ball has a single number on it, so that if we put the balls in a bowl, mix them thoroughly, and then draw one out, only one number can occur. If the ball we select has a 3 on it, for example, it cannot simultaneously have any other number. Some events are *not* mutually exclusive. For example, suppose a group of individuals is asked whether they used public transportation or some other means of getting to work yesterday. Some might say they had used a combination of both, such as driving their car to the railroad station and taking a train from there. Use of one form of transportation on the way to work did not preclude use of other forms. Similarly, each of our five balls could have more than one number on it from the set of numbers from 1 to 5. If we wished to classify each ball as having the property 1, 2, 3, 4, or 5, a ball having the numbers 2 and 3, for example, could be simultaneously classified as having the property 2 and 3; the set of properties in this instance is not mutually exclusive.

Returning to our mutually exclusive set of five numbered balls—that is, the balls with single numbers on them—suppose we went many, many times through the procedures of drawing a ball, replacing it, mixing up the balls, and drawing again. How frequently would we expect each ball to be selected? Before answering, we should note an additional characteristic of this procedure that is critical: *On any occasion, each element is equally likely to occur.* That is, except for the numbers, the balls are identical; and if we scramble them thoroughly no ball is any more likely than another to come to hand when we draw. It can therefore confidently be stated that over the long run, each ball will be selected $\frac{1}{5}$ of the time. It can also be said that on the average each ball will be selected 20 percent of the time, or that the average *proportion* of the time it will be selected is .20.

Now if we draw a single ball, how likely (how probable) is it that a given number will occur? What is the probability that the ball is, say, a 4? The answer follows from our conclusion that each ball will be selected $\frac{1}{5}$ of the time. That is, over repeated occasions, we expect a 4 to occur $\frac{1}{5}$ of the time. For a single draw, therefore, there is one chance in five of getting a 4; alternately, we can say that a 4 will occur 20 percent of the time on the average, or that it has a *probability* of occurrence of .20. What is the probability that in a single draw from our set of five balls we will pick a ball with an *even* number—that is, a ball from the *subset* of balls numbered 2 and 4? We have 2 chances out of 5 of drawing an even number, so the answer is $\frac{2}{5}$ or .40. What is the probability of *not* drawing an even number? We have divided the set into two nonoverlapping subsets, two balls with even numbers and three with odd numbers. The probability of drawing noneven (that is, odd) numbered balls is therefore $\frac{3}{5}$ or .60. What about the probability of a ball with either an odd number *or* an even number? Since the question encompasses all elements of the set, the probability has to be unity: $\frac{5}{5}$ or 1.00.

We offer one more example. Suppose we have ten balls, five of them numbered 1, three of them 2, and two of them 3. Each of the individual *balls*—that is, each of the elements of the set, has an equal probability of being selected on a single draw $(\frac{1}{10})$, but what about each *number*? Since there are five balls numbered 1, $\frac{5}{10}$ or half of them fall into this "1" category; the probability of drawing a 1 therefore is .50. Similarly, the probability of a 2 is $\frac{3}{10}$ or .30, and of a 3, $\frac{2}{10}$ or .20.

Probability Defined

With these examples before us we can now offer a definition of probability *for equally likely events*. When a single, random observation from a set of mutually exclusive events is to be made, the *probability of an event from a specific subset of events is the ratio of the number of events belonging to that subset to the total number of possible events*. This definition can be expressed in a simple formula, in which n_A stands for the number of events belonging to subset A and N for the total number of all possible events:

$$p(A) = \frac{n_A}{N} \tag{7.1}$$

To illustrate the application of this formula, suppose we had six objects, three of them yellow, two of them red, and one of them blue. In this example, N is 6. What is the probability of selecting a red object on a single draw? There are two red objects, so n_A is 2, and the probability of selecting a red one is $\frac{2}{6}$ = .33.

We can also express N as $n_A + n_{\bar{A}}$, where $\bar{A}$ stands for all categories of the elements in the set that are *not* in the A category. $\bar{A}$ is technically called the *complement* of A. Thus, we can also write:

$$p(A) = \frac{n_A}{n_A + n_{\bar{A}}} \tag{7.2}$$

The probability that A will not occur—that is, $p(\overline{A})$—is given by:

$$p(\overline{A}) = \frac{n_{\overline{A}}}{n_A + n_{\overline{A}}} \qquad (7.3)$$

The two probabilities, $p(A)$ and $p(\overline{A})$, must add up to 1.00. In our example of six objects, the probability of selecting one of the two red (R) objects or $p(R)$ is $\frac{2}{6}$ and the probability of *not* selecting a red object or $p(\overline{R})$ is $\frac{4}{6}$, and $p(R) + p(\overline{R})$ is $\frac{2}{6} + \frac{4}{6} = 1.00$.

Notice that the definition above is of the probability of *equally likely events*. The condition of equal likelihood does not always hold. Take the simple case of a "loaded" coin—a coin that does not have perfect balance so that H and T do *not* come up equally often. For example, a coin may be biased in a manner that results in 60 percent H and 40 percent T over an infinite number of tosses. The likelihood of H on a given occasion thus is greater than the likelihood of T, specifically .60 versus .40. We may give a more general definition of probability that holds for this as well as for the equal likelihood case. *The probability of a random event of a certain type occurring on a given occasion is the proportion of times it would occur on an unlimited number of occasions.* Our subsequent discussion of the principles of probability, however, will be limited to examples of equally likely events.

SOME PRINCIPLES OF PROBABILITY

The Addition Rule

We have a deck of 52 well-shuffled cards and pick one at random. The probability of selecting an ace of hearts is $\frac{1}{52}$; similarly, the probability of selecting a king of spades is $\frac{1}{52}$. What is the probability that the card will be either an ace of hearts *or* a king of spades? The answer is the sum of the probabilities of the two individual events, $\frac{1}{52} + \frac{1}{52}$, or $\frac{1}{26}$. Similarly, what is the probability of selecting an ace or a king of any suit? Since there are four aces and four kings in the deck, the answer is $\frac{4}{52} + \frac{4}{52}$ or $\frac{2}{13}$. These examples are illustrations of the *addition rule: In a set of mutually exclusive random events the probability of occurrence of either one event or another event (or subset of events) is the sum of their individual probabilities.* For the two-event case, this rule may be expressed as:

$$p(A \text{ or } B) = p(A) + p(B) \qquad (7.4)$$

The formula may be expanded to include any number of events. For example, to determine the probability of one of three events (or subsets of events), A or B or C, we would add $p(C)$ to the right-hand side of the equation. The probability that a card drawn at random would be an ace, king, or jack therefore is $\frac{4}{52} + \frac{4}{52} + \frac{4}{52} = \frac{3}{13}$.

The Multiplication Rule

In the previous section we were concerned with specifying the probability that on a single occasion one of two or more mutually exclusive events or subsets of events will occur—for example, an ace *or* a king. We now consider the probability of occurrence of a *series* of events. Suppose we toss a fair coin two times; what is the probability that on the first toss a head will turn up and on the second a tail? Observe first that in two tosses four equally likely patterns are possible: HH, HT, TH, TT; over the long run, two tosses of the coin will produce our specified pair of events, HT, $\frac{1}{4}$ of the time. The probability of obtaining HT on a single pair of tosses therefore is $\frac{1}{4}$ or .25. Similarly, what is the probability of obtaining HTH, in that order, in three tosses? With three tosses eight different patterns are possible: HHH, HTH, HHT, HTT, THH, TTH, THT, TTT. The probability of HTH is therefore $\frac{1}{8}$. For four tosses there are 16 possibilities, so that the probability of any one of them is $\frac{1}{16}$, and so on.

Application of the *multiplication rule* allows us to determine these probabilities without having to figure out and then count all the possibilities. This rule states that *the probability of two or more independent events occurring on separate occasions is the product of their individual probabilities*. For the case of two independent events, this rule can be expressed as:

$$p(A, B) = p(A) \times p(B) \tag{7.5}$$

where $p(A, B)$ is to be read as the probability of both A and B occurring.

Let's see how the rule works. In the case of a coin, H and T each have a probability of $\frac{1}{2}$ on a single toss. The pattern HT about which we inquired should thus occur $\frac{1}{2} \times \frac{1}{2} = \frac{1}{4}$ of the time. For three tosses the probability of a given pattern is $\frac{1}{2} \times \frac{1}{2} \times \frac{1}{2} = \frac{1}{8}$, and so on. How about this example: What is the probability of drawing from a deck of 52 cards a 2-spot, followed (after replacing our first card in the deck) by a heart? Since a 2 has a probability of $\frac{4}{52}$ or $\frac{1}{13}$ and a heart a probability of $\frac{13}{52}$ or $\frac{1}{4}$, the answer is $\frac{1}{13} \times \frac{1}{4} = \frac{1}{52}$.

Note that our definition of the multiplication rule states that the events whose probability of occurrence we are specifying must be *independent*. By independent we mean that the occurrence of one event does not change the probability of the occurrence of any other event. Tossing a fair coin, for example, and observing a head does not mean that on the next toss a second head is any more or any less likely to occur; its probability remains $\frac{1}{2}$.[1] This independence can be demonstrated by tossing a coin many times and comparing the proportion of heads following heads and following tails. The proportion of H following both H and T, you will discover, remains $\frac{1}{2}$.

---

[1] We are here denying what is known as the *gambler's fallacy*. Many gamblers believe, for example, that if a head has occurred with unusual frequency, a tail is "overdue" and therefore has a higher probability of occurrence than .50. In actual fact, no matter how many heads happen to come up in a row—5, 10, or 200—the probability of a tail on the next toss of a fair coin stubbornly remains at .5.

Conditional Probability

Independent Events. Suppose we have one hundred balls, numbered 1 through 100. We draw a ball numbered 3 and replace it. We now inquire about the probability that on the next draw we will select the ball numbered 68. This type of question concerns a *conditional probability*: the likelihood that an event will occur, given the fact that another event or series of events *has already occurred*. A conditional probability is indicated by the expression $p(B|A)$, which is read as "the probability of B given A." In our example, 3 is event A and 68 is event B.

The answer to our question above is that $p(B|A) = \frac{1}{100}$. With replacement of each ball before drawing another to keep the N at 100, the probability of selecting a specific ball remains $\frac{1}{100}$, no matter what ball or how many balls have previously been selected. That is, as we discussed in the section above, the events are independent; when we replace each ball before drawing another, the probability of the individual event remains constant whatever prior events have occurred. When events are independent, you notice, the probability of B given A $[p(B|A)]$ reduces to $p(B)$. When events are *not* independent, $p(B|A)$ is *not* equal to $p(B)$, as we shall see in the next section.

Nonindependent Events. We now turn to the conditional probabilities of nonindependent events. Suppose, to relieve the tedium of our previous examples, we imagine that six men on a hunting trip find themselves trapped in a mountain cabin by a snowstorm. Three of them happen to be lawyers, two are physicians, and one is an architect. One of them, they decide, should go for help, and they decide to draw lots for the honor. The man chosen is one of the lawyers (the most likely event, since the probability of selection of the architect is $\frac{1}{6}$, of one of the physicians $\frac{2}{6}$, and of one of the lawyers $\frac{3}{6}$). After days of waiting, the five remaining men decide to send out a second messenger. What is the probability that the person selected by their second lottery will be a lawyer? With one of the lawyers gone and the group diminished to 5, the probability of another lawyer's being selected as the second man is $\frac{2}{5}$. If, instead, the *architect* had gone first, the probability of a lawyer's going second would be $\frac{3}{5}$. Or suppose that we were concerned with the probability that a *physician* went second, given that a physician had also gone first. In this instance the probability would be $\frac{1}{5}$.

These are examples of *conditional probabilities for nonindependent events*. Questions are being asked about $p(B|A)$, the likelihood of B given A, in a situation in which event A is not replaced in the original set before selection of B (the first man didn't come back). Without replacement, the probability of B *is* influenced by what has gone before. The reasoning by which $p(B|A)$ is determined is no different from that by which $p(A)$ is determined, except that we must take into account the change in the set brought about by prior events that have not been replaced. For example, if the balls numbered 2 and 4 are removed from a set numbered 1 through 100 and are not replaced, what is the probability that any particular ball of the remaining 98 will be selected on the next draw? The answer is $\frac{1}{98}$. Since the balls that were removed were both even, the probability of an *even* number on the next draw is $\frac{48}{98}$, and so on.

In the example of the hunters above, what is the probability that the first and the second men will both be lawyers? In a set of balls numbered 1 through 10, what is the probability, assuming nonreplacement, that the first ball will be 2, the second 4, and the third 5? These questions about a series of events are subject to the same sort of *multiplication rule* governing independent events. This rule, we recall, states that the joint probability of a series of events is the product of the probabilities of the individual events. Our previous statement of this rule, however, has to be modified to take into account that the probability of event B is conditional upon event A, C is conditional upon A and B, and so on. Thus, for three events (A, B, C) we have: $p(A, B) = p(A) \times p(B|A)$, so that $p(A, B, C) = p(A, B) \times p(C|A, B)$ and:

$$p(A, B, C) = p(A) \times p(B|A) \times p(C|A, B) \tag{7.6}$$

where $p(A)$ is the probability of A, $p(B|A)$ is the probability of B given A, and $p(C|A, B)$ is the probability of C given A and B in that order.

The answer to our question about the two hunters being lawyers is therefore $(\frac{3}{6})(\frac{2}{5}) = \frac{1}{5}$ or .20. The probability of drawing balls 2, 4, and 5 in that order without replacement from the set of 10 is $(\frac{1}{10})(\frac{1}{9})(\frac{1}{8}) = \frac{1}{720}$ (or .0014), and so on.

Conditional Probabilities and Joint Events

We now extend our discussion to consider conditional probabilities in situations in which each member of a group has been categorized according to two characteristics rather than one. Suppose, for example, that you are trying out as a ship's recreation director on a weekend cruise for "singles." You have been told that if the cruise is a success, you can have the job permanently. Quickly you size up the 80 men and 120 women who are on the cruise and assign each to one of three groups: Swingers, who are going to have a good time anyway, Duds, for whom little can be done, and Eagers, who are ready to enjoy themselves, given a little encouragement. You decide to concentrate your efforts on the Eagers. For convenience, we will assume the validity of your judgments. We will also assume that for your purposes—getting the job—the group of 200 constitutes a complete *population* of individuals. The results of your assignments are shown in Table 7.1.

Table 7.1   Number of Men and Women Classified in the Three Enjoyment-Potential Groups

	Swingers (A₁)	Eagers (A₂)	Duds (A₃)	Total
Men (B₁)	25	30	25	80
Women (B₂)	25	60	35	120
Total	50	90	60	200

Table 7.2 Proportion of the 200 Individuals Falling into Each Category

	Swingers (A_1)	Eagers (A_2)	Duds (A_3)	Total
Men (B_1)	.125	.150	.125	.400
Women (B_2)	.125	.300	.175	.600
Total	.250	.450	.300	1.000

Members of the group, you will notice, have been classified according to *two* characteristics: first, their enjoyment potential, in which each of the 200 individuals is assigned to one of three mutually exclusive categories (A_1, A_2, A_3), and second, their sex, in which each is assigned to one of two mutually exclusive categories (B_1, B_2). The entries in the cells of the table indicate the number of individuals exhibiting each pair of characteristics. The $B_1 A_3$ cell, for example, shows that 25 of the 80 men are classified as Duds. These numbers can also be expressed as the *proportion* of the total group exhibiting each pair of characteristics. The frequency data for the group of 200 men and women just reported, when translated into proportions, yield the values shown in Table 7.2.

Table 7.2 is also a *probability table*. For example, the probability that an individual drawn at random will be a woman (fall in the B_2 category) is .60, the sum of the three A entries for B_2. Similarly, the probability that an individual chosen at random is a Swinger (A_1) is .25, the sum of the two B entries of A_1. We could also inquire about the probability that an individual will jointly exhibit a given pair of characteristics. For example, what is the probability that a randomly selected individual will be both female *and* a Swinger (will be classified as $A_1 B_2$)? We can read this probability directly from the table by locating the entry in the appropriate cell; we find it is .125.

Now we ask a somewhat different question. Given that an individual selected from the population of 200 is a *woman* (belongs to category B_2), what is the probability that she is an Eager (belongs to category A_2)? You will recognize this question as inquiring about a *conditional* probability, in this instance $p(A_2 | B_2)$ or the probability of A_2 given B_2. In attempting to determine this conditional probability, let us first go back to the frequencies in Table 7.1. There are 120 women in the group, 60 (50 percent) of whom are classified as Eager. Given that an individual is a woman, the probability that she falls into the Eager category is therefore .50. Among the males, the probability that a man is a Dud is $\frac{25}{80}$, and so on.

This type of conditional probability can also be determined by the following formula, using the probability values shown in Table 7.2:

$$p(A|B) = \frac{p(A, B)}{p(B)} \tag{7.7}$$

Note that this formula is a rearrangement of the multiplication law: $p(A, B) = p(B) \times p(A|B)$, obtained by dividing both sides of the latter equation by $p(B)$. As an illustration of the application of this formula, we again consider the probability of being Eager, given an individual who is a woman, $p(A_2|B_2)$. We first find, by inspecting Table 7.2, the probability that the individual is both A_2 *and* B_2. This $p(A_2, B_2)$ value we find to be .30. Next we determine, from examining the column at the far right of the table, that the probability of B_2 is .60. This $p(A_2, B_2)$ value is then divided by $p(B_2)$, giving us $\frac{.30}{.60}$ or .50, the same figure for the conditional probability $p(A_2|B_2)$ that we obtained earlier.

What about $p(A_3|B_1)$, the probability of being a *Dud*, given a *man*? The joint probability of A_3 and B_1, Table 7.2 shows, is .125, and the probability of B_1 is .40; the conditional probability of A_3 given B_1 therefore is $\frac{.125}{.40}$ or .3125. Application of this formula to all combinations of $A|B$—that is, probabilities of the three Enjoyment categories given the individual's sex—yields the conditional probabilities shown in Table 7.3. Note that for each B category the probabilities total 1.00; each man or each woman falls into one of the Enjoyment categories, so that the probabilities in the three Enjoyment groups must add up to unity in each sex.

Conditional probabilities can also be found for B, given A, by a parallel formula:

$$p(B|A) = \frac{p(A, B)}{p(A)} \tag{7.8}$$

For example, given an individual who is a Swinger (A_1), what is the probability that the individual is male (B_1)? From the joint probabilities in Table 7.2 we determine that the $A_1 B_1$ probability is .125 and that $p(A_1)$ is .25; $p(B_1|A_1)$ is therefore .125/.25 or .50. The complete set of $p(B|A)$ values is shown in Table 7.4. Again you will see that for a given class of A (enjoyment potential) the two categories of B (sex) account for all cases of that A; the two B probabilities therefore total 1.00 for each A.

PERMUTATIONS AND COMBINATIONS

Permutations

A budding art collector has four paintings, which he displays on his living room walls. The wall hooks are fixed, but he regularly changes the places where each painting hangs. How many different spatial arrangements or orders of the four paintings are possible? This question concerns the number of different sequences or *permutations* a set of discrete objects can take. With two objects two orders or permutations are possible: AB and BA; with three objects, A, B, and C, six permutations are possible: ABC, ACB, BAC, BCA, CAB, and CBA. With four objects there are 24 permutations—an answer we could have figured out by jotting down and counting all the possibilities. We can determine the answer more easily by recognizing that the number of permutations (P) a set of N objects can take is equal to the product of the integers from N to 1. The product is called N

Table 7.3   Conditional Probabilities, $p(A|B)$, of the Individual's Being in
Each Enjoyment-Potential Category, Given the Individual's Sex

	Swingers (A_1)	Eagers (A_2)	Duds (A_3)	Total
Men (B_1)	.3125	.3750	.3125	1.00
Women (B_2)	.2083	.5000	.2917	1.00

factorial or N!.[2] Thus:

$$P_N = N! = N(N-1)(N-2) \cdots (1) \tag{7.9}$$

Application of this formula to three objects gives us $3 \times 2 \times 1 = 6$, and to four objects $4 \times 3 \times 2 \times 1 = 24$.

The reason that the number of orders N objects may take is equal to N! is as follows. There are N ways to choose the first object or event. For example, in a set of four objects, A, B, C, and D, any one of the four may be chosen first. For each of these ways, there are N - 1 ways to choose the second object. For example, if B is chosen first from the set of four, A, C, or D may be chosen second. This gives N(N - 1) possible orders so far. For each of these possibilities, there are then N - 2 ways to choose the third object and N(N - 1)(N - 2) possible orders. Continuing this argument to its end yields N! ways to arrange the set of N things.

In Formula 7.9, you will note that P has a subscript, N. This subscript indicates that we are trying to determine the number of possible permutations for N objects taken N at a time. But we may also be interested in N objects taken *r* at a time, where *r* is some number less than N. For example, our art collector may have increased his holdings to

Table 7.4   Conditional Probabilities, $p(B|A)$, of the Individual's Being a Man
or Woman, Given the Individual's Enjoyment Potential

	Swingers (A_1)	Eagers (A_2)	Duds (A_3)
Men (B_1)	.50	.33	.417
Women (B_2)	.50	.67	.583
Total	1.00	1.00	1.000

---

[2]Occasionally the number may turn out to be 0 or 1. Both 0! and 1! take the value 1.

nine paintings but have room to display only five of them at one time. How many arrangements or permutations of these nine paintings, hung five at a time, are possible? The formula that permits us to answer this question is:

$$P_r^N = \frac{N!}{(N-r)!} \qquad (7.10)$$

where the subscript *r* attached to P indicates that we are seeking the number of permutations for *r* objects at a time, rather than the entire set. On the right side of the equation the numerator is the same N factorial as above, while the denominator indicates that *r* is to be subtracted from N and the factorial of the resulting number found—that is, the product of the integers from (N - *r*) to 1. If six objects are taken two at a time, then the number of permutations is determined by 6!/(6 - 2)! or (6 × 5 × 4 × 3 × 2 × 1)/ (4 × 3 × 2 × 1). The result of these calculations is 30.

You may understand Formula 7.10 better if you again notice that in this example of 6 objects taken two at a time, there are 6 ways of choosing the first object and, for each of these possibilities, 5 ways of choosing the second object, thus yielding 6 × 5 or 30.

Combinations

Our art collector with nine paintings that he wants to hang in all possible orders of five at a time faces an almost impossible task, since 9!/(9 - 5)! yields 15,120 permutations. He would have a better chance of succeeding if he contented himself with all possible *combinations* of his paintings. Combinations refer to the number of different subsets of *r* objects each that we can select from a set of N objects *without regard to sequence*. For example, three objects taken two at a time permit six sequential arrangements or permutations: AB and BA, AC and CA, BC and CB. If we disregard order (for example, treat AB and BA as the same), we see that there are three different subsets of objects; therefore the number of unordered subsets or *combinations* of three objects taken two at a time is 3. The formula for determining this number is as follows:

$$C_r^N = \frac{N!}{r!(N-r)!} \qquad (7.11)$$

The number of ways that nine paintings can be displayed five at a time, ignoring the particular arrangement of a given subset of five on the walls, is therefore 9!/5!(9 - 5)!. This equals 126, a more manageable number than 15,120.

Knowledge of permutations and combinations also allows us to make probability statements about sets of events. One example will suffice. Four sisters were left 8 pieces of jewelry by their grandmother, her will stating that to avoid quarrels, each was to receive 2 pieces to be determined at random. One of the sisters had her eye on a topaz ring and a sapphire pin. What is the probability she will get her wish? Application of Formula

7.11 shows that there are 28 combinations of 8 things taken two at a time. She has only 1 chance out of 28, or a probability of only .036, to get both the pieces she covets.

The principles of probability will also help us to understand binomial probability distributions and the normal probability distribution, discussed in the next chapter.

DEFINITIONS OF TERMS AND SYMBOLS

Set. A collection of objects or events. For example, a deck of cards represents a set of 52 objects.

Elements. The individual members of a set of objects or events.

Equally likely events. The probability of occurrence of a single element on any given occasion is equally likely for each element in a set.

Mutually exclusive events. Two or more events that cannot possibly occur together. For example, a person is either alive or dead and cannot be both.

Independent events. The probability of an event remains unchanged whether or not some other event occurs.

Nonindependent events. The probability of an event is different when some other event occurs than when it does not occur.

Probability of equally likely events. In a set of N mutually exclusive events that are equally likely to occur, the likelihood of occurrence of any one event on a single occasion is 1/N. The likelihood of occurrence of a subset of events (n) is the ratio of the number of events belonging to the subset to N, the total number of events: n/N.

Addition rule. In a set of mutually exclusive random events, the probability of occurrence of either one event or another event (or subset of events) is the sum of their individual probabilities. In the two-event case, this rule is expressed as p(A or B) = p(A) + p(B).

Multiplication rule for independent events. The probability of two or more independent events or subsets of events occurring on separate occasions is the product of their individual probabilities. In the two-event case, this rule is expressed as p(A, B) = p(A) × p(B).

Conditional probability of successive events. The probability that an event will occur, given the fact that another event or series of events has already occurred. In the instance of the probability of B occurring, given that A has already occurred, this probability is expressed as p(B$|$A).

Multiplication rule for dependent events. The probability of two or more dependent events occurring is the product of their individual probabilities, with the probability of each event after the first being a conditional probability. In the case of the probability of event A followed by event B, this rule is expressed as $p(A, B) = p(A) \times p(B|A)$.

Joint events. Events are classified according to more than one set of mutually exclusive categories. For example, in a group of N people, each person could be classified by sex (male or female) and by marital status (married or not married). The probability of an individual having a given combination of characteristics (e.g., married and female) is the ratio of the number of individuals in the group having that combination to the total number in the group.

N factorial (N!). Multiplication of every integer from 1 to N by every other integer. Thus, $N! = N(N-1)(N-2)\cdots 1$

Permutations of N things (P_N). The number of orders or sequences that N discrete objects or events can take. $P_N = N!$

Permutations of *r* things out of N (P_r^N). The number of possible orders of a specified number (*r*) of objects selected from a larger set of N objects. $P_r^N = N!/(N-r)!$

Combinations (C_r^N). The number of different subsets of *r* objects that can be selected from N objects without regard to sequence. $C_r^N = N!/r!(N-r)!$

PROBLEMS

1. In each of the following, indicate whether the individuals or events have been classified into mutually exclusive (ME) or non-mutually exclusive (NME) categories. Explain your choice.
   (a) The recorded crimes committed in Travis County last year are categorized according to the type of offense (for example, murder, armed robbery, attempted rape, and so forth), and the totals for each category of offense are reported.
   (b) The number of undergraduate students in a state university who are enrolled in the College of Liberal Arts, College of Sciences, College of Business, and College of Engineering.
   (c) The instructor in a course in freshman English determines the number of students who were given course grades of A, B, C, D, and F.

2. In each of the following, indicate whether the samples or series of events being described are independent (I) or nonindependent (NI).
   (a) In a study of the effects of two types of diet on young rats, an investigator selects two males from each of 20 litters of pups. One pup in each pair is selected at random and assigned to one diet condition and the other pup is assigned to the second diet condition.
   (b) Students' grade point averages over three successive semesters.
   (c) The number of aces in each of a person's bridge hands over the course of three rubbers.

3. In a group of 200 individuals, 60 do not wear glasses, 50 wear glasses all the time, and 90 wear glasses only for reading. For purposes of this question, identify these subgroups as A, B, and C, respectively.
   (a) What is the probability that an individual randomly selected from the group of 200 never wears glasses?
   (b) What is the probability of selecting someone who wears glasses either all or some of the time?
   (c) If the answer to (a) is identified as $p(A)$, how can the answer to (b) be identified?
   (d) What is the probability of randomly selecting a person who wears glasses only for reading, followed by a person who wears glasses all the time, *assuming replacement*?
   (e) What is the probability of randomly selecting three people in a row who never wear glasses, again assuming replacement?

4. A nationally distributed magazine stages a prize lottery in which a contestant is first chosen from each of the 50 states. From these 50, three are chosen at random, the first drawn receiving the grand prize, the second drawn the second prize, and the third the third prize. No contestant may receive more than one prize.
   (a) What is the probability that the grand prize winner will be from Alaska or Hawaii?
   (b) What is the probability that the grand prize winner will come from one of the original 13 states?
   (c) Suppose that the grand prize winner comes from Texas. What is the probability that the second prize winner will come from California?
   (d) What is the probability that the first prize winner will come from Iowa, the second from Maine, and the third from Wyoming?

5. In a study of the development of the principle of conservation, a psychologist studies groups of 3-year-old, 5-year-old, and 8-year-old children. Each child is shown a glass half full of colored water. Then the child watches as the water is poured into a taller glass of smaller diameter. The child is next asked whether the amount of water in the second glass is less than, the same as, or more than the amount that was in the first glass. The children's responses are shown below.

	Less	Same	More	Total
3-year-olds	1	2	6	9
5-year-olds	0	5	5	10
8-year-olds	0	9	2	11
Total	1	16	13	30

   (a) Construct a table, similar to the one above, in which the *proportions* of the total group of 30 are shown instead of frequencies.
   (b) What is the probability that a child responded "same"?
   (c) Set up a conditional probability table of the child's being in each response category, given the child's age.
   (d) Given a 5-year-old child, what is the probability that the child responded "same"?
   (e) Given a 3-year-old, what is the probability that the child responded "more"?

6. A psychologist intends to give subjects a battery of four personality tests. She is uncertain whether the order in which the tests are given will influence subjects' responses and decides to vary order among subjects. How many possible orders are there?

7. In a psychophysical experiment, an investigator prepares 10 cylinders that are identical in appearance but form a finely graded series in weight. Each subject is to be presented two cylinders at a time, one placed in the right hand and the other placed in the left hand, and asked to judge which one is heavier.

    (a) If subjects were presented with all possible permutations of pairs, how many pairs would be involved?

    (b) If subjects were presented with all possible combinations of pairs, how many pairs would be involved?

The Binomial and the Normal Distributions

CHAPTER 8

THE BINOMIAL DISTRIBUTION

A wine-loving friend of yours claims his palate is so sensitive that he can discriminate between two wines of the same type produced in the same year by neighboring vintners. He happens to have two bottles of wine that meet this description, and you challenge him to prove his boast. On each of ten "trials" you have him sip from each of two unmarked glasses and then announce which wine is which. On eight of the ten trials his identification is correct. At first you're impressed; eight out of ten is a pretty good hit rate, you think to yourself. But then it occurs to you that if he couldn't discriminate at all, he would be expected to be "right" half the time by chance alone. Was it pure luck he guessed correctly 80 percent of the time, or is it likely that he really can discriminate above the chance level?

This example illustrates a *binomial experiment.* In the binomial situation, one of two possible outcomes occurs on each of a number of independent occasions. The wine-taster was given 10 independent opportunities to judge which of the two wines was which and on each occasion was either *right or wrong.* To answer the question the experiment was designed to test, we must know something about a theoretical distribution called the *binomial probability distribution.*

In the binomial situation, we have said, one of two possible outcomes occurs on each occasion. Assume, for purposes of illustration, that each outcome has a probability of occurrence of $\frac{1}{2}$. For example, we might be tossing a fair coin in which H and T have an equal probability of turning up. Or the subject in our wine-tasting experiment might be unable to discriminate between the two wines and therefore be equally likely to be right as wrong. If we call the occurrence of a head or of a correct guess a "hit," and a tail or a wrong guess a "miss," we can determine by applying the multiplication rule the number of possible patterns of hits and misses on a given number of occasions and the probability of occurrence of each. If we toss a coin three times, for example, eight equally likely patterns of hits and misses are possible (HHH, HHM, HMM, HMH, MHH, MHM, MMH, MMM) and the probability of each is $(\frac{1}{2})^3$. That is, the probability of each is $\frac{1}{2} \times \frac{1}{2} \times \frac{1}{2}$ or $\frac{1}{8}$. Similarly, with five occasions there are 32 possible outcomes, each having a probability of $(\frac{1}{2})^5$ or $\frac{1}{32}$.

Now suppose that for the five-trial case we write out the 32 possible patterns of hits and misses. For each pattern we then count the number of *hits*, ranging from five out of five to zero. We find that five hits occur in only *one* of the 32 possible outcomes, but there are five patterns in which four hits occur (HHHHM, HHHMH, HHMHH, HMHHH, MHHHH), ten patterns in which three hits occur, and so on. A complete frequency distribution of the various numbers of hits, from five to zero, is shown in Table 8.1. In the final column of the table is the probability of occurrence of each number of hits. This list of probabilities is known as a *binomial probability distribution.*

The binomial probabilities in the table were obtained by applying the addition rule, discussed in Chapter 7. That is, each of the *individual* patterns of hits and misses has a probability of $\frac{1}{32}$ or .03125. Five hits occur only once among the 32 patterns, so its probability is .03125. Four hits occur in *five* of the patterns; adding .03125 together five times gives .15625, and so on.

Table 8.1 Theoretical Frequency Distribution of Number of "Hits," Where Number of Occasions = 5 and p = .50

Hits	Frequency	prob (Hits)
5	1	.03125
4	5	.15625
3	10	.31250
2	10	.31250
1	5	.15625
0	1	.03125
	$\sum f = \overline{32}$	$\sum prob\ (Hits) = \overline{1.00000}$

The Binomial Expansion

In explaining the binomial distribution above, we suggested writing down all the possible patterns of the two events that could occur on N occasions and then counting the number of patterns containing N of the specified ("favored") event, N - 1 of the favored event, and so forth down to (N - N) or 0. We can determine the distribution more efficiently by expanding the binomial:

$$(p + q)^N$$

where p is the probability of obtaining the favored event on a given occasion, q is the probability of the nonfavored event, and N is the number of independent occasions or observations. The expansion takes the general form:

$$(p + q)^N = p^N + Np^{N-1}q + \frac{N(N-1)}{1 \times 2} p^{N-2}q^2$$

$$+ \frac{N(N-1)(N-2)}{1 \times 2 \times 3} p^{N-3}q^3 + \cdots + q^N \tag{8.1}$$

For $p = \frac{1}{2}$ and N = 5 we thus have:

$$\left(\frac{1}{2} + \frac{1}{2}\right)^5 = \left(\frac{1}{2}\right)^5 + 5 \left(\frac{1}{2}\right)^4 \left(\frac{1}{2}\right) + 10 \left(\frac{1}{2}\right)^3 \left(\frac{1}{2}\right)^2$$

$$+ 10 \left(\frac{1}{2}\right)^2 \left(\frac{1}{2}\right)^3 + 5 \left(\frac{1}{2}\right) \left(\frac{1}{2}\right)^4 + \left(\frac{1}{2}\right)^5$$

Note the coefficients in each term in the expansion. The coefficient for the first term, $(\frac{1}{2})^5$, is understood to be 1, and the remainder are 5, 10, 10, 5, and 1. These coefficients correspond to the frequencies in Table 8.1, while the terms themselves correspond to the probabilities listed in the table. For example, the probability of four hits is given by the second term, $5(\frac{1}{2})^4(\frac{1}{2})$; the value of this term equals .15625, the same figure we obtained earlier by the addition rule.

Each term in the binomial expansion can also be obtained by:

$$C_r^N p^r q^{N-r}$$

where C_r^N is the combination of N things taken r at a time, as we discussed in Chapter 7. Since, according to Formula 7.11, C_r^N is equal to $N!/r!(N-r)!$, we can write:

$$C_r^N p^r q^{N-r} = \frac{N!}{r!(N-r)!} p^r q^{N-r} \tag{8.2}$$

As an example of the application of this equation, we determine the probability of obtaining three hits in a sample of five observations, where $p = \frac{1}{2}$:

$$C_3^5 p^3 q^2 = \frac{5!}{3!(5-3)!} \left(\frac{1}{2}\right)^3 \left(\frac{1}{2}\right)^2 = 10\left(\frac{1}{32}\right) = .3125$$

The coefficient (10) and the probability (.3125) agree, you will observe, with the term for three hits in the expansion of the binomial equation and the entries in Table 8.1.

Let us consider one more illustration of the application of the equation immediately above. If we rolled four dice, we could determine the probability of getting various numbers of, let us say, six-spots. Sixes could come up on all four dice, on only three of them, two of them, and so on. Since the probability of rolling a six-spot in a single die is $\frac{1}{6}$ and of rolling any other number $\frac{5}{6}$, we could determine the probability of these various numbers of six-spots in four rolls by expanding the binomial $(p+q)^N = (\frac{1}{6} + \frac{5}{6})^4$. What if we wanted to know only the probability of obtaining *two* sixes among the four dice? We could expand the binomial and use the appropriate term, or we could apply Formula 8.2. The latter procedure gives us:

$$C_2^4 p^2 q^2 = \frac{4!}{2!(4-2)!} \left(\frac{1}{6}\right)^2 \left(\frac{5}{6}\right)^2 = 6\left(\frac{25}{1296}\right) = .116$$

Before concluding this section, we return briefly to the results of the wine-tasting experiment with which we started. The would-be expert, you recall, succeeded in guessing correctly on eight of ten occasions. Assume that he cannot discriminate at all—that is, assume that the probability of a hit on each trial is $\frac{1}{2}$. Expansion of the binomial with $p = \frac{1}{2}$ shows that the probability is only about .05 that on 10 occasions he would make *eight or more* correct guesses by chance alone. This probability was obtained by first expanding the binomial for $p = \frac{1}{2}$ and N = 10. (In Figure 8.1, this expansion is shown in graphic rather than in tabular form.) Then the frequencies of eight, nine, and ten hits are added and the sum divided by the total number of possible outcomes. These frequencies are 45, 10, and 1, respectively, and the total number of outcomes (or Σf) is 1024; thus $\frac{56}{1024} = .0547$.

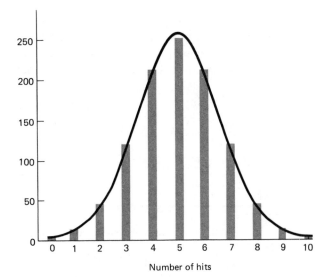

Figure 8.1 A binomial frequency distribution in which $p = .50$ and $N = 10$ and a superimposed normal distribution with the same mean and σ.

With a probability as small as this, it is unlikely that your friend guessed correctly as often as he did by pure dumb luck; very probably he *can* discriminate above the level of chance. The kind of reasoning we have just gone through involves setting up and testing a hypothesis about some state of affairs, and then coming to some conclusion about the validity of the hypothesis. In the next chapter, we will discuss the topic of hypothesis testing in more detail. We return now to describing the binomial distribution.

The Mean and Standard Deviation of the Binomial Distribution

It is possible to calculate the mean and standard deviation of a binomial distribution, just as we learned to do for any other kind of distribution. If you look at Table 8.1, for example, where $p = .50$ and $N = 5$, what is the mean number of hits? Since the distribution is perfectly symmetrical, the mean falls at the midpoint of the range, which we observe to be 2.5. We could also calculate the mean in the ordinary way by multiplying each hit by the corresponding p (which is essentially percent frequency), summing these products, and dividing by Σp. Σp, of course, equals 1. If you were to carry out these operations, you would find that the mean turns out to be 2.5, just as it should be. The formula for the mean of the binomial can be reduced to a simpler form that considerably lessens our calculational labors:

$$\mu_b = Np \tag{8.3}$$

where N is the number of observations and p the probability of the favored event. In our example μ_b therefore equals $5(.5)$ or 2.5. You will observe that we are dealing with a theoretical population of values, so the symbol we use for the mean is μ(mu). The subscript, b, indicates that we are referring to the mean of the binomial distribution.

We can determine the variance and the standard deviation of a binomial distribution in the usual way by finding the squared deviations of the "scores" from μ_b, finding the mean of these squared deviations, and so on. But just as in the case of μ_b, simpler formulas can be used. The *variance* of the binomial is given by:

$$\sigma_b{}^2 = Npq \tag{8.4}$$

and the *standard deviation* by:

$$\sigma_b = \sqrt{Npq} \tag{8.5}$$

where $q = 1 - p$.

Thus, in our example, we have for the standard deviation:

$$\sigma_b = \sqrt{(5)(.5)(.5)} = \sqrt{1.25} = 1.118$$

Shape of the Binomial Distribution When $p = \frac{1}{2}$

When $p = \frac{1}{2}$, the binomial distribution is always symmetrical. As N, the number of possible occurrences, becomes larger and larger, the shape of the binomial distribution with $p = \frac{1}{2}$ becomes increasingly similar to another theoretical distribution. This is the *normal probability distribution,* a particular member of the family of bell-shaped curves based on continuous rather than discrete measures. The similarity is illustrated in Figure 8.1, in which $p = \frac{1}{2}$ and N = 10. On the baseline we have indicated the number of "hits"—for example, number of heads coming up in a toss of ten coins. On the Y axis is shown the frequency, out of the 1024 possible patterns of hits and misses, with which each *number* of hits could be expected to occur. These frequencies, you recall, correspond to the coefficients of the terms in the binomial expansion. The frequencies have been plotted as bars to emphasize that the measures are discrete. That is, they take the values 0, 1, 2, 3, and so on; intermediate values such as 1.54 hits or 3.791 do not occur. Superimposed on the bar graph for the binomial distribution is a normal probability curve whose mean and standard deviation have the same values as those of the binomial. Since continuous measures (those that can take any value) are represented, the curve is unbroken.

You will observe that the two shapes are highly similar. As the N in a binomial increases toward infinity, the size of the "steps" between adjacent bars becomes smaller and smaller, and the distribution increasingly resembles the continuous normal distribution. Thus, we can state that the *limit of the binomial distribution*, where $p = \frac{1}{2}$ and N is infinitely large, is the *normal probability distribution.*

The similarity between the two distributions often leads us to substitute the normal distribution for the binomial distribution when we are evaluating binomial data like those in our wine-tasting experiment. When $p = \frac{1}{2}$, this substitution may safely be made when N is relatively small—as low as 10, let us say. As p departs from $\frac{1}{2}$ toward 0 or 1, the normal curve may provide a satisfactory approximation but N must be increasingly larger for the normal curve to be used. We shall discuss this use of the normal curve more specifically in the next chapter. We will devote the rest of this chapter to describing the properties of the normal curve.

THE NORMAL PROBABILITY DISTRIBUTION

Like the binomial distribution, the normal probability distribution is a theoretical mathematical model, based on a population of measures that is infinite in size. But there is an infinite variety of curves, and by choosing the appropriate mathematical equation we can plot and describe the properties of any of them we wish. Why, then, are we giving so much emphasis to this particular curve? The primary reason has to do with statistical theory. We have already seen that the binomial distribution approaches this form as N increases, and we shall see in later chapters that other theoretical sampling distributions also take this form.

A second reason why the normal curve is of interest is that many empirical distributions—those obtained by actually measuring some characteristic—are approximately normal in shape. That is, if we attempt to fit an idealized mathematical shape to the data we obtain, the equation for the normal distribution is an appropriate model. By applying our knowledge of the properties of the normal curve to an empirical distribution, we can answer many types of questions about the distribution.

In explaining the properties of normal curves, we will use only empirical distributions for illustrative purposes, deferring discussion of the curve's theoretical uses until later chapters.

Characteristics of the Normal Curve

The normal curve is shown in Figure 8.2. Since it is a plot of a theoretical distribution and not of actual data, the units of measurement along the baseline are expressed in z scores. We should also recall that the distribution is based on an infinite number of cases, which can take any z-score value. The curve therefore is not dropped to the baseline at either end and can be understood to reach it only at z values of $\pm\infty$.

The most obvious characteristic of the normal curve is its shape, somewhat like a bell, rising to a rounded peak in the middle and tapering off symmetrically at both tails.

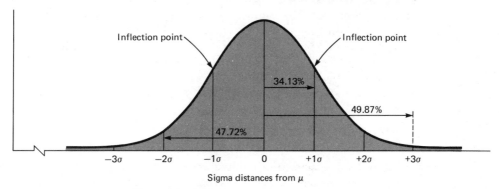

Figure 8.2 The normal distribution curve.

The half of the curve below μ has an $/$ shape and the half above μ a reversed $/$. The vertical lines erected from the baseline at distances of $+1\sigma$ and -1σ both intersect the curve at its "inflection point"—the point at which the $/$ and its mirror image each change directions, from being concave to being convex.

The total *area* under the curve—the area between the curve and the baseline—is taken as 100 percent. We can also mark off portions of the curve and describe their areas as a percent of the total. In Figure 8.2, for example, we have dropped lines from the curve to the baseline at μ and at one σ above μ. Now in *any* curve, normal or any other shape, the portion of the total area under the curve between μ and the point that is one σ above μ is always the same for *that particular curve*. In other words, for any single curve, regardless of its shape, there is a fixed relationship between distance along the baseline measured in z scores, and area under the curve. In the *normal* curve the area under the curve between μ and $+1\sigma$ turns out to be 34.13 percent of the total area. *The percent area is the same as the percent frequency;* therefore, we know immediately that in the normal curve 34.13 percent of the total frequency is included between μ and the point that is one σ above μ. *When distance along the baseline is measured in σ units, percent area under the normal curve between μ and any other point is the same as percent frequency.* Recall that σ units are z scores; the score that is one σ above the mean is the score where $z = +1.00$. Thus, we can restate what we have just said in z-score terms: In the normal curve the percent frequency between μ and the point where $z = +1.00$ is 34.13.

Since the normal curve is symmetrical, another 34.13 percent of the frequency lies between μ and the point one σ below μ, between μ and $z = -1.00$. Therefore, 68.26 percent of the cases fall between $z = -1.00$ and $z = +1.00$. From μ to the point two σ's from μ is 47.72 percent of the frequency, so 95.44 percent fall between $z = -2.00$ and $z = +2.00$. Between μ and three σ's is 49.87 percent. Thus, three σ's above and below μ, six σ's in all, include 99.74 percent, or almost all, of the total frequency.

Let us illustrate what we have stated. If the weight distribution for a sample of 10,000 men were exactly normal, 3,413 men (34.13 percent of N) would fall between the mean weight and the weight that was one σ above the mean. If the mean were 170 and σ 15, then 68.26 percent or 6,826 of them could be expected to weigh between 155 and 185 (170 ± 15, the scores where $z = \pm1.00$).

As a final illustration, consider the distribution of measurements of chest circumference of American women. Since all women have not been measured, the actual μ and σ are not known. But the results of several studies permit us to estimate that μ of this distribution is about 35 inches, with $\sigma = 3$ inches. Probably the distribution is sufficiently normal to permit us to predict that about 95 percent of American women would achieve, naturally, scores between 29 and 41 inches ($\mu - 2\sigma$ and $\mu + 2\sigma$). Almost all women would score between 26 inches ($35 - 9$) and 44 inches ($35 + 9$)—that is, between $z = -3.00$ and $z = +3.00$.

Probability and Percent Area

We have discussed the percent of the total area in a normal curve that lies between representative points on the baseline, and the relationship between these percent areas and percent frequencies. For example, approximately 34 percent of the area, and hence about

34 percent of the total number of cases, fall in the interval between μ and one σ above μ. These percentages could instead be expressed as proportions. For example, the proportion of cases falling between μ and $z = 1.00$ is approximately .34.

These area values also reflect *probabilities*. For example, the probability that an individual selected at random from a normal distribution will fall between μ and one σ above μ (will have a z score ranging from 0 to +1.00) is .34.

When we were discussing equally likely discrete events in the last chapter, such as the outcome of coin tossing or drawing cards from a deck, we defined probability as the ratio of the number of times the specified event can occur to the total number of possible events. The normal curve, however, involves continuous rather than discrete variables. Therefore, we must employ a somewhat different definition of probability for events randomly sampled from a normal distribution. This definition states that probability in the normal curve is the *ratio of the area under the specified portion of the curve to the total area*. This definition, you will observe, states formally what we have said in the paragraph above.

The Normal Curve Table

The percent area under the normal curve, and therefore the percent frequency, between μ and various σ distances from μ is shown in Table B at the end of the book. This table will be used repeatedly from now on. Let us examine the table and determine how to read it. The left-hand column, labeled x/σ, gives deviations from μ in σ or z-score units.

The values in the x/σ column in Table B are given to one decimal place; the second decimal place is shown in the top row of the table. The values in the body of the table are the *percent area* or *percent frequency* between μ and the point x/σ distant from μ. For example, if we wished to determine the percentage of the total frequency between μ and the point where $z = 1.96$, we would first find the value of 1.9 in the x/σ column, then go along that row to the column headed .06 and read off the percent frequency, which in this case is 47.50 percent.

Note that the table gives percentages only on *one* side of μ—that is, for only one-half of the normal curve—so none of the values in the table is larger than 50 percent. Because the normal curve is symmetrical, it is not necessary to print another table of the percentages for the other half of the curve; the percent frequency between μ and any x/σ value is the same whether the x/σ value is above or below μ, whether the z score is plus or minus.

Using the Normal Curve Table to
Solve Problems

Let us now take up some problems involving real data and show how the normal curve table can be used to solve them. To avoid terminological confusion, we will assume that each group that has been measured constitutes a population of some sort, even if only a highly artificial one. Therefore, we will assume that the means and standard deviations of the obtained distributions are population values and refer to them, respectively, by the symbols μ and σ.

It may be helpful to state our general procedure first. If our empirical distribution is normal, or approximately so, the relation shown in Table B between percent frequency and z scores in the normal curve also holds for the empirical distribution. But the raw scores in an empirical distribution may be in any kind of units, such as feet, seconds, letters canceled per four minutes, and the like. In the normal curve the units are z scores, x/σ values. Therefore, we shall have to convert the raw score from the empirical distribution, given us in the problem, into a z score or x/σ value, and then immediately Table B becomes applicable. Let us see how this works.

In a study dealing with various athletic skills, 90 nine-year-old boys were measured as to how far they could throw a softball.[1] The mean was 79.06 feet, the standard deviation was 19.87 feet, and the distribution sufficiently approximated the normal distribution to allow us to assume normality. With this information, we are in a position to answer all sorts of questions. For example, what percentage of the boys threw 109 feet or more? How many threw 45 feet or less? Any boy who qualifies among the top 10 percent must be able to throw at least how far? What is the probability that a boy picked at random from the group threw somewhere between 59 and 99 feet? What distances are so extreme that they were thrown by only 1 percent of the boys? What extreme distances were thrown by only 5 percent?

Since it will not affect the method we are illustrating, we will make the data more convenient to handle by rounding off μ to 79 feet and σ to 20 feet, although we would not do this in actual practice. We also will assume, for purposes of illustration, that the data were normally distributed. The first question we asked concerning this distribution was, What percentage of the boys could throw 109 feet or more? This problem is shown graphically in Figure 8.3. We are interested in determining the percentage of the group falling at or above 109 feet in their scores, the shaded upper portion of the curve. To determine this percentage, the raw score of 109 feet must first be converted to a z-score value:

$$z = \frac{X - \mu}{\sigma} = \frac{x}{\sigma} = \frac{109 - 79}{20} = \frac{30}{20} = 1.50$$

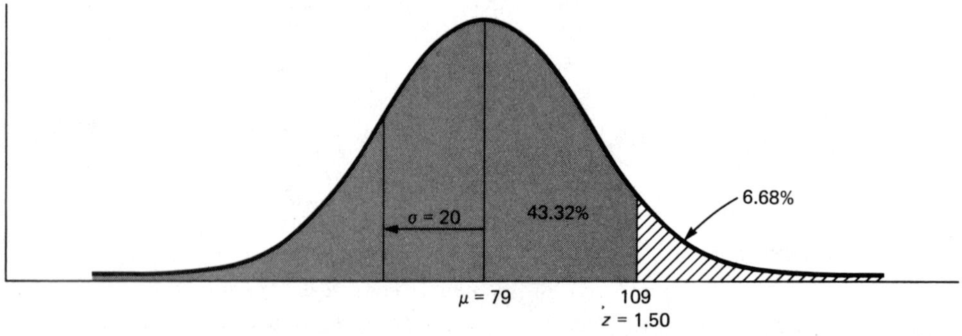

Figure 8.3 Percentage of boys throwing 109 feet or more, assuming normality of distribution.

---

[1] Data courtesy of Dr. Robert Malina.

From Table B, we find that 43.32 percent of the frequency in a normal distribution falls between μ and $z = 1.50$. We are interested, however, in the percentage that falls *beyond* $z = 1.50$. Since the total percent frequency above μ is 50 percent, we determine that $50 - 43.32 = 6.68$ percent of the boys throw 109 feet or more.

The second question asked was, *How many* threw 45 feet or less? Note carefully that this question asks about the *lower* extremes of the curve, scores of 45 feet or *less*, and about the raw frequency rather than percent frequency. Although we will not show this problem graphically, it might be helpful for you to draw a normal curve and mark off the portion of the curve in which we are interested. Our first step is to convert 45 feet to a z score: $z = (45 - 79)/20 = -34/20 = -1.70$. Table B shows us that 45.54 percent of the area lies between μ and $z = -1.70$; $50 - 45.54$ or 4.46 percent of the area lies below this z score. Since the questions ask about raw frequency, we must now find the number of boys, out of 90, that correspond to the percentage we have just determined. We find that $(4.46)(90/100)$ gives us 4.014. Rounding off this number, 4 boys threw 45 feet or less.

The third question asks, Any boy who qualifies among the top 10 percent must be able to throw at least how far? Note the difference between this question and the previous ones. Instead of being given a raw score and ending up with a percent frequency or a raw frequency, we are here given a percent frequency and asked for a raw score. We must go through the same steps as before, but in the opposite direction. Remember that a total of 50 percent of the area of the normal curve lies at or above μ. We want to find the z score that divides the percentage area above μ into two portions: 40 percent between μ and the z score and 10 percent beyond the z score. Since Table B gives the percent area between μ and a given z score, we look in the *body* of the table for 40 percent; this will give us the z score *beyond* which 10 percent remains. The closest value to 40 percent we can find in Table B is 39.97 percent, which corresponds to $z = 1.28$. We now know that the top 10 percent of the boys have z scores of $+1.28$ or more. But the question asks for a raw score, so we must convert the z of 1.28 into a raw score. Recall that the formula for this conversion is $X = \mu + z(\sigma)$ and that in our problem, $\mu = 79$ and $\sigma = 20$. Thus $X = 79 + 1.28(20) = 79 + 25.6 = 104.6$. The top 10 percent of the boys threw 104.6 feet or more.

The fourth question asks, What is the probability that a boy threw somewhere between 59 and 99 feet? This problem has been shown pictorially in Figure 8.4. Observe that 59 feet lies below μ and 99 feet lies above μ and that the question inquires about the probability of falling between these two values, the shaded portion of Figure 8.4. To find this probability, we must first convert both raw scores to z scores. As it turns out, both fall one σ unit from μ: $z = (59 - 79)/20 = -1.00$ and $z = (99 - 79)/20 = +1.00$. Our next task, then, is to find the percent frequency between $z = \pm1.00$. Looking at Table B, we find that 34.13 percent of the frequency falls between μ and $z = +1.00$ and the same percent between μ and $z = -1.00$. Thus 34.13×2 or 68.26 percent of the boys fall between z scores of ±1.00 or between raw scores of 59 and 99 feet. This percentage can also be expressed as a proportion by moving the decimal points two places to the left: The proportion of the boys falling between the raw scores of 59 and 99 feet is approximately .68. This proportion reflects the *probability* that the score of a single boy, selected at random,

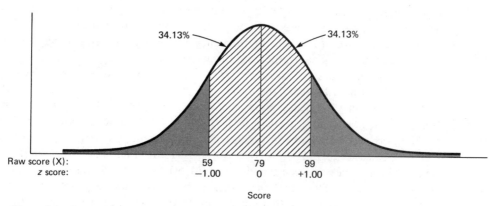

Raw score (X):   59   79   99
   z score: −1.00   0   +1.00

Score

Figure 8.4 Percentage of boys throwing between 59 and 99 feet, assuming normality of distribution.

fell between these two values, which is what our question asked. The answer to our question is thus .68.

The fifth question asks, What distances are so extreme that they are thrown by only 1 percent of the boys? This question, unlike the previous questions, does not specify whether it is the upper or the lower extreme that is of concern. Therefore, it should be assumed that the question refers to *both* extremes, the top $\frac{1}{2}$ percent plus the bottom $\frac{1}{2}$ percent, for a total of 1 percent. The problem we are working on may be clearer if we restate the question as follows: What scores (distances of throw) deviate from μ in either direction to such an extent that the *total* frequency of scores beyond these values is 1 percent? The problem is also shown graphically in Figure 8.5.

Our first step is to find the z score *above* which $\frac{1}{2}$ percent of the cases fall, the z score that divides the upper half of the curve into two portions with respect to area or percent frequency: 49.5 percent between μ and the z-score point, and .5 percent beyond this point. Looking in the body of Table B, we find that 49.5 percent lies midway be-

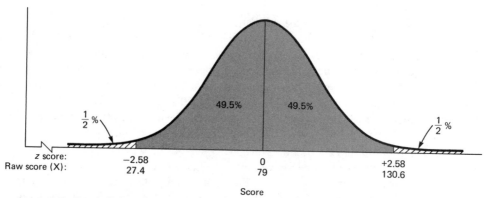

z score: −2.58   0   +2.58
Raw score (X):   27.4   79   130.6

Score

Figure 8.5 Baseball-throwing scores that occur 1 percent of the time or less, assuming normality of distribution.

tween $z = 2.57$ and 2.58, so we may choose either one. We decide to select 2.58; thus we may state that $\frac{1}{2}$ percent of the cases lie above $z = +2.58$. Correspondingly, $\frac{1}{2}$ percent lie below $z = -2.58$. We now have the answer to our question in z-score terms: the extreme 1 percent of the cases have z scores of ± 2.58 or greater. Our final step is to convert these two z-score values into their raw-score equivalents, using the formulas $\mu + z(\sigma)$ and $\mu - z(\sigma)$. Filling in the appropriate values, we thus have $79 + 2.58(20) = 79 + 51.6 = 130.6$, and $79 - 51.6 = 27.4$. The extreme 1 percent of the cases, then, have scores of 27.4 or less and 130.6 or greater.

The problem we have just solved was chosen to illustrate something we will now emphasize. The problem asked for the deviation beyond which a certain percent frequency occurs without specifying the direction of the deviation. As we will find in later chapters, this type of problem occurs often. We may state as a general rule that whenever the direction is *not* specified in a problem, you always assume the deviation in both directions is meant. In other words, when you are given a percent frequency and have to find the deviation corresponding to it, *unless it specifically states otherwise* the problem means that the deviation be such that half the percent frequency remains beyond it at each end of the curve.

Another reason why the above problem was chosen is that in the next chapter, when we take up the important problem of testing hypotheses, we will find that the frequency of 1 percent, one-half percent under each tail of the curve, has special significance. You should memorize the fact that 1 percent of the frequency in a normal distribution lies beyond $z = \pm 2.58$.

The final question asked the distances that could be thrown by the extreme 5 percent of the boys. This question is of the same type as the one we have just answered. It inquires about both extremes of the distribution, 2.5 percent at one end and 2.5 percent at the other. We must look in Table B for the z score that includes 47.5 percent of the area between μ and that z. (We obtain 47.5 by subtracting 2.5 from 50.) Doing so, we find the z of 1.96; 2.5 percent of the cases have a z of $+1.96$ or greater and 2.5 percent of the cases have a z of -1.96 or less. We next convert these z scores to raw scores: $79 \pm 1.96(20) = 39.8$ and 118.2. The extreme 5 percent of the cases have scores of 39.8 or less or 118.2 or more. The frequency of 5 percent—2.5 percent under each tail of the curve—also has special significance when we test hypotheses about populations. You should memorize that 5 percent of the cases in a normal distribution equal or exceed z scores of ± 1.96.

Percentiles from the Normal Curve

Table B is also useful for determining percentile ranks for raw scores from normally distributed data. It must be emphasized, however, that percentile ranks determined from Table B will not be accurate unless the raw scores are distributed normally or very nearly so. Recall that the percentile rank of a given score is the percentage of the total number of cases, the percent frequency, lying *below* that score. To illustrate, assume a normal distribution with a mean of 121 and a standard deviation of 20. What is the percentile rank of the raw score 141? As usual, we have to get the z score corresponding to the raw score; in this case $z = +1.00$. From Table B, 34.13 percent fall between this z-score value and μ. Therefore, the raw score 141 is above this 34.13 percent of the cases plus the 50 percent that are below μ, so the percentile rank is 84 (rounded). As you can see, we have merely found the total percent frequency falling below the given raw score. In other words, to use Table B to find the percentile rank for a raw score from normally distributed data, we have only to perform the usual step of converting the raw score to a z-score value and then, using Table B, find the total percent frequency falling below the z.

Let us also illustrate how to find the raw score corresponding to a known percentile rank; for example, what score in the distribution above (where $\mu = 121$ and $\sigma = 20$) corresponds to the 31st percentile? This must be the score below which 31 percent of the cases fall, which means 19 percent of the cases are between that score and μ. From Table B we find that .50 is the z score most closely corresponding to 19 percent. We now know that the raw score we are looking for is $(.50)(20)$ score units below μ; therefore the 31st percentile corresponds to a score of 111.

We have now considered a variety of normal curve problems; before going further it may be helpful to point out how similar they are. You should find the solution of normal curve problems easier if you keep in mind that all problems give some information expressed in one kind of unit, and that to answer the problem the given information will have to be translated into some other kind of unit. Specifically, the problem will give us at least one of the following units: a raw score, a z score or x/σ value, a percent frequency, or a raw frequency. Answering the problem will require translating from one to another of these units. Perhaps the diagram below will make this clearer. The top row shows the general procedure in all problems: We are given one kind of unit and we translate into another kind of unit. The two-headed arrows indicate that we can translate in either direction. Note, however, that whether we are translating from left to right or right to left, *we cannot skip a step.* To get from X to %f we have to go from X to z, then from z to %f; to get from f to X, we must first convert f to %f, then %f to z, then z to X.

$$X \longleftrightarrow z \longleftrightarrow \%f \longleftrightarrow f$$

$$\frac{X - \mu}{\sigma} = z \longrightarrow \text{Table B} \longrightarrow \frac{\%f(N)}{100} = f$$

$$X = z\sigma + \mu \longleftarrow \text{Table B} \longleftarrow \%f = \frac{f}{N}(100)$$

The other two rows of the diagram show the specific steps to take. Thus, suppose a problem gives a raw score (X) and asks for the percent frequency (%f), or raw frequency (f), above or below that score. For example, how many boys scored below 50 feet in the baseball-throwing distribution? We would proceed from *left to right* in the top row of the diagram and we would also go through the steps shown in the *second* row.

On the other hand, suppose the problem gives a percent frequency or raw frequency and asks for either a z score or raw score. As an example, the top 30 boys are above what score in the baseball distribution? In this case we would go from *right to left* in the top row of the diagram and would also go through the steps shown in the *bottom* row.

The point we want to make is that there are really not several different kinds of normal curve problems. All normal curve problems are essentially the same; they only look different because they can be stated in so many different ways. One more thing: we strongly advise drawing a graph for each problem. You often will find that if you express graphically the information given in a problem, it will help you get started answering it.

DEFINITIONS OF TERMS AND SYMBOLS

Binomial probability distribution. Theoretical distribution giving the probability of each possible number of hits (or favored event) and misses (or nonfavored event) when N independent attempts are made. The probability of a hit is p, and the probability of a miss is $q = 1 - p$.

Binomial expansion. Expansion of the binomial $(p + q)^N$ to give the probability of each number of hits, r. This probability is given by

$$\frac{N!}{r!\,(N - r)!}\, p^r q^{N - r}.$$

Mean of binomial (μ_b). Mean of the binomial distribution. $\mu_b = Np$, where N is the number of observations and p is the probability of the favored event.

Standard deviation of binomial (σ_b). This standard deviation is given by $\sigma_b = \sqrt{Npq}$, where N is the number of observations, p is the probability of the favored event, and q is the probability of the favored event not occurring, with $q = 1 - p$.

Normal probability distribution. A particular theoretical probability distribution, based on continuous measures, of an infinite number of scores. The shape of the normal distribution is symmetrical and somewhat like a bell, rising to a rounded peak in the middle and tapering off symmetrically at both tails. The percentage of the area under the normal curve between μ and each sigma (z score) distance from μ is fixed and unique to the normal curve.

Inflection point. Point on a curve such as a normal curve where the curve changes direction from being concave to convex or vice versa. In a normal curve, the inflection points occur at $z = \pm 1$.

PROBLEMS

1. As a promotional gimmick, a grocery store gives each customer at the checkout counter a card with four coated lines on it. The customer is to pick one line and scrape off the coating. Behind one of the lines is the name of an item of merchandise that the customer wins if he or she picks that line. Nothing is printed on the other lines, and the customer who chooses one of these lines "loses." The cards have been

printed so that one-fourth of them have the "winner" on each line. One customer accumulates three cards before making her choice of a line for each card.

(a) On these three cards, how many *different* patterns of "winners" and "losers" are possible and what is the probability of each?

(b) Expand the binomial (Formula 8.1 in text) to find the probabilities of obtaining 0, 1, 2, and 3 winners on the three cards.

(c) Determine the probability of getting two losers and one winner by Formula 8.2. Check to see if your answer agrees with the probability value you calculated for one winner in (b).

(d) What is the probability of having two *or* three winners?

(e) What is the mean and the standard deviation of the distribution of number of losers in the set of three cards?

2. A test measuring knowledge of the U.S. Constitution has been given to a national sample of 5,000 high school seniors. The distribution of scores closely approximates the normal curve and has a mean of 75 and a standard deviation of 12.

(a) What percentage of those tested scored 63 or below?

(b) What percentage scored between 63 and 87?

(c) *How many* of the 5,000 individuals scored between these score values?

(d) What percentage scored between 81 and 93?

(e) A raw score of 66 has a percentile rank of what?

(f) What is the raw-score equivalent of the 67th percentile?

3. The 5,000 high school seniors were also given a test of their knowledge of current events. This distribution also closely approximated the normal curve and had a mean of 58 and a standard deviation of 10.

(a) What is the probability that a senior chosen at random would earn a score of 62 or below?

(b) What is the probability that a randomly chosen student would have a score that deviated 5 or more score units *in either direction* from the mean?

(c) The extreme 5 percent of the seniors (2.5 percent at either end) earned raw scores of what and above or what and below?

Sampling Distributions: Single Samples

In Chapter 8 we discussed the binomial distribution–the distribution expected from a large number of samples of events in situations in which each event has one of two possible outcomes. For example, in Figure 8.1 we plotted the relative frequency, in an infinite number of samples of 10 coin tosses each, with which samples with 0 to 10 heads would theoretically occur. Although we did not then describe it as such, Figure 8.1 represents what is called a *sampling distribution*. A sampling distribution is a theoretical distribution, based on a mathematical model, that represents the *distribution of values of a statistic that would be obtained from an infinite number of random samples of a given size.*

In this chapter we will extend our discussion of sampling distributions to include the distribution of sample *means*, based on the measurement of continuous variables, and will consider the use of sampling distributions in making inferences about population characteristics.

SAMPLING DISTRIBUTION OF MEANS

When we measure the amount of some characteristic that each member of a random sample exhibits, one of the statistics we compute from the sample data is $\overline{X}$. Our purpose in determining $\overline{X}$ is not merely to describe the individuals in the sample, but also to get an idea about the value of the population mean. The likelihood is very small that this $\overline{X}$ has *exactly* the same value as μ of the population from which the sample was selected. That is, if we measured a large number of samples from the population, we would discover that the values of the sample $\overline{X}$'s were varied and that few were exactly the same as μ. If we were to use any given $\overline{X}$ as an estimate of μ, then, our estimate would be likely to contain a certain degree of inaccuracy. This fact is known as the *sampling error* of $\overline{X}$'s with respect to μ.

In Chapter 4 we said that the value of $\overline{X}$ was the best *single* guess we could make about the value of μ. But if $\overline{X}$ is not expected to be the same as μ, how can we use the sample $\overline{X}$ as indicative of the population mean? The best answer is, first, that sample means show no systematic tendency to be either larger or smaller than μ and hence can serve as an unbiased estimate of μ. Second, the mean of a truly random sample rarely deviates far from the population mean, particularly if the sample is large, so that $\overline{X}$ is usually a fairly close approximation of μ. In practice, however, we seldom use $\overline{X}$ to assign a precise value to μ; instead we use the sample data to test some specific hypothesis about the value of μ or to make estimates about a range of values that might reasonably be expected to contain μ. We will go on now with further elaboration of this important problem.

MEANS OF SUCCESSIVE RANDOM SAMPLES

Strictly speaking, the steps in reasoning we are going through now are based upon the assumption that we are drawing an infinite number of samples from an infinite population. However, it is quite satisfactory for illustrative purposes to go through the statistical logic with a finite population and a finite number of samples. We will follow this procedure for the clarity it affords.

Suppose, then, we take successive random samples from a large, known population. We draw one sample of subjects and measure them on some aspect of behavior; then we replace the sample and draw another of the same size, measure the subjects in it, and so on. Let us say we draw 5000 samples in this way, each sample consisting of 225 cases. We then calculate $\overline{X}$ for each of the samples. The result is a set of 5000 numbers, each of which represents the mean of a sample. We can describe these 5000 numbers in the same way we describe raw scores. That is, the mean of the 5000 $\overline{X}$'s can be determined in the ordinary way by adding them up and dividing by 5000. Similarly, the same procedures used with raw scores (X's) can be used to determine the standard deviation of the set of 5000 $\overline{X}$'s. And, of course, we can order the 5000 $\overline{X}$'s and cast them in the form of a frequency distribution. This frequency distribution is given a special name: *the sampling distribution of $\overline{X}$.*

The sampling distribution of $\overline{X}$'s can also be presented graphically as a frequency polygon, the values of $\overline{X}$'s being placed along the baseline (X axis) and the frequencies on the ordinate (Y axis). When the size of each of the samples is relatively large, the frequency distribution of $\overline{X}$'s approximates a *normal distribution* even if the distribution of measures (X's) in the population is not normal.

Suppose further that the 5000 random samples we have obtained, each composed of 225 cases, come from a population whose μ and σ are known. Let us say that these values are 90 and 7.5, respectively. The distribution of the 5000 $\overline{X}$'s will be approximately normal, as shown in Figure 9.1. What do we expect the mean of this distribution of $\overline{X}$'s to be? Recall that $\overline{X}$'s are *unbiased estimates* of μ. That is, overall the $\overline{X}$'s are not systematically larger or smaller than μ. Further, errors of overestimation of a given size occur equally as often as errors of underestimation of μ. The mean of a sampling distribution of $\overline{X}$'s, therefore, can theoretically be expected to take the *same value as μ for the*

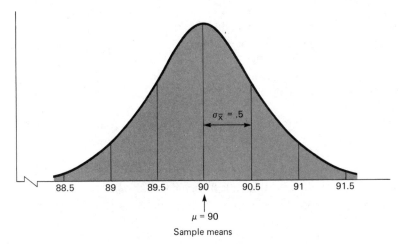

Figure 9.1   Sampling distribution of $\overline{X}$'s of 225 cases each from a population in which μ = 90 and σ = 7.5.

population. In our illustration, we have said that μ is 90. In Figure 9.1, the mean of the sampling distribution of $\overline{X}$'s can be seen to take this same value. Thus another way of defining lack of bias in $\overline{X}$ as an estimate of μ is to say that the average $\overline{X}$ in a sampling distribution is equal to μ, that is, the population parameter that is parallel to the statistic obtained for each sample.

In addition to calculating the mean of the sampling distribution of $\overline{X}$'s, we could also calculate its standard deviation by treating each $\overline{X}$ as if it were a raw score and applying the procedures described in Chapter 5. The standard deviation of the sampling distribution of $\overline{X}$'s is given a special name, the *standard error of the mean*, and is symbolized as $\sigma_{\overline{X}}$. The word "error" is used to label this particular standard deviation because each $\overline{X}$ in the distribution that differs from μ incorporates a sampling error and is not an accurate estimate of μ.

Although the standard deviation of the sampling distribution of $\overline{X}$'s—that is, the standard error or $\sigma_{\overline{X}}$—could be calculated directly, an alternate method is available. Statisticians have demonstrated that the value of the standard error of the mean, $\sigma_{\overline{X}}$, has a constant relationship to the standard deviation of the population from which the samples were obtained and to the size of each of the samples. The exact relationship is given by the formula:

$$\sigma_{\overline{X}} = \frac{\sigma}{\sqrt{N}} \tag{9.1}$$

where N = sample size

We specified earlier that the σ of our population is known to be 7.5. Since each sample N is 225, $\sigma_{\overline{X}}$ can be calculated as:

$$\sigma_{\overline{X}} = \frac{7.5}{\sqrt{225}} = .5$$

The sampling distribution in Figure 9.1 therefore has a mean of 90, the same value as μ, and a standard deviation or standard error of .5.

Inspection of Formula 9.1 shows that as sample N's grow larger, $\sigma_{\overline{X}}$ grows smaller. This reflects the fact that as the size of the individual samples is increased, the $\overline{X}$'s of the samples are increasingly likely to be close to μ and hence to the mean of the sampling distribution.

You can see why this is so by considering the following example. Assume that the mean height of women in the United States is 5 feet 6 inches. If a series of random samples were obtained, each containing 500 women, the $\overline{X}$'s would all be very close to the population value. But if the samples consisted of only two women each, it could easily happen that an occasional sample happened by chance to be composed of two shorties of 5 feet or less or of two 6-footers. With small samples, considerable variability among $\overline{X}$'s can be expected.

To repeat, as sample size is increased, $\overline{X}$'s cluster closer and closer around μ. As sample size is increased, the distribution of $\overline{X}$'s also increasingly resembles the normal probability distribution. These relationships with sample size comprise what is known as the *central limit theorem.*

Knowing the shape of the sampling distribution and its mean and standard deviation, we are in a position to make probability statements about $\overline{X}$'s. In Chapter 8, we discovered that a case drawn at random from a normal distribution has a probability of approximately .68 of falling between the points established by $+1\sigma$ and -1σ—that is, between z scores of ± 1. Similarly, we determined from Table B that an individual has a probability of .05 of falling at or beyond the point established by a z of $+1.65$. Since for all practical purposes our sampling distribution of $\overline{X}$'s can be assumed to be normal when N is large, we can make similar probability statements about a single sample $\overline{X}$. Thus, for example, the probability that an $\overline{X}$ will fall within the interval bounded by a z of ± 1 in the sampling distribution is .68.

We now ask a more concrete question about the distribution in Figure 9.1. What is the probability that a $\overline{X}$ will deviate from the μ of 90, *in either direction*, by 1 or more points? That is, what is the probability of obtaining $\overline{X}$'s of 91 or greater or of 89 or less? We first must convert this deviation into a z score, defined, you recall, as the score minus the mean of the distribution, divided by its standard deviation.

The "score" in this instance is $\overline{X}$; the mean of the sampling distribution takes the value of the μ of the population; and the distribution's standard deviation is $\sigma_{\overline{X}}$. Thus:

$$z = \frac{\overline{X} - \mu}{\sigma_{\overline{X}}} = \frac{x}{\sigma_{\overline{X}}} \tag{9.2}$$

Filling in the values for our example, $\pm 1/.5$ is equal to a z of ± 2.00. Consultation with Table B shows that approximately 2.28 percent of the area under the normal curve lies at or beyond a z of $+2.00$. The same percent lies at or beyond a z of -2.00. Thus 4.56 percent of the area lies at or beyond $z = \pm 2.00$. The probability of drawing a random sample whose $\overline{X}$ is 91 or greater or 89 or less is therefore .0456.

Our aim, however, is not to learn how to make probability statements about obtaining sample $\overline{X}$'s of certain values, given the values of the population μ and σ. If we had this information about the population, there would be little purpose in measuring samples. Our goal is the reverse: to make inferences about the unknown μ, given data from a single random sample. If we knew the standard error of the sampling distribution of $\overline{X}$'s, we would be able to make these inferences about μ. Unfortunately, when we do not know the population μ, we also do not know the standard error of the mean, $\sigma_{\overline{X}}$. We can, however, calculate an *estimate* of $\sigma_{\overline{X}}$ from the data of a sample. Recall from Chapter 5 that estimating a single value for a population parameter from the data of a sample is known as *point estimation.*

A Point Estimate of the Standard Error of the Mean

The $\sigma_{\overline{X}}$ of a sampling distribution, we have seen, can be determined by dividing σ, the standard deviation of the population, by the square root of N, the number in each sample. This formula is used as a basis for estimating the standard error of the mean from the data of a single sample by substituting for the unknown σ an *estimate* of σ. Recall first from Chapter 5 that sample variances tend to be smaller than the variance of their population. However, an unbiased estimate of σ^2, symbolized as s^2, can be obtained by dividing Σx^2 by $N - 1$ rather than N. The square root of the resulting number, symbolized as s, yields a corrected estimate of σ. Thus:

$$s = \sqrt{\Sigma x^2/(N - 1)}$$

Our estimate of the standard error of the mean, symbolized as $s_{\bar{X}}$, is obtained by substituting s for σ in Formula 9.1. Thus:

$$s_{\bar{X}} = \frac{s}{\sqrt{N}} \qquad\qquad (9.3)$$

$$\text{where} \quad s = \sqrt{\Sigma x^2/(N - 1)}$$

$$N = \text{sample size}$$

Before proceeding further, let us state once more what $s_{\bar{X}}$ represents. It is an *estimate*, based on the data from a single sample, of the *standard deviation of a distribution of randomly drawn sample means*. This standard deviation is given a special label, the *standard error of the mean*. In other words, $s_{\bar{X}}$ is an estimate of the *variability* of the sampling distribution of $\bar{X}$'s and thus reflects the *errors* in the distribution of $\bar{X}$'s in estimating μ.

We can now demonstrate the calculation of $s_{\bar{X}}$. We have a random sample of 25 cases with s of 15. Thus:

$$s_{\bar{X}} = \frac{s}{\sqrt{N}} = \frac{15}{\sqrt{25}} = 3.00$$

Suppose that we increased our sample N from 25 to 100 cases, but again found $s = 15$. In this instance, we calculate $s_{\bar{X}}$ to be $15/\sqrt{100}$ or 1.5. This value is considerably smaller than the one calculated above; in fact, it happens to be just half the size of $s_{\bar{X}}$ where $N = 25$. This example illustrates the implications of the central limit theorem: as sample N's increases, $\bar{X}$'s in a sampling distribution become less and less variable, clustering closer and closer around μ.

We now have before us the statistical logic that will permit us to make inferences about population μ's from sample data. Several approaches are open to us. Sometimes we wish to test a hypothesis that the μ of a population is some particular value and collect data from a sample of individuals so that we may evaluate how likely it is that the hypothesis is true. On other occasions we have no prior hypothesis about the value of μ. We could make a point estimate, specifying that the $\bar{X}$ of the sample we have measured is our best estimate of μ. However, when we have no prior hypothesis, we typically make what is known as an *interval estimate*, in which we establish a range or interval of values within which μ probably falls. Interval estimation will be discussed further in the section on confidence intervals later in the chapter.

TESTING HYPOTHESES ABOUT μ: LARGE SAMPLES

We conduct a study to determine the beliefs of American middle-class adults about people who are impoverished. One of the questions asks for a rating, on a 7-point scale, of the relative interest of the poor and nonpoor in having material things. A scale point of 1 indicates that the poor are very much *less* materialistic than the nonpoor, and 7 that the poor are very much *more* materialistic; a scale point of 4 indicates that the two are equally materialistic.

Before we test a random group of middle-class adults, we set up two rival hypotheses. The first hypothesis we choose to set up is that in the population as a whole, the mean scale point that would be chosen is 4.00—the hypothesis that in the population of middle-class adults, the poor and nonpoor are believed to be equal. (From a statistical point of view we could have selected any value from 1 through 7 for our hypothesis, but for other reasons we have chosen 4.) This hypothesis, which *always specifies an exact value for* μ, is known as the *null hypothesis* or H_0. The second hypothesis, called the *alternative hypothesis* and symbolized as H_1, is typically more general and specifies that μ is some value other than the one specified in the null hypothesis.[1]

The H_1, or alternative, hypothesis can be either *bidirectional or unidirectional.* That is, we may hypothesize merely that the true value of μ is not equal to the value specified in the null hypothesis ($\mu \neq$ value in H_0). This is called a bidirectional hypothesis because it specifies that μ is some value either above or below the value specified in H_0. Or we may specify the *direction* in which we expect the difference; for example, μ is some value *less,* but not greater, than the one specified in H_0 ($\mu <$ value in H_0). This is called a unidirectional hypothesis. The importance of the distinction between a unidirectional and a bidirectional alternative hypothesis will become clear later when we discuss the procedures used to evaluate H_0. We will see that we normally use a so-called *two-tailed* test when H_1 is bidirectional but may use a *one-tailed* test when H_1 is unidirectional.

Returning to our example, let us suppose that if the poor and nonpoor are *not* believed to be equally materialistic by middle-class adults—that is, if H_0 that $\mu = 4.00$ is not true—we have no sound reason for guessing which group would be judged to be more materialistic. Our H_1 would therefore be bidirectional. Our two hypotheses are therefore:

H_0: $\mu = 4.00$

H_1: $\mu \neq 4.00$

---

[1]The alternative hypothesis could be a specific value, let us say in our example a value of 2.00. However, it is more common to set up a more general H_1, specifying that μ may be *any* value that differs from H_0 in a given direction (unidirectional hypothesis) or in either direction (bidirectional). For example, if H_0 states that $\mu = 64$, a unidirectional H_1 might be that μ is some value greater than 64; the bidirectional H_1 would state that μ is some value greater *or* smaller than 64.

After stating H_0 and H_1, we test 121 individuals and find that the mean scale point chosen is 3.91 and that the s of the sample ratings is .70. We now proceed to evaluate the null hypothesis that $\mu = 4.00$.

What we want to do first is to determine what the sampling distribution of $\overline{X}$'s would look like if $\mu = 4.00$ and we had tested not one but a very large number of random samples of 121 cases each. We know that if $\mu = 4.00$, then the mean of the sampling distribution of $\overline{X}$'s would also equal 4.00. We must now inquire about the shape of the sampling distribution and the value of the standard error of the distribution, $\sigma_{\overline{X}}$. Since we do not know the value of σ, we have no way of calculating $\sigma_{\overline{X}}$. But we may obtain an *estimate* of the standard error, which we have agreed to symbolize as $s_{\overline{X}}$, by using the sample s given above:

$$s_{\overline{X}} = \frac{s}{\sqrt{N}} = \frac{.70}{\sqrt{121}} = .06$$

The standard error of the sampling distribution, we thus estimate, is .06.

With a sample as large as the one we have obtained, the *shape* of our distribution of sample means is close enough to normal to permit us to assume the normal distribution. Our hypothetical sampling distribution of $\overline{X}$'s under H_0, then, is normal in shape, has a mean of 4.00, and an estimated standard error of .06. This hypothetical distribution is shown in Figure 9.2. The figure also shows where our obtained $\overline{X}$ of 3.91 falls in this sampling distribution.

Ordinarily, our next task would be to find the z score of our obtained $\overline{X}$ value by dividing the quantity $(\overline{X} - \mu)$ by the standard error of the distribution. However, since

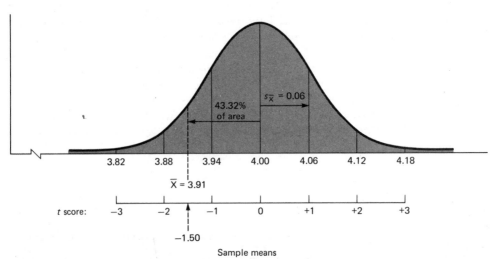

Figure 9.2 Visual test of the hypothesis that $\mu = 4.00$ when $\overline{X} = 3.91$ and $s_{\overline{X}} = .06$.

we can only estimate the standard error, we can only estimate z. The estimated z is identified as *Student's t* or simply t. ("Student" was the nom de plume of the statistician who did the original work on the distribution of estimated z or t.) The formula for this estimated z or t score is as follows:

$$t = \frac{\overline{X} - \mu}{s_{\overline{X}}} \tag{9.4}$$

This formula, of course, is the same one we used earlier for z (see Formula 9.2), except that we have substituted for the unknown standard error, $\sigma_{\overline{X}}$, our estimate of its value, $s_{\overline{X}}$. Applying this t formula to our data, we find:

$$t = \frac{\overline{X} - \mu}{s_{\overline{X}}} = \frac{3.91 - 4.00}{.06} = -1.50$$

Our obtained $\overline{X}$, then, falls 1.50 $s_{\overline{X}}$ units below μ. Since with a sample as large as ours the sampling distribution of t can be assumed to be normal, we can treat this value as if it were a z score and determine from Table B the percentage of the area lying between μ and $z = -1.50$. Table B indicates that 43.32 percent of the area under the curve lies between μ and -1.50, and that 6.68 percent of the area lies *beyond* -1.50. Since our alternative hypothesis, H_1, is bidirectional, we must be concerned with the other end of the distribution as well. We therefore state that if the null hypothesis, H_0, were correct, the probability of obtaining by chance a sample $\overline{X}$ that deviated from μ *this much or more* in either direction would be .134 (.0668 $\times$ 2 = .1336). In other words, if the study were repeated over and over and over again, sample $\overline{X}$'s that deviated this much or more in either direction from $\mu = 4.00$ would be expected approximately 13 percent of the time.

Significance Levels

We now face a decision: is H_0, the hypothesis that $\mu = 4.00$, a reasonable one? Or should we conclude that it is unreasonable and that H_1, the bidirectional hypothesis that μ is some value other than 4, is more likely to be correct? We need some decison rule, some cutoff point, that will allow us to reach a conclusion. One of two probability levels, .01 or .05, is typically used in scientific research for this purpose. We have determined in our example that if H_0 were true, the probability is .13 that we would obtain a $\overline{X}$ that deviated as much or more from the hypothesized μ as our obtained $\overline{X}$. Since this probability is greater than either .01 or .05, we conclude that our null hypothesis that $\mu = 4.00$ is a *reasonable* one.

When used to evaluate null hypotheses, the .01 and .05 probability levels are commonly called the .01 (or 1%) and .05 (or 5%) *significance levels*. The particular significance level an investigator decides to use in reaching a decision about the null hypothesis is called the *alpha* (α) level.

If the investigator decides to use the 1% or .01 significance level as the cutoff point in evaluating H_0, $\alpha = .01$. It is stated that if $\overline{X}$ would be obtained only 1% or *less* of the time by chance alone if H_0 were true, H_0 will be *rejected* as being unreasonable. In other words, instead of concluding that H_0 is correct and that this unusual ~~sur~~ $\overline{X}$ has occurred by chance, the investigator decides that H_0 is probably false and that H_1, the hypothesis that μ is some other value, is more reasonable. If, on the other hand, $\overline{X}$ could be obtained *more* than 1 percent of the time if H_0 were correct, the investigator does not reject H_0, deciding instead to retain it as a reasonable hypothesis about the value of μ. In stating a conclusion about whether H_0 is accepted or rejected, the investigator should also indicate the α level that was used to reach a decision.

The 5% significance level sets a less stringent criterion than the 1% level for rejecting the null hypothesis about μ. If investigators set α at .05, they will *reject H_0* if a difference as great as or greater than the difference between the obtained $\overline{X}$ and μ would occur by chance only 5 percent of the time or less (if H_0 were true.) Conversely, if a $\overline{X}$ this extreme would occur *more* than 5 percent of the time by chance, the investigators will retain H_0 as a reasonable hypothesis at the 5% significance level.

Now we take a slightly different approach to our study of beliefs about poverty. Suppose we had decided beforehand to use the 5% significance level in reaching a decision about the hypothesis that $\mu = 4.00$. We therefore will reject as unlikely any null hypothesis that would result in a sample $\overline{X}$ as different from μ as the one we have obtained 5 percent of the time or less by chance alone. We remind you that we have specified that H_1 is bidirectional; that is, we are concerned about deviations that occur 5 percent or less of the time in *either* direction from μ. In a normal distribution, 5 percent of the cases are expected to fall at or beyond a z of ± 1.96 with 2.5 percent at either extreme. With large samples, the distribution of estimated z scores or t's can be assumed to be normal. When the α level is .05 and H_1 is a bidirectional hypothesis, the critical values of t that we use to assess H_0 are ± 1.96. If the t we have computed for our obtained $\overline{X}$ is greater in absolute value than the critical value of 1.96, we *reject H_0*. We reject H_0 because there are only 5 chances in 100 or less that we would get a $\overline{X}$ that deviated this much or more from μ if H_0 were correct. Conversely, if our computed t has an absolute value that is less than 1.96, we retain H_0. In our example, t was determined to be -1.50. Absolutely, this t value is less than 1.96. We therefore do not reject, at the 5% significance level, the null hypothesis that $\mu = 4.00$.

These procedures are illustrated in Figure 9.3, which shows a t distribution based on large samples and which we therefore have assumed to be normal. In this figure a center portion of the curve, bounded by t's of $+1.96$ and -1.96, is marked off. If the value of any $\overline{X}$, expressed in $s_{\overline{x}}$ units (t score), falls within this area, we do not reject H_0; we accept H_0 as reasonable. If, on the other hand, the t score for $\overline{X}$ falls in either tail, we reject H_0. We saw above that our sample $\overline{X}$ fell only -1.50 $s_{\overline{x}}$ units from μ. We see in Figure 9.3 that this result falls within the "don't reject" area. We therefore retain, at the 5% significance level, the null hypothesis that $\mu = 4.00$; we conclude that H_0 is a reasonable one.

We could, of course, prepare a figure similar to Figure 9.3 for evaluating a bidirectional H_1 using the 1% alpha level. The extreme one-half percent of the area at either end of a normal curve falls at or beyond ± 2.58. The two tails of the distribution at or beyond t's of ± 2.58 thus define the area of rejection.

When H_1 is bidirectional and we consider both extremes of the z or t distribution in

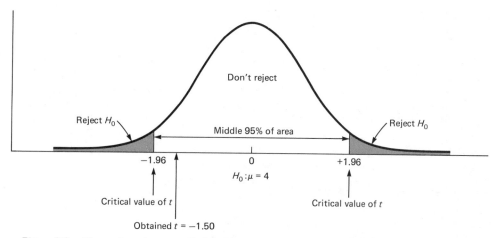

Figure 9.3 Illustration of a two-tailed significance test for μ at $\alpha = .05$ with normal curve values.

evaluating H_0, we are using a *two-tailed* test. When we have good reason to believe before collecting the sample data that any deviation of $\overline{X}$ from the null hypothesis will be in a specific direction, we may elect to set up an H_1 that is unidirectional and to employ a *one-tailed* test, one that considers the probability of $\overline{X}$'s falling at only one extreme of the distribution. For example, suppose that in investigating beliefs about poverty we also ask our sample of 121 to compare the ability of the poor and the nonpoor to manage their money effectively. Again, our subjects use a 7-point scale, with 1 indicating that the poor are much less able to manage their money, 4 that the poor and nonpoor are equal, and so on. We also choose again to test the null hypothesis that $\mu = 4.00$. We suspect, however, that the poor will be rated as *less* able to manage money than the nonpoor; that is, we believe μ will be less than 4. Our hypotheses are therefore:

H_0: $\mu = 4.00$

H_1: $\mu < 4.00$

(where $<$ means less than).

Suppose that the $\overline{X}$ turned out to be 3.75 and s to be .81. The $\overline{X}$, you will notice, is in the direction specified by H_1. Suppose also that we had decided to use $\alpha = .01$ instead of .05 as in the example above. The extreme 1 percent of the cases falling at *one* end of a normal distribution, we find in Table B, lie at or beyond a z of 2.33. Since H_1 specifies that $\mu < 4.00$, the critical value of t that is used to assess our null hypothesis is -2.33. We calculate t for our $\overline{X}$ of 3.75 as follows:

$$t = \frac{\overline{X} - \mu}{s_{\overline{X}}} = \frac{3.75 - 4.00}{.074} = -3.38$$

where $s_{\overline{X}} = s/\sqrt{N} = .81/\sqrt{121} = .074$

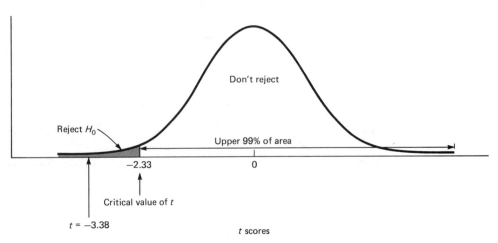

Figure 9.4   Illustration of a one-tailed significance test for μ at $\alpha = .01$ with normal curve values.

The t for our $\overline{X}$ falls far beyond -2.33. If H_0 were true, we would obtain an $\overline{X}$ this different from μ less than 1 percent of the time. As is illustrated in Figure 9.4, we therefore *reject* the null hypothesis that $\mu = 4.00$ and accept as more reasonable the unidirectional alternate hypothesis that μ is less than 4.[2]

Statistical vs. Experimental Hypotheses

The example above may have puzzled you. If we had good reason to expect that middle-class individuals believe that the poor manage their money less well than the nonpoor, why did we test the particular null hypothesis that we did—that they are judged, by middle-class individuals as a group, to be *equally* capable? That is, why did we test the hypothesis that $\mu = 4$ rather than a value less than 4? The answer is that we suspected that μ has *some* value less than 4 but had no basis for predicting how much less. The null hypothesis, however, must be assigned an exact value. We therefore chose to select 4.00 (no difference between poor and nonpoor) as the null hypothesis to test, with the hope that we could *reject* this hypothesis and accept, instead, the more general H_1 hypothesis that μ is *some* value less than 4.00.

The null hypothesis that we test, then, may sometimes be picked only for statistical reasons. Our actual *experimental* hypothesis—our guess about the true state of affairs that we conducted the study to evaluate—may not always correspond to the null hypothesis. Of course, there are instances in which our experimental hypothesis specifies an exact value for μ, rather than a range of values, such as less than 4. In these instances the statistical null hypothesis we test is the same as our experimental hypothesis.

---

[2]What if the difference between $\overline{X}$ and the μ specified in H_0 turns out to be in the *opposite* direction as specified in a unidirectional H_1? Obviously, H_1 can be rejected without further ado. Unfortunately, it is not legitimate to test H_0 by changing H_1 to a bidirectional form after the fact and performing a two-tailed significance test or to change the direction of H_1 to correspond to the results.

One-Tailed vs. Two-Tailed Tests. When the null hypothesis an investigator tests statistically does not coincide with the experimental hypothesis, the investigator typically has predicted beforehand the direction in which $\overline{X}$ is expected to differ from the μ specified in H_0. There is some controversy among statisticians as to whether a unidirectional H_1 should be set up and H_0 evaluated by means of a one-tailed test when the investigator has predicted the outcome or whether all tests of H_0 should be two-tailed. We will not attempt here to discuss the reasons for the controversy. We will say only that one-tailed tests are sometimes used when the investigator has good reason to set up a unidirectional H_1. Despite the fact that investigators usually have unidirectional experimental hypotheses, bidirectional *statistical* hypotheses are much more common. It typically is assumed that H_1 is bidirectional and that a two-tailed significance test has been performed, unless a unidirectional H_1 and a one-tailed test are specifically mentioned. (We will follow this practice in all subsequent discussions.) In reporting the results of our statistical analyses, then, we should always specifically mention the use of a one-tailed probability figure if we have employed this procedure.

Unidirectional H_1's could be discussed in connection with other kinds of significance tests introduced in later chapters. We have chosen not to do so, however, both because no new principles are involved and because they are used less frequently than bidirectional H_1's.

TESTING HYPOTHESES ABOUT μ: SMALL SAMPLES

The central limit theorem states that as samples increase in size, the shape of the sampling distribution of $\overline{X}$'s approaches the normal curve. When the sample N is large, the sampling distribution of t so closely approximates the normal curve that for practical purposes we may treat it as if it were normal. In our previous examples, all of which involved large samples, we therefore used the z score values found in the normal curve table in testing hypotheses about μ using the 1% and 5% significance levels.

The t values obtained from small samples are *not* normally distributed. Assuming that the measured characteristic is normally distributed in the population from which the samples were drawn, the distribution of small sample t's is, however, *symmetrical* and its exact shape varies systematically with the degrees of freedom associated with the samples. Degrees of freedom (df), as explained in Chapter 5, refers to the number of values in a distribution of measures that are free to vary and still yield the original value of $\overline{X}$. The df associated with t for single samples is N - 1, the number of cases in the sample minus one.

$$df = N - 1 \tag{9.5}$$

As sample df's become smaller, the shape of the sampling distribution of t becomes increasingly flatter than the normal curve, with the peak being lower and the tails of the distribution being "pulled out" farther along the baseline. This increasing flatness means that we must go farther and farther along the t scale on the baseline to encompass the same area.

The upshot of all of this is that when samples are small, we cannot use normal curve values in assessing hypotheses about μ. In making decisions about H_0 at a given α level, we must instead use the critical values of t at $\alpha = .05$ or $.01$ that are associated with the t distribution for the particular df associated with our sample. Special tables have been prepared that list these critical values; one such is Table C in Appendix II. At the left of Table C is a column labeled df (degrees of freedom). In the other columns are listed the t values for each df that are required at the 10%, 5%, 2%, and 1% significance levels for a two-tailed test, and at the 5%, 2.5%, 1%, and 0.5% significance levels for a one-tailed test.

As we described earlier, as sample N and therefore sample df decrease, the critical value of t associated with a given α level increases. Inspection of the second page of Table C shows, for example, that with $df = 120$, the value of t at $\alpha = .05$ for a two-tailed test is 1.9799; this figure is very close to the normal curve value of 1.9600. With $df = 50$, the corresponding value is 2.0086 and with $df = 10$, it is 2.2281.

As an example of the use of Table C, assume that an investigator has hypothesized that μ is 80 and has specified a bidirectional H_1. To test this hypothesis, the investigator has measured 25 individuals, finding $\overline{X} = 84$ and $s = 8$. Thus:

$$H_0: \mu = 80$$

$$H_1: \mu \neq 80$$

$$t = \frac{\overline{X} - \mu}{s_{\overline{X}}} = \frac{84 - 80}{1.6} = 2.50$$

where $\quad s_{\overline{X}} = s/\sqrt{N} = 8/5 = 1.60$

$$df = N - 1 = 25 - 1 = 24$$

Consultation of Table C shows that the two-tailed values of t needed at $\alpha = .05$ and $.01$ for $df = 24$ are 2.06 and 2.80. Since the calculated t is 2.50, H_0 can be rejected at the 5% but not at the 1% significance level.

How Small Is Small? We have emphasized that when the sample N is small, we must use the values in the t table associated with the sample df in evaluating the significance of a calculated t. But we have not stated how large a sample must be before we may reasonably assume the normal distribution and use the values in Table B or, conversely, how small a sample must be before we use the values listed in Table C. There is no precise answer to this question, but inspection of the bottom entries in Table C shows that with $df = 120$, the critical value of t at a given α level is very similar to the parallel value in a normal curve, obtained with an infinite df. Thus we have chosen to assume the normal distribution in all sampling distributions of t in which df is greater than 120.

Steps in Testing a Hypothesis about μ

It may be helpful at this point to pause and list the steps that are followed in testing a hypothesis about μ.

1. Set up a null hypothesis about the value of μ and an alternate hypothesis:

   H_0: μ = some specific value

   H_1: $\mu \neq$ value in H_0 (bidirectional hypothesis)

   OR

   H_1: $\mu >$ value in H_0 (unidirectional hypothesis)

   OR

   H_1: $\mu <$ value in H_0 (another unidirectional hypothesis)

2. After testing a random sample of individuals and determining $\overline{X}$ and s, compute t and df.

   $$t = \frac{\overline{X} - \mu}{s_{\overline{X}}} \qquad df = N - 1$$

   where $\quad s_{\overline{X}} = s/\sqrt{N}$

   μ = value specified in H_0

3. Look up the critical values of $t_{.05}$ and $t_{.01}$ in Table C for the calculated df. If H_1 is bidirectional, these are the values associated with a two-tailed test. If H_1 is unidirectional, these may be the values associated with either a one-tailed or a two-tailed test.

4. Compare the calculated t with the critical value of t at a given α level. When a two-tailed test is being conducted and the calculated t is *larger* in absolute value than the critical value, reject H_0 at that α level. If the calculated t is *smaller* in absolute value than the critical value, retain H_0.

   If a one-tailed test is being conducted, H_0 is rejected when the calculated t is larger than the critical value of t in the direction specified by H_1; that is, the obtained t must be *negative* in sign when H_1 specifies that μ is *less than* the value stated in H_1 or must be *positive* in sign when H_1 specifies that μ is *greater than* the value stated in H_0. If the obtained t has the sign predicted by H_1 but is less than the critical value at the given α level, *or*, whatever its absolute value, if t has the sign opposite to that predicted in H_1, H_1 is rejected and there is no choice but to retain H_0.

5. In stating the conclusion about H_0, the α level and df should be specified. If a one-tailed test has been used, this should be stated. It will otherwise be assumed that a two-tailed test was conducted.

Two Types of Error

Let us repeat once more what we have done when we reject a null hypothesis about μ at a given α level. For example, we reject a hypothesis at the 5% level. We say that if H_0 is correct, a difference between $\overline{X}$ and the hypothesized μ as large as or larger than the one we have obtained would occur by chance 5 percent of the time or less. Rather than concluding that by chance our sample happened to deviate this much from μ, we decide instead that the null hypothesis we tested is faulty. But we could be wrong. In fact, we know that the probability of being wrong is .05. If H_0 were *true* and we repeated our study over and over again, we would get $\overline{X}$'s that differed this much or more from μ 5 times out of 100 purely by chance. Thus, in rejecting the null hypothesis on a given occasion, we run the risk of labeling as false a hypothesis that is in fact true. Rejecting a true null hypothesis is known as a *Type I error*. The Type I error also is identified as an α error.

How might we reduce the likelihood of making a Type I error? The answer is simple enough: we set a more stringent α level for rejecting H_0. If we used $\alpha = .01$, we would demand that a deviation between $\overline{X}$ and μ occur by chance no more than 1 time in 100, thus cutting down the probability of rejecting a true hypothesis to .01. We would be even surer of avoiding a Type I error by using the .001 α level (only 1 chance in 1000 of obtaining the results by chance), and so on. However, while we reduce the risk of a Type I error by lowering our α level, we simultaneously *increase* the likelihood of making what is called a *Type II error*. In a Type II error (also called a *beta* or β error), we retain (fail to reject) a hypothesis that is actually *false*. We can lower the risk of a Type II in exactly the *opposite* way as we minimized a Type I error: by *raising* the α level. If we use the 5% level of significance, for example, we will reduce Type II error—that is, fail to reject false hypotheses less easily than if we used $\alpha = 1\%$—and reduce this error even more if we raise α to 10 or 20%.

Although changing our α level has opposite effects on the likelihood of Type I and II errors, the probabilities of the two errors are not mathematically the exact complement of each other. That is, if α is 5%, the probability of a Type I error is .05 when we reject a hypothesis but may be *higher* than .05 for a Type II error when we accept a hypothesis. You may begin to appreciate the reason why Type II errors may be more frequent when you consider the following. Suppose we postulate μ to be 52 and find $\overline{X}$ to be 53. After testing this H_0, we conclude at a given α level that it is tenable. But we could have set H_0 as $\mu = 53$ or $\mu = 54$, among other values, and we would have concluded that these hypotheses are also tenable. In short, a whole *range* of null hypotheses would be acceptable at a given α level. When we retain a null hypothesis, all we are saying is that it is *one* of a number of acceptable hypotheses. For this reason, investigators prefer not to speak of "accepting" the null hypothesis; they are more likely to be cautious and say they have *failed to reject* the null hypothesis.

As noted above in the section on statistical versus experimental hypotheses, investigators often set up null hypotheses to evaluate statistically that they hope to be able to reject. As a result, investigators have a special concern with the ability of their statistical tests to detect false null hypotheses. This ability is known as the *power* of the test. The topic of power will be discussed in detail in the next chapter. We will note here only that one of the factors affecting the power of a test is the sample N. That is, we are more likely to detect null hypotheses that are false, thus reducing the incidence of Type II errors, when the sample size is large than when it is small. This relationship can be understood by recalling that the formula for $s_{\overline{X}}$ is $s/\sqrt{N}$.

Assume, for example, that we have incorrectly hypothesized that $\mu = 122$ when it actually has some other value. Testing a sample of two, we obtain $\overline{X}$ of 119 and s of 10. Thus, $s_{\overline{X}}$ would be 7.07 and t under $H_0 : \mu = 122$ would be 3/7.07 or .424. We would thus *fail* to reject the false hypothesis that $\mu = 122$. Increasing N to 100, on the other hand, and continuing to assume $\overline{X} = 119$ and $s = 10$, $s_{\overline{X}}$ is calculated to be 1.00 and t is calculated to be 3.00; thus with N = 100, we would *reject* the false null hypothesis that $\mu = 122$ beyond the 1% significance level.

CONFIDENCE INTERVALS

In working with research data, we often do not have hypotheses about specific values of μ that we wish to test. Suppose, for example, that we were interested in the satisfaction of assembly line workers with their jobs and were able to persuade a random sample of 144 workers to take a test that measured job satisfaction. We have no prior hypothesis about how satisfied the average worker is and, in fact, have done the study to find out what workers' feelings are like. It would be possible to use the $\overline{X}$ of the sample as an exact estimate of the population μ. However, rather than making this kind of point estimate, it is more common to make what is known as an *interval estimate*. That is, we establish a range of values within which we may reasonably assert that μ lies. The assertion that μ falls somewhere within a specific range or interval of values has a certain probability of being correct; we make it with a certain degree of confidence. When we establish such a range, it is commonly spoken of as a *confidence interval*. In research reports two confidence intervals are used most frequently: the *99% confidence interval* and the *95% confidence interval*.

Ninety-Nine Percent Confidence Interval

You will remember that if we were testing a hypothesis about μ using $\alpha = .01$ (and a bidirectional H_1), we would reject any H_0 that would make an absolute difference between μ and $\overline{X}$ as large as or larger than the one we have obtained occur 1 percent of the time or less by chance alone (that is, a positive difference between μ and $\overline{X}$ of the size we have obtained or larger would occur by chance one-half percent of the time or less and a negative difference of this size or larger one-half percent of the time or less). Conversely, we would retain as reasonable *any* H_0 that would make the difference between $\overline{X}$ and μ occur by chance more than 1 percent of the time. As we indicated in an earlier discussion, there is a whole range of reasonable or acceptable hypotheses at a given α level. If $\alpha = 1\%$, the range of acceptable hypotheses for μ establishes what is known as the *99% confidence interval* for μ. That is, we assert that μ falls someplace within this range of values, 99% of such statements being correct, and conversely, 1% of such statements being incorrect. Let us now demonstrate how to go about establishing the 99% confidence interval.

If we were testing a bidirectional hypothesis about μ, we would convert $\overline{X}$ into a t score and using $\alpha = .01$, we would reject the hypothesis if the absolute value of our obtained t were greater than the critical value of t associated with the 1% significance level for the sample *df*. Conversely, if our obtained *t* were smaller than this critical value, we would accept the hypothesis.

Given $\overline{X}$ and *s* computed from our sample, how large could our hypothesis about μ be and still be accepted at $\alpha = .01$? We could answer this question by setting *t* at the value associated with $\alpha = .01$ for our *df* and then entering our obtained $\overline{X}$ and $s_{\overline{X}}$ into the right-hand side of the *t* equation. Finally, we solve for μ:

$$t_{.01} = \frac{\overline{X} - \mu}{s_{\overline{X}}}$$

and by rearrangement:

$$\mu = \overline{X} + t_{.01}(s_{\overline{X}})$$

Suppose that for the 144 workers to whom we gave a test of job satisfaction, $\overline{X}$ was 73 and *s* was 18, yielding $s_{\overline{X}} = 1.5$. With a sample this large, we use the normal curve figure, 2.58, as the critical value of $t_{.01}$. Thus the largest acceptable hypothesis about μ at $\alpha = .01$ is:

$$\mu = 73 + 2.58(1.5) = 73 + 3.87 = 76.87$$

We can find the smallest acceptable hypothesis about μ by setting $t_{.01}$ at a minus value and again solving for μ:

$$\mu = \overline{X} - t_{.01}(s_{\overline{X}})$$

In our example, the smallest value of μ that would be acceptable at $\alpha = .01$ is therefore:

$$\mu = 73 - 3.87 = 69.13$$

If we generalize from this series of steps, we see that the range of acceptable values for μ at $\alpha = .01$, which we identify as the 99% confidence interval, is given by:

$$99\% \text{ CI: } \overline{X} \pm t_{.01}(s_{\overline{X}}) \tag{9.6}$$

In our example, the 99% confidence interval is thus 73 ± 3.87 or 69.13 to 76.87. When we assert that the unknown μ falls within this range of values, 99% of such assertions will be correct.

One more example. We have tested 36 cases, and have calculated $\overline{X}$ to be 105 and $s_{\overline{X}}$ to be 1.78. Before we can determine the 99% confidence interval for μ, we must find the critical value of *t* for a two-tailed test at $\alpha = .01$. The *df* of our sample is 35—that is, $36 - 1$. In Table C, the value of *t* associated with *df* = 35 is 2.72. Therefore:

$$99\% \text{ CI: } \overline{X} \pm t_{.01}(s_{\overline{X}}) = 105 \pm (2.72)(1.78) = 105 \pm 4.84 = 100.16 \text{ to } 109.84$$

Ninety-Five Percent Confidence Interval

The 95% confidence interval establishes the range of acceptable hypotheses for μ with $\alpha = .05$. We determine the 95% confidence interval exactly as we did the 99% interval, except that we now use the value of t associated with $\alpha = .05$. The formula for the 95% confidence interval is:

$$95\% \text{ CI: } \overline{X} \pm t_{.05}(s_{\overline{X}}) \tag{9.7}$$

In our example in which $\overline{X}$ for 36 cases was 105 and $s_{\overline{X}}$ was 1.78, we first determine from Table C the value of t associated with 35 df for a two-tailed test. This value is 2.03. Thus:

$$95\% \text{ CI: } 105 \pm (2.03)(1.78) =$$
$$105 \pm 3.61 = 101.39 \text{ to } 108.61$$

You will note that the range of values established for the 95% confidence interval is narrower than the range of values established for the 99% confidence interval, which we calculated to be 100.16 to 109.84.

In using the 95% confidence interval to specify a range of values within which the mean of a population may lie we are less sure of including the true μ than with the 99% interval, but for the sake of greater specificity many research workers are willing to take the chance that μ does not fall beyond the limits set up by the 95% interval. The 95% confidence level, however, is usually considered the minimum probability value that should be used in setting up a range of values for μ. We could, for example, set up the 90% confidence interval. But then there would be 10 chances in 100, or 1 in 10, that μ lies outside this interval, and this probability of error is generally agreed to be too high. One could set up confidence intervals that fall between the 99 and 95% intervals, such as a 97% confidence interval. As a matter of custom, however, the 99 and 95% confidence intervals are commonly used as reference values. Similar remarks can be made about the use of the 5% and 1% significance levels in testing hypotheses about μ. Under ordinary circumstances the maximum α and thus the minimal significance level that should be used is 5%. If we go much beyond this level—for example, to the 10% level—we make the risk unacceptably high that we will commit a Type I error; that is, we will too frequently reject hypotheses about μ that are true.

The use of significance levels, which allow us to reject or not reject hypotheses with a stated probability, simplifies our writing of research reports. We merely state which one of the two α levels we are using and reject any hypothesis that exceeds this level. It is important to note once more that we must always state whether we are using the 5% or 1% α level in deciding to reject or not reject a hypothesis, so that there will be no misunderstanding. When we wish to establish a confidence interval for μ rather than to test a specific hypothesis about its value, we must also be sure to state whether it is the 99% or 95% interval we have determined.

TESTING HYPOTHESES ABOUT POPULATION PROPORTIONS

Formulas are available that allow us to obtain standard errors for the sampling distributions of other statistics than the mean and hence to test hypotheses about their population parameters. We will discuss here only one such technique, a test of a hypothesis about a population proportion.

Suppose we are investigating the biases that may influence choice behavior. It has been shown, for example, that in predicting heads or tails in the toss of a coin, people tend to pick heads; in guessing what number from 1 to 10 someone is thinking of, they tend to pick an even number; in a choice between right and left, they tend to choose right, and so on. We give a random sample of 50 individuals a novel series of tests, one of which involves asking them to guess whether the nonsense statement, "The sagem is bep," is true or false. If there is no bias, we expect that 50 percent of the subjects would choose true and 50 percent false. The null hypothesis (H_0) that we decide to test, then, is that the percentage choosing true is 50. We find that the actual numbers are 32 choosing true and 18 choosing false—percentages of 64 and 36. What should we conclude?

By this time we are wise enough not to jump to the conclusion that H_0 is false and that there is a bias for choosing true simply because more than half the sample chose this alternative. There are sampling errors involved, and the question we really need to ask ourselves is this: "If there is no bias—that is, if the population proportion choosing true is .50—how likely is it that we would obtain a proportion as high as .64 with 50 cases?"

It is critical that you recognize we are no longer dealing with measurement data—that is, the property we have observed in the individual case does not differ in amount. Rather, each individual has been assigned to one of two mutually exclusive categories according to the answer, true or false, to the test item. More particularly, we are discussing a binomial sampling distribution in which we determine, in each sample, the proportion of individuals choosing true, which we will identify as p, and the proportion of individuals choosing false, which we will identify as q, where $p + q = 1.00$.

In Chapter 8 we saw that when $p = .5$, the normal distribution provides a good approximation of the binomial distribution with moderate N's or greater, say 10 or more. It would be somewhat more precise to utilize the binomial distribution to answer our question, but it is usually more convenient to use the normal distribution.

If we drew a very large number of random samples of the same size from a population in which $p = .5$ and for each obtained a proportion, we may thus assume, if N is large enough, that these sample proportions would be distributed normally. If we calculated the standard deviation of this distribution, we could make statements of probability about where any sample would fall. (If these statements have a familiar ring, it is because we used the same language in discussing the sampling distribution of $\overline{X}$.) Thus, we need to know the standard deviation of the distribution of sample proportions—that is, the standard error of the proportion, σ_{prop}.

We recall from Chapter 8 that in a binomial sampling distribution in which we specify the frequency of a certain event out of a total of N occasions, the standard deviation, which is the standard error of the distribution of frequencies or σ_f, is given by $\sqrt{Npq}$. However, we are looking for the standard error of a distribution of sample *proportions*.

Since proportion is determined by dividing frequency by N, σ_{prop} may be obtained by dividing the expression for σ_f by N: $\sqrt{Npq}/N$. This may be reduced to $\sqrt{pq/N}$. Thus:

$$\sigma_{prop} = \sqrt{\frac{p_t q_t}{N}} \tag{9.8}$$

where p_t = true population proportion

$q_t = 1 - p_t$

N = sample size

It is important to note that p in the formula for σ_{prop} is not the obtained *sample* proportion; it is the *true* proportion in a known or hypothesized population.

The question we ask in our bias experiment, we repeat, is, "If the true population proportion is .50, is it likely that our obtained proportion, .64, occurred by chance?" In answering the question, we first calculate σ_{prop}:

$$\sigma_{prop} = \sqrt{\frac{pq}{N}} = \sqrt{\frac{(.5)(.5)}{50}} = \sqrt{\frac{.25}{50}} = .071$$

We can visualize our hypothetical sampling distribution and determine where our obtained proportion falls. This has been done in Figure 9.5. We can see by inspection that it would be quite unusual to get a sample proportion of .64 if the true proportion were .50. However, we need a more precise probability statement and must therefore convert our sample proportion (p_s) into a z score. Since we know σ_{prop} of the distribution, note that we can calculate the actual z rather than an estimate of it. The mean of the sampling distribution of sample proportions takes the same value as the value of the population proportion specified in our null hypothesis. The standard deviation or stan-

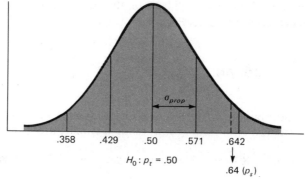

Figure 9.5 Visual test of the hypothesis that the true proportion is .50 when the obtained sample proportion is .64.

dard error of the sampling distribution is σ_{prop}. Thus the z formula may be written:

$$z = \frac{p_s - p_t}{\sigma_{prop}} \qquad\qquad (9.9)$$

where p_s = sample proportion and p_t = hypothetical true proportion specified in H_0. In our example, we thus have:

$$z = \frac{.64 - .50}{.071} = 1.97$$

If we had set α at .05 and our H_1 were bidirectional—that is, our alternate hypothesis to H_0 is $H_1: p_t \neq .50$—we would reject any null hypothesis that would result in a z score, for our p_s, of ± 1.96 or greater. We use ± 1.96 as cutoff points, of course, because these are the values associated with the extreme 5 percent of the area in a normal distribution, 2.5 percent at each tail. Our z was calculated to be 1.97. We therefore reject the null hypothesis that $p_t = .50$; we conclude that the difference between p_s and p_t is significant at the .05 level. More concretely, we conclude that people are probably biased in favor of "true."

It might seem that for any problem in which we obtain a proportion we could test any hypothesis about the true proportion for a given N, using the kinds of procedures we have just illustrated. A word of caution is in order, however. The binomial distribution, we remind you, generally approaches the normal distribution as N grows larger, doing so most rapidly when p_t is .50. We therefore should not test hypotheses about p_t based on the normal curve assumption unless N is some reasonable size, 10 at a minimum. As our hypothesis about p_t departs from .5, the sample N should be even larger than this minimum. If we test extreme hypotheses about true proportions, about .80 or greater or .20 or smaller, use of the normal curve may lead to serious errors of interpretation, when Ns are not large, because the sampling distributions will be markedly skewed. Under this circumstance we can, however, fall back on the less convenient but more exact binomial distribution discussed in Chapter 8.

DEFINITIONS OF TERMS AND SYMBOLS

Sampling distribution.   A theoretical distribution of values of a statistic such as $\overline{X}$, obtained from an infinite number of random samples of a given size.

Standard error of $\overline{X}$ ($\sigma_{\overline{X}}$).   The standard deviation of a sampling distribution of $\overline{X}$'s.

Estimated standard error of $\overline{X}$ ($s_{\overline{X}}$).   An estimate of $\sigma_{\overline{X}}$ based on the data from a single random sample. $s_{\overline{X}} = s/\sqrt{N}$.

Central limit theorem. The theorem that as sample size increases, the shape of the sampling distribution of $\overline{X}$'s increasingly resembles the normal probability distribution and the values of the $\overline{X}$'s cluster closer and closer to μ—that is, $\sigma_{\overline{X}}$ becomes increasingly smaller.

Null hypothesis (H_0). The hypothesis that a specified population parameter, such as the mean of a population (μ), takes some specified value. The null hypothesis is evaluated by the data from a single random sample.

Alternative hypothesis (H_1). The hypothesis that a population parameter, such as μ, has some value other than the value specified in H_0. A bidirectional H_1 merely states that the population value is unequal to the value stated in H_0. A unidirectional H_1 states the direction of the difference: either that the population parameter has some value greater than the one specified in H_0 or that it has some value less than the value specified in H_0.

Student's t (t ratio). The estimated value of z for a sample statistic in a sampling distribution of that statistic under a given H_0 about the parallel population parameter. In the case of a sampling distribution of $\overline{X}$'s, the t for an observed sample $\overline{X}$ is given by $t = (\overline{X} - \mu)/s_{\overline{X}}$.

Significance level (α level). The probability level that an investigator decides to use as a critical value in evaluating H_0. Investigators typically used either $\alpha = .05$ or $\alpha = .01$.

Two-tailed test. Both extremes of the t or z distribution are considered in evaluating H_0. A two-tailed test must be used to test H_0 when H_1 is bidirectional.

One-tailed test. Only one extreme of the t or z distribution is considered in evaluating H_0. A one-tailed test can be used only when H_1 is unidirectional. The extreme to be considered (upper or lower) is specified in H_1.

Type I (α) error. Rejecting a null hypothesis (H_0) as false when it is true. The Type I error is also identified as an *alpha* (α) error.

Type II (β) error. Failing to reject H_0 when it is false. The Type II error is also identified as a *beta* (β) error.

Confidence interval. The range of acceptable values for a population parameter at a given α level, determined from the data of a single sample. When $\alpha = .01$, the 99% confidence interval is established. When $\alpha = .05$, the 95% confidence interval is established.

Confidence interval for μ. The 95% confidence interval for μ is given by $\overline{X} \pm t_{.05}(s_{\overline{X}})$. The 99% confidence interval for μ is given by $\overline{X} \pm t_{.01}(s_{\overline{X}})$.

Standard error of proportion (σ_{prop}). The standard deviation of a binomial sampling distribution of sample proportions. $\sigma_{\text{prop}} = \sqrt{p_t q_t / N}$ where p is the population proportion of the favored event and q is $1 - p$.

PROBLEMS

1. Scores are available for all individuals who have taken a particular Civil Service examination over the past five years. These scores can be said to constitute a population of measures. The distribution of scores is found to be approximately normal. The mean and the standard deviation have the following values: $\mu = 78$, $\sigma = 10$.

   (a) If an infinite number of random samples of 25 cases each were drawn from the population and $\overline{X}$ determined for each sample, what would the resulting distribution of $\overline{X}$'s be called?

   (b) What numerical value would the mean of the distribution of $\overline{X}$'s be expected to take?

   (c) What is the standard deviation of the distribution of $\overline{X}$'s called? With samples of 25 cases each, what value would this standard deviation be expected to take?

   (d) If, instead, an infinite number of random samples of 49 cases each were drawn from the population and their $\overline{X}$'s determined, what numerical values would the mean and the standard deviation of this sampling distribution of $\overline{X}$'s be expected to take?

   (e) What would the standard deviation be if the samples consisted of 100 cases each?

   (f) Using the calculations you performed above, sketch three normal distributions of sample $\overline{X}$'s for 25, 49, and 100 cases each in a *single* graph. Show values of $\overline{X}$'s along the baseline and the mean and standard deviation in each distribution. (Suggestion: mark off the baseline in one-score units ranging from 72 to 84.)

   (g) After inspecting the figure in (f), describe how the variability in the distribution of $\overline{X}$'s changes as sample N's increase. As sample N's increase, $\sigma_{\overline{X}}$ approaches what value?

2. For a normally distributed population of scores, $\mu = 150$ and $\sigma = 16$. A single sample of 64 cases is drawn at random from the population.

   (a) What is the probability that the $\overline{X}$ of this sample of 64 cases is 153 or *greater?* (In answering this and the following questions, it might be helpful to draw first the sampling distribution of $\overline{X}$'s to be expected for an infinite number of samples with N = 64.)

   (b) What is the probability that this $\overline{X}$ deviates 4 or more units *in either direction* from the population μ?

   (c) The probability is .05 that the $\overline{X}$ of this single sample will have a value of what or above?

   (d) Suppose that the population μ and σ were unknown to you and that for a sample of 64 cases, $s = 17$. Using the sample data, estimate the standard error of the mean. Where does the discrepancy between $s_{\overline{X}}$ and the true $\sigma_{\overline{X}}$ come from?

3. A study of sex stereotypes about occupations is conducted in which 144 sixteen-year-old girls are given a list of occupations, each accompanied by a 7-point rating scale. Point 1 is labeled "much more appropriate for men," point 4 is labeled "equally appropriate for men and women," and point 7 is labeled "much more appropriate for women." For the occupation "salesclerk," the following data were obtained: $\overline{X} = 4.15$, $s = 1.20$, N = 144.

   (a) What is $s_{\overline{X}}$ for the sampling distribution of $\overline{X}$'s?

   (b) A sex stereotype is said to exist if it is believed that an occupation is more appropriate for one sex than another. In determining whether there is a sex stereotype among 16-year-old girls about salesclerks, what null hypothesis should you test? After specifying H_0, set up a bidirectional alternative hypothesis.

(c) Test H_0, using $\alpha = .05$. What do you conclude about this hypothesis? State your conclusions specifically, in terms of the stereotype about salesclerks.

(d) Your conclusion in (c) could be in error. Describe this error verbally and identify which type (I or II) it is.

(e) If you wanted to minimize this type of error, would $\alpha = .05$ or $\alpha = .01$ be preferable?

4. In the study of stereotypes described in problem 3, the following data were obtained from 25 sixteen-year-old boys for the occupation of lawyer: $\overline{X} = 3.12$, $s = 1.50$, $N = 25$.

(a) Specify H_0 and a unidirectional H_1. Explain your choice of H_1.

(b) Test H_0 against H_1, using $\alpha = .01$. What do you conclude? Be specific.

5. Assuming that the shape of the distribution in the population from which samples were drawn is normal, describe the shape of the sampling distribution of t as df's based on the sample N's go from low to high.

6. A manufacturer claims that its lightbulbs will last for an average of 500 hours. A consumer organization tests this claim, with the following results: $\overline{X} = 480$, $s = 50$.

(a) If the sample had consisted of 9 bulbs, what would you conclude about the validity of the manufacturer's claim? In conducting your test, set up a bidirectional H_1 and use $\alpha = .05$.

(b) If the sample had consisted of 16 bulbs and $\overline{X}$ and s were the same as above, what would you conclude?

(c) What would you conclude if the sample consisted of 49 bulbs?

(d) After examining your answers in (a), (b), and (c), state how sample size affects the power of the t test to detect a false null hypothesis at a given α level.

7. The waist circumferences of 144 twenty-year-old women are measured in centimeters, with the following results: $\overline{X} = 78.4$ cm, $s = 7.5$.

(a) What are the values that define the limits of the 95% confidence interval for the μ of the population of 20-year-old women?

(b) What are the values that define the limits of the 99% confidence interval?

8. An investigator is planning a study in which subjects are to be given pairs of words and to learn which arbitrarily chosen member of each pair has been designated as "correct." The investigator wants to select word pairs for which there is no initial selection bias among subjects and does a pilot study in which 40 subjects are shown a number of pairs and for each asked to guess which is "correct." On one pair, "grass-table," 22 pick "grass" and 18 pick "table." Perform a statistical test that will allow the investigator to conclude whether there is equal or unequal initial preference for the two words. Be sure to specify the α level you have used in reaching your conclusion. (Be sure to note in answering this question that the data involve two mutually exclusive categories and not measurement of some property.)

9. Each child in a sample of 20 five-year-old children was given a set of 9 objects consisting of 3 squares, 3 circles, and 3 triangles. For each shape, one object was yellow, one was blue, and one was red. The child was instructed to sort the objects, putting those together that belonged together. Of the 20 children, 15 sorted by color and 5 sorted by form. Is there an equal or unequal preference for form versus color?

Sampling Distributions: Independent Samples from Two Populations

CHAPTER **10**

Psychologists have begun to explore the possibility that under some circumstances, giving people tangible rewards for doing tasks they find enjoyable may lessen their interest in these tasks and may result in poorer task performance. An investigator who is interested in this topic gives second-grade children an attractive jigsaw puzzle to put together. The children, each of whom is seen individually, are randomly assigned to one of two conditions. In one condition, the child is given a bag of candy after completing the puzzle. In the other condition, no reward is given. Then each child is given a second puzzle, and the time the child takes to complete it is recorded.[1] The mean time taken to complete the second puzzle is 73 seconds for the children in the rewarded group and 62 seconds for the children in the nonrewarded group. Should the investigator conclude that reward leads to poorer performance? In this chapter, we consider techniques for answering this question.

Before getting to the statistics of the matter, we will discuss some of the features of the study just described and introduce some terminology. The study is an illustration of a basic two-group experiment. In such an experiment, the investigator manipulates the conditions under which subjects are tested, treating members of one group in one way and members of another group in another. In our example, there was an *experimental group* given a reward and a *control group* given no reward.

Although we will discuss in this chapter only two-group experiments, more than two conditions can be studied in the same investigation. For example, in our illustrative study, the investigation might include, as a second experimental condition, a group of children who are praised after completing the first puzzle for having done it so well. Techniques for analyzing experiments in which more than two conditions are manipulated are discussed in the next chapter.

The conditions the investigator manipulates are identified as the *independent variable*. The observations the investigator gathers from each subject are known as the *dependent variable*. The purpose of the experiment is to determine whether the variations in experimental condition, the independent variable, bring about variations in the dependent variable.[2] In our example, the investigator wished to find out whether the independent variable, reward condition, results in differences in the dependent variable, time to completion on the second puzzle.

To reach a reasonable conclusion, investigators must attempt to make the values of all variables that could affect performance equal in the two groups, except for the vari-

---

[1] Although this study is hypothetical, it is modeled after a number of actual experiments showing that under certain conditions, tangible rewards often have deleterious effects. See M. R. Lepper & D. Greene (Eds.), *The hidden costs of rewards: New perspectives on the psychology of human motivation* (Hillsdale, N.J.: Lawrence Erlebaum, 1978).

[2] In this section we are discussing true experiments—that is, research in which the investigator systematically varies the conditions under which groups of subjects are observed in order to determine whether the conditions have a differential effect on some measure. Investigators may also be interested in comparing on some dependent variable groups of individuals who differ in some preexisting property—for example, sex, age, occupation, or personality characteristics. The statistical techniques discussed in this chapter are appropriate for both types of research.

able under study. In our example, the children were all tested in the same room, given the same instructions, treated the same by the experimenter except for the reward, and so forth. Investigators must also try to obtain initially comparable groups—that is, groups that, if treated alike, would have behaved alike. In our example, an effort was made to obtain comparable groups by randomly assigning the children to the two conditions. This technique of forming groups was earlier identified as randomization.

When we say that experimental conditions have differential effects on the dependent variable, we imply that the μ's of the hypothetical populations would not be the same. For example, if giving children a candy reward actually does lead to poorer performance on an interesting puzzle than no reward, the μ of the hypothetical population of all children who could be tested under the reward condition would be less than the μ of the population of all children who could be tested under the nonreward condition. When μ's differ, it is also likely that the $\overline{X}$'s of samples of children tested under the two conditions would show a similar difference.

On the other hand, when experimental conditions have *no* differential effects on the dependent variable, the μ's of the hypothetical populations from which the samples were drawn would be equal. If investigators were able to obtain perfectly comparable samples and, with the exception of the dependent variable, tested the two samples under exactly the same conditions, then not only the means of the populations but also the means of the samples would be the same. However, these goals are rarely if ever perfectly achieved, and as a result, the $\overline{X}$'s may differ purely because of chance factors the investigator is unable to identify or control. For this reason, a statistical technique is needed that allows investigators to come to a reasonable conclusion about the difference between the $\overline{X}$'s of their experimental groups—to decide whether the difference is due simply to sampling error or whether, instead, the population μ's differ and the obtained difference between $\overline{X}$'s reflects this fact. The conclusion that μ's differ implies that the experimental conditions *do* differentially influence the dependent variable.

The procedures used to assess these rival hypotheses—that the conditions are and are not differential in their effects—involve an extension of the logic used to test a hypothesis about the μ of a single population from the data of a single sample. We learned in Chapter 9 that if an infinite number of random samples were measured and the $\overline{X}$ of each determined, the mean of the sampling distribution of $\overline{X}$'s would take the same value as the mean of the population, μ. We also learned how to estimate the standard error of the mean—that is, the standard deviation of the sampling distribution of $\overline{X}$'s. With this estimate and information about the shape of the sampling distribution, we were able to test a hypothesis about μ. Similarly, an infinite number of *pairs* of samples could theoretically be measured, one of the samples in each pair under one condition and the second under another condition. After finding $\overline{X}$ for each sample, we could find the difference between each pair of $\overline{X}$'s. The resulting series of differences we identify as the *sampling distribution of the differences between means.* The mean of the sampling distribution of differences could be determined and would be found to take the same value as the *difference between μ's.* The standard deviation of the sampling distribution, which is identified as the *standard error of the difference between means*, could also be determined.

It is humanly impossible, of course, to measure an infinite number of pairs of samples, and typically we have available to us the data from only a single pair of samples. Our goal is to use the information from the pair of sample $\overline{X}$'s to test a hypothesis about the value of the difference between μ's of the parent populations. With a way to estimate the standard deviation of the sampling distribution of the difference between pairs of sample $\overline{X}$'s—that is, the standard error of the difference between means—and information about the shape of the sampling distribution, we are in a position to test such hypotheses about the difference between μ's.

With this brief overview before us, we will discuss in more detail the nature of the sampling distribution of the difference between $\overline{X}$'s and then proceed to a description of how to test a null hypothesis about the difference between μ's.

SAMPLING DISTRIBUTION OF THE DIFFERENCE BETWEEN MEANS

In discussing the sampling distribution of the difference between means, we must keep *two* populations in mind—not a single population as in Chapter 9. To illustrate, suppose we are interested in whether or not men and women differ in their memory for faces. We select two random samples of equal size, one composed of men and the other of women. After giving the subjects a memory test, we find the mean performance of each group. We next determine the difference between $\overline{X}$'s, let us say the women's $\overline{X}$ ($\overline{X}_W$) minus the men's ($\overline{X}_M$). We could repeat this procedure very many times—technically, an infinite number—on each occasion finding the difference between the pair of the $\overline{X}$'s in the same order, $\overline{X}_W - \overline{X}_M$. The result is a series of numbers, each representing the difference between a pair of sample $\overline{X}$'s. These numbers can be arranged to form a frequency distribution, in the ordinary way. Assuming that each of the populations from which the samples were drawn are normally distributed or that the size of the samples is relatively large, we expect the shape of this sampling distribution of differences between $\overline{X}$'s to approximate a normal distribution, just as in the case of the sampling distribution of $\overline{X}$'s from a single population.

The mean of the sampling distribution can be obtained by summing the difference scores and dividing by their number. The mean of the distribution of differences between pairs of sample $\overline{X}$'s will take the same value as the difference between the two population μ's—in our illustration, $\mu_W - \mu_M$, or more generally, $\mu_1 - \mu_2$. The standard deviation of the distribution of differences could also be computed. The standard deviation, called the standard error of the difference between means, is symbolized as $\sigma_{\overline{X}_1 - \overline{X}_2}$. With this information about the shape of the sampling distribution, and its mean and standard deviation, we are able to determine the probability of obtaining any given difference between $\overline{X}$'s of a given size or larger.

Standard Error of the Difference Between Means

We noted that the standard error of the difference, $\sigma_{\overline{X}_1 - \overline{X}_2}$, could be computed directly from the actual distribution of differences between $\overline{X}$'s. However, a constant relationship has been demonstrated between this standard error and the standard error of the mean ($\sigma_{\bar{X}}$) of each of the populations from which the samples are drawn; $\sigma_{\bar{X}}$ is in turn related to the σ of the population and the sample N, as we learned earlier. This relationship is shown in the following formula:

$$\sigma_{\bar{X}_1 - \bar{X}_2} = \sqrt{\sigma_{\bar{X}_1}^2 + \sigma_{\bar{X}_2}^2} \qquad\qquad (10.1)$$

where

$$\sigma_{\bar{X}_1}^2 = \frac{\sigma_1^2}{N_1} \qquad \text{and} \qquad \sigma_{\bar{X}_2}^2 = \frac{\sigma_2^2}{N_2}$$

Using Formula 10.1, we can determine $\sigma_{\bar{X}_1 - \bar{X}_2}$, the standard error of the difference between $\bar{X}$'s, if we know the σ of each population and the sample sizes, N_1 and N_2.

Suppose that in our illustration about women's memory for faces vs. men's the population difference ($\mu_W - \mu_M$) is 3.00 and that, for a given sample N, $\sigma_{\bar{X}_1 - \bar{X}_2}$ is 1.5. The sampling distribution of differences between $\bar{X}$'s is illustrated in Figure 10.1. We now measure a single pair of random samples of N cases each and find the difference between $\bar{X}$'s ($\bar{X}_W - \bar{X}_M$) to be 1.6.

The discrepancy between the value of $\mu_1 - \mu_2$ and the difference between $\bar{X}$'s in this pair of samples is thus 3.0 − 1.6 = 1.4. How common is this event? That is, how probable is it that in a single pair of $\bar{X}$'s, we would obtain a difference between $\bar{X}$'s that deviates 1.4 or more units in either direction from the mean of the sampling distribution? (The mean of the sampling distribution, you recall, takes the same value as $\mu_1 - \mu_2$.) We can see from Figure 10.1 that our obtained ($\bar{X}_W - \bar{X}_M$) value lies quite close to the mean of the distribution; a deviation of ±1.4 score units or more would be a quite probable

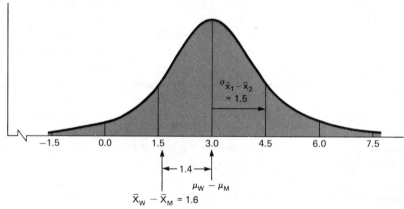

Figure 10.1 Hypothetical distribution of differences between sample $\bar{X}$'s for women's and men's memory for faces when the difference between population μ's ($\mu_W - \mu_M$) is 3.00 and the standard error is 1.5. For a pair of samples in which $\bar{X}_W - \bar{X}_M$ = 1.6, the deviation from the mean of the distribution is 1.4 score units.

event. We can determine the probability more exactly by computing a z score for our obtained difference between $\overline{X}$'s and by then consulting the normal curve table. We define z as a score minus the mean of the distribution from which it came, divided by the standard deviation of the distribution. Thus the formula for determining z in sampling distributions of differences between means is:

$$z = \frac{(\overline{X}_1 - \overline{X}_2) - (\mu_1 - \mu_2)}{\sigma_{\overline{X}_1 - \overline{X}_2}} \tag{10.2}$$

thus yielding for our example:

$$z = \frac{1.6 - 3.0}{1.5} = \frac{-1.4}{1.5} = -.93$$

Looking at Table B, we find that 32.38 percent of the area in a normal curve falls between the mean and a z of .93 and 17.62 percent beyond this point; z scores exceeding ±.93 therefore will occur approximately 35 percent of the time (17.62% × 2).

Estimating the Standard Error of the Difference

We rarely know μ and σ for a single population, let alone for two populations. Therefore, we cannot obtain either $(\mu_1 - \mu_2)$ or $\sigma_{\overline{X}_1 - \overline{X}_2}$. However, if we had some way to estimate $\sigma_{\overline{X}_1 - \overline{X}_2}$, we would be able to test hypotheses about the difference between μ's, based on the data from a single pair of samples. Fortunately, we can arrive at such an estimate by using, in place of each $\sigma_{\overline{X}}$, estimated standard errors of the mean ($s_{\overline{X}}$), which we learned to calculate in Chapter 9.

The formula for estimating the standard error of the difference between $\overline{X}$'s parallels Formula 10.1 given above for determining $\sigma_{\overline{X}_1 - \overline{X}_2}$. Thus:

$$s_{\overline{X}_1 - \overline{X}_2} = \sqrt{s_{\overline{X}_1}^2 + s_{\overline{X}_2}^2} \tag{10.3}$$

where

$$s_{\overline{X}_1}^2 = \frac{s_1^2}{N_1} \qquad \text{and} \qquad s_{\overline{X}_2}^2 = \frac{s_2^2}{N_2}$$

The calculation of $s_{\overline{X}_1 - \overline{X}_2}$ is illustrated in the section below.

TESTING A HYPOTHESIS ABOUT $\mu_1 - \mu_2$: t RATIO

We perform an experiment with laboratory animals to determine whether a diet deficient in protein affects their general activity level. One group of 60 animals is fed a low-protein diet for six months; another group of 60 is given a normal amount of protein. Near the end of this interval members of both groups are tested for activity level. The $\overline{X}$ of the normal group turns out to be 97.51, and s is 13.7; for the deficient group $\overline{X}$ is 91.36 and s is 14.2.

Our purpose in testing these samples is to make an inference about *populations*— populations of animals given diets that are normal or deficient in protein. We decide to test the hypothesis that the μ's of the two populations are the *same;* in other words, our null hypothesis for this study is $\mu_1 - \mu_2 = 0$. Our alternate hypothesis is the bidirectional hypothesis that the difference is some value other than zero. Thus:

$$H_0: \mu_1 - \mu_2 = 0$$
$$H_1: \mu_1 - \mu_2 \neq 0$$

Which population is identified as 1 and which is 2 is purely arbitrary; in our example, 1 will refer to the normal diet condition and 2 to the deficient.

To test the hypothesis, we first must calculate the estimated standard error of the difference:

$$s_{\overline{X}_1 - \overline{X}_2} = \sqrt{s_{\overline{X}_1}^2 + s_{\overline{X}_2}^2} = \sqrt{\frac{s_1^2}{N_1} + \frac{s_2^2}{N_2}}$$

$$= \sqrt{\frac{(13.7)^2}{60} + \frac{(14.2)^2}{60}} = \sqrt{3.13 + 3.36}$$

$$= \sqrt{6.49} = 2.55$$

If our null hypothesis that $\mu_1 - \mu_2 = 0$ is true, we know that the mean of the sampling distribution of differences between $\overline{X}$'s is also zero. The standard deviation of the distribution, we estimate, is 2.55. The difference between $\overline{X}$'s in the two samples we actually measured is $97.51 - 91.36 = 6.15$. Now we ask ourselves the following question. If $\mu_1 - \mu_2 = 0$, how probable is it that the difference between a pair of sample $\overline{X}$'s would deviate from zero as much as or more in either direction as occurred in our samples? In order to answer this question, we must first determine the estimated z score of our obtained difference between $\overline{X}$'s. We recall from Chapter 9 that the estimated z is identified as t. To obtain the estimated z or t, we simply substitute $s_{\overline{X}_1 - \overline{X}_2}$ for $\sigma_{\overline{X}_1 - \overline{X}_2}$ in Formula 10.2. Thus, the t formula for the difference between means is:

$$t = \frac{(\overline{X}_1 - \overline{X}_2) - (\mu_1 - \mu_2)}{s_{\overline{X}_1 - \overline{X}_2}} \tag{10.4}$$

In our example we proposed the null hypothesis that $\mu_1 - \mu_2 = 0$. When this particular H_0 is being tested, the expression $(\mu_1 - \mu_2)$ can be omitted from the numerator, so that the formula for t becomes:

$$t = \frac{\overline{X}_1 - \overline{X}_2}{s_{\overline{X}_1 - \overline{X}_2}} \tag{10.5}$$

where H_0: $\mu_1 - \mu_2 = 0$.

Application of Formula 10.5 to the data from our groups of laboratory animals gives us:

$$t = \frac{97.51 - 91.36}{2.55} = \frac{6.15}{2.55} = 2.41$$

Significance Levels of t.   As we described in Chapter 9 when discussing the testing of hypotheses about μ, we evaluate the null hypothesis about the difference between means by comparing our obtained t with the t associated with a predetermined α level. As before, α typically is either the 5% or the 1% level. When the obtained t equals or exceeds the t associated with our specified α level, we conclude that our t is *significant*. We *reject* the null hypothesis that we have set up about the difference between μ's and conclude that our alternate hypothesis is more reasonable.

When the sample N's are large, the sampling distribution of t may be assumed to be normal. The values associated with the 5% and 1% significance levels are therefore those found in the normal curve table. We also remind you that when H_1 is bidirectional, these values are 1.96 and 2.58 for the 5% and 1% levels, respectively.

In our example, the samples of 60 each are large enough to permit us to assume the normal distribution. If α were set at the 5% level, we would reject the null hypothesis that $\mu_1 - \mu_2 = 0$, since the calculated t of 2.41 is greater than the critical value of 1.96, as illustrated in Figure 10.2. We would conclude, at the 5% level, that our obtained difference between $\overline{X}$'s was significant—that in the population as a whole, prolonged protein deficiency leads to lowered activity level.

Choice of the Null Hypothesis

If investigators anticipate that the experimental conditions they study have no differential effects on the dependent variable, they assume that the two population μ's are the same. Therefore, it is understandable that the null hypothesis that is tested is $\mu_1 - \mu_2 = 0$. In most experiments, however, investigators expect that the conditions they study *do* differentially affect the dependent variable; they conduct the study because they suspect the population μ's are *not* equal.

If investigators have reason to hypothesize that the population μ's differ by some *specific* amount, they will probably choose to test this hypothesis. For example, an investigator may have a theoretical model that leads to the prediction that $\mu_1 - \mu_2 = 6$ and therefore sets H_0 at this value. In computing t, the investigator uses Formula 10.4, inserting 6 as the value of $\mu_1 - \mu_2$.

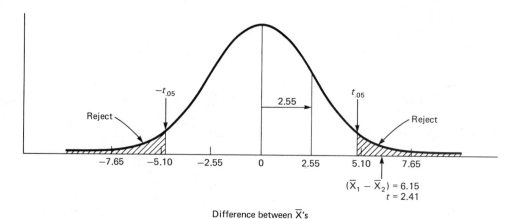

Figure 10.2 Illustration of significance test for H_0: $\mu_1 - \mu_2 = 0$, and H_1: $\mu_1 - \mu_2 \neq 0$ at $\alpha = .05$ where $(\bar{X}_1 - \bar{X}_2) = 6.15$, $s_{\bar{X}_1 - \bar{X}_2} = 2.55$, and $t = 2.41$.

However, even when investigators have firm grounds for postulating not only a difference between μ's but also the direction of the difference, they rarely have a basis for predicting the *exact* value of this difference. But the logic of hypothesis testing, we noted previously, demands that H_0 be assigned a precise value. The solution is to approach the matter backwards, as it were. For *statistical* purposes, the investigator sets up the null hypothesis that $\mu_1 - \mu_2 = 0$, thus specifying an exact value for the difference. The investigator then sets up a more general alternative hypothesis (H_1) that states that the difference between μ's is some unspecified value other than zero. The investigator's expectation is that it will be possible to *reject* H_0 and to accept instead the alternative hypothesis that the μ's are different, that the experimental conditions *do* have a different effect.

Thus, whatever the investigators' experimental hypothesis, the statistical hypothesis they almost always test is H_0: $\mu_1 - \mu_2 = 0$. This procedure is so common that the expression "the null hypothesis" often is used synonymously with the specific hypothesis that the difference between μ's is zero, and the formula for t is frequently the one shown in Formula 10.5, in which the term $(\mu_1 - \mu_2)$ is omitted.

We have said that the null hypothesis an investigator sets up to test statistically may or may not correspond to the *experimental* hypothesis. That is, in some instances, investigators may predict beforehand that the null hypothesis will be shown to be reasonable and thus will be retained. In other instances, they may predict that it will be shown to be unreasonable and thus will be rejected. Whatever their intent, investigators are left in more doubt when they accept the null hypothesis than when they reject it. Perhaps the simplest way to explain this statement is to take a different approach to the difference between $\bar{X}$'s obtained in a single pair of samples; to use the sample data to set up a confidence interval for $\mu_1 - \mu_2$, in the same manner as we set up a confidence interval for μ in

Chapter 9. For $\alpha = .05$, this confidence interval is defined as

$$95\% \text{ CI: } (\overline{X}_1 - \overline{X}_2) \pm t_{.05}(s_{\overline{X}_1 - \overline{X}_2})$$

Now suppose that $(\overline{X}_1 - \overline{X}_2)$ for a pair of large samples is 7.0 and $s_{\overline{X}_1 - \overline{X}_2}$ is calculated to be 3.0. Since the samples are large, $t_{.05}$ takes the normal curve value of 1.96. Thus:

$$95\% \text{ CI: } 7 \pm 1.96(3) = 7 \pm 5.88 = 1.12 \text{ to } 12.88$$

At the 95% confidence level, we assert that the value of $\mu_1 - \mu_2$ falls someplace within the interval of 1.12 to 12.88. Suppose instead that we had used the sample data to test $H_0: \mu_1 - \mu_2 = 0$. We find $t = \frac{7}{3} = 2.3$ and, with $\alpha = .05$, reject H_0. This conclusion corresponds to what is indicated by the 95% confidence interval: the range of values established for $\mu_1 - \mu_2$ *does not* include zero.

But now suppose instead that $(\overline{X}_1 - \overline{X}_2)$ had turned out to be 2. Assuming $s_{\overline{X}_1 - \overline{X}_2}$ to be the same, the 95% confidence interval for turns out to be:

$$95\% \text{ CI: } 2 \pm 1.96(3) = 2 \pm 5.88 = -3.88 \text{ to } +7.88$$

At the 95% confidence level, we assert that $\mu_1 - \mu_2$ falls someplace in the interval -3.88 to $+7.88$. If we had used the sample data to test $H_0: \mu_1 - \mu_2 = 0$, we would find that $t = \frac{2}{3} = .67$ and would accept H_0 at the 5% level. But inspection of the range of values established for the 95% confidence interval shows that the hypothesis $\mu_1 - \mu_2 = 0$ is but one of many hypotheses that would be acceptable at $\alpha = .05$. In fact, *any* H_0 between -3.88 and $+7.88$ would be acceptable at this α level.

As these examples illustrate, rejecting H_0 with a statistical test is a relatively clean process, but accepting H_0 leaves investigators in greater uncertainty. For this reason, as we have noted before, investigators are reluctant to describe a null hypothesis as having been "accepted" because it connotes a positive conclusion. When the computed t falls short of the critical value of t at a specified α level, it is preferable to be cautious and to conclude instead that H_0 has *not been rejected.*

THE t RATIO AND SMALL SAMPLES

Standard Error of Difference and Unequal N's

In our discussion of t thus far, we have dealt only with relatively large samples—samples that have 60 to 70 members or more. We make several modifications in our procedures when we wish to determine the significance of the difference between the means of small samples. First, we should use Formula 10.3 for $s_{\overline{X}_1 - \overline{X}_2}$ *only if the N's for the two samples are equal.* To repeat this formula and the conditions under which it is used in comparing small samples:

$$s_{\bar{X}_1 - \bar{X}_2} = \sqrt{s_{\bar{X}_1}{}^2 + s_{\bar{X}_2}{}^2}$$

where $s_{\bar{X}_1}{}^2 = \dfrac{s_1{}^2}{N_1}$ and $s_{\bar{X}_2}{}^2 = \dfrac{s_2{}^2}{N_2}$ (small samples with equal N's)

In this formula, observe that the $s_{\bar{X}}{}^2$ of each sample, which in turn is based on the s^2 of each sample, contributes equally to the estimate of the standard error of the difference. When the sample N's are not equal, we must amend our formula so that the contribution of the data from each sample to our estimate of the standard error of the difference is weighted according to the sample N. This modification, shown below, *must* be used for small samples with unequal N's but also may be used when N's are the same, since in this instance it yields the same value as Formula 10.3.

$$\text{\Large ✳} \quad s_{\bar{X}_1 - \bar{X}_2} = \sqrt{\frac{(N_1 - 1)\,s_1{}^2 + (N_2 - 1)\,s_2{}^2}{(N_1 + N_2 - 2)} \left(\frac{1}{N_1} + \frac{1}{N_2} \right)} \qquad \textbf{(10.6)}$$

(small samples with equal or unequal N's)

A raw-score formula also may be used:

$$s_{\bar{X}_1 - \bar{X}_2} = \sqrt{\frac{\left(\sum X_1{}^2 + \sum X_2{}^2 \right) - \left(N_1 \bar{X}_1{}^2 + N_2 \bar{X}_2{}^2 \right)}{(N_1 + N_2 - 2)} \left(\frac{1}{N_1} + \frac{1}{N_2} \right)} \qquad \textbf{(10.7)}$$

Note that when sample N's *are* equal, these formulas produce exactly the same answer as does the simpler formula based on the addition of the two $s_{\bar{X}}{}^2$'s. Thus, as we have noted, the use of these modified formulas always is appropriate, whether the sample N's are the same or different.

Use of the *t* Table

A second difference in the procedures we employ with small samples has to do with interpreting the t that we have calculated. As we discussed earlier, the t values obtained from large samples approximate the normal distribution. Thus, in interpreting large-sample t's, we use Table B, the normal curve table. You will recall from Chapter 9 that the t values obtained from small samples are *not* normally distributed. However, assuming that the populations from which the samples were drawn are normally distributed, the distributions of small sample t's are symmetrical, their shape being related to the degrees of freedom (df) associated with the samples.

The df associated with a single sample, it will be recalled, is equal to $N - 1$. In evaluating a hypothesis about the difference between μ's, the calculated t is based on the data from two samples, each with $N - 1$ df. The df that is used in evaluating t is therefore the sum of the *df*'s for the two samples:

$$df = (N_1 - 1) + (N_2 - 1) = N_1 + N_2 - 2 \tag{10.8}$$

If we draw two independent samples where N is 10 each, the *df* is 18 (10 + 10 - 2). If we have two samples, one of 7 and the other of 9 cases, the *df* is 14.

Having calculated *t*, we consult the *t* table, Table C, to find the critical values of *t* at the 5% and 1% significance levels that are associated with the *df* for our pair of samples. If the calculated *t* exceeds the critical value of *t* at a given α level, we reject the null hypothesis that $\mu_1 - \mu_2 = 0$. If, on the other hand, the calculated *t* falls short of the tabled value, we conclude we cannot reject the null hypothesis at the given α level.

Two Examples

We can now demonstrate how to go about testing the significance of the difference between $\overline{X}$'s of small samples. Suppose we test five thirsty rats in a maze and give them a drink of water after each trial. Another group of five is given a saccharine solution to drink after each trial. Our experimental question is whether learning the maze will be more rapid with one reward than with the other. The null hypothesis we test is that the rewards have equal effects on performance; the alternate hypothesis is that they do not. Therefore:

$$H_0: \mu_1 - \mu_2 = 0$$

$$H_1: \mu_1 - \mu_2 \neq 0$$

We have reported below the mean number of trials taken by members of each group to go through the maze without error and the s^2 that we have calculated from the data of each sample. Since the two groups have the same N's, we use the simpler equal-N Formula 10.3 for computing $s_{\overline{X}_1 - \overline{X}_2}$.

Water	*Saccharine*
$N_1 = 5$	$N_2 = 5$
$\overline{X}_1 = 18.40$	$\overline{X}_2 = 15.60$
$s_1^2 = 9.95$	$s_2^2 = 8.30$
$s_{\overline{X}_1}^2 = \dfrac{s_1^2}{N_1}$	$s_{\overline{X}_2}^2 = \dfrac{s_2^2}{N_2}$
$= \dfrac{9.95}{5}$	$= \dfrac{8.30}{5}$
$= 1.99$	$= 1.66$

$$s_{\bar{X}_1 - \bar{X}_2} = \sqrt{s_{\bar{X}_1}^2 + s_{\bar{X}_2}^2} = \sqrt{1.99 + 1.66}$$

$$= \sqrt{3.65} = 1.91$$

$$t = \frac{\bar{X}_1 - \bar{X}_2}{s_{\bar{X}_1 - \bar{X}_2}} = \frac{18.40 - 15.60}{1.91} = \frac{2.80}{1.91}$$

$$= 1.47$$

$$df = N_1 + N_2 - 2 = 5 + 5 - 2 = 8$$

Having determined t and df, we now look at Table C. Assuming that we have chosen $\alpha = 5\%$, we determine that with 8 df we need a t of 2.31 to be significant at the 5% level. Since our t was only 1.47, we cannot assert with any confidence that the two rewards produced any difference in the speed of acquiring the maze. In short, we cannot reject the null hypothesis that $\mu_1 - \mu_2 = 0$.

Our second example comes from a study in which the perceptual-motor performance of individuals with a certain type of brain injury was tested. We wish to compare the mean performance of a group of 12 persons whose injury was unilateral—on only one side of the cortex—with the performance of a group of 5 whose injury was bilateral. You will notice that, since the groups do not have the same number of cases, we must use the formula appropriate for unequal N's when computing $s_{\bar{X}_1 - \bar{X}_2}$.

Unilateral	Bilateral
$N_1 = 12$	$N_2 = 5$
$\bar{X}_1 = 49.19$	$\bar{X}_2 = 38.43$
$s_1^2 = 74.65$	$s_2^2 = 49.28$

$H_0: \mu_{Bi} - \mu_{Uni} = 0$

$H_1: \mu_{Bi} - \mu_{Uni} \neq 0$

$$s_{\bar{X}_1 - \bar{X}_2} = \sqrt{\frac{(N_1 - 1)\, s_1^2 + (N_2 - 1)\, s_2^2}{(N_1 + N_2 - 2)} \left(\frac{1}{N_1} + \frac{1}{N_2} \right)}$$

$$= \sqrt{\frac{11(74.65) + 4(49.28)}{12 + 5 - 2} \left(\frac{1}{12} + \frac{1}{5} \right)}$$

$$= \sqrt{\frac{821.15 + 197.12}{15} (.083 + .200)} = \sqrt{19.21}$$

$$s_{\overline{X}_1 - \overline{X}_2} = 4.38$$

$$t = \frac{\overline{X}_1 - \overline{X}_2}{s_{\overline{X}_1 - \overline{X}_2}} = \frac{49.19 - 38.43}{4.38}$$

$$= \frac{10.76}{4.38} = 2.46$$

$$df = 12 + 5 - 2 = 15$$

For $df = 15$, $t_{.05} = 2.13$ and $t_{.01} = 2.95$ (Table C). Conclusion: Assuming that α was set beforehand at .05, H_0: $\mu_1 - \mu_2 = 0$ is rejected at the 5% level, since the calculated t of 2.46 is greater than the critical value of 2.13. Assuming that low scores reflect poor performance, it is concluded at the 5% level that bilateral brain injury results in poorer perceptual-motor performance than unilateral injury. However, if α had been set at .01, the null hypothesis would *not* be rejected, since $t = 2.46$ is less than $t_{.01} = 2.95$.

Assumptions Underlying *t*

We mentioned earlier, but did not emphasize, that the theoretically derived t distributions are based on the assumption that the characteristic being measured is normally distributed in the populations from which the samples were drawn. Use of the t ratio to test a hypothesis about the difference between population means also assumes that the population σs are the same. It has been demonstrated, however, that t tests may be used without noticeably distorting our conclusions even when these assumptions are not met, as long as the σ's are not markedly different or the distributions do not depart radically from normal. If one or more of these assumptions is seriously violated, one procedure that is open to us is to use an alternate type of statistical test that makes no assumptions about population parameters such as distribution shapes or σ's. Some of these *nonparametric techniques*, as they are called, are discussed in Chapter 15.

In the procedures for calculating and evaluating t that have been discussed in this chapter, a final assumption is that the two samples are both randomly drawn from their respective populations and are independent of each other. Techniques are also available for calculating t from the data of samples that have not been independently selected but instead have been deliberately matched. This topic will be discussed in Chapter 11.

In practice, investigators almost never have access to all the members of a population and hence are unable to select and test random samples in their experiments. As we noted in Chapter 1, the most common procedure is to use *randomized* groups. That is, a pool of individuals is available, sometimes called a *sample of convenience*, and individuals in the pool are randomly assigned to experimental conditions. Strictly speaking, statistical tests based on the assumption of randomized assignment of subjects to groups should be used in analyzing the data from experiments using this type of design. However, statistical theory is far better developed for situations involving random sampling than for those involving randomization. Primarily for this reason, the results of investigations using the randomization design are almost always treated as though the groups had been obtained by random sampling and analyzed by statistical tests, such as t, based on the random-sampling assumption. Fortunately, the significance level given by t appears to be quite close to that yielded by a more exact test when both are applied to randomized data.

CHOICE OF SIGNIFICANCE LEVEL

Statisticians advise investigators to specify beforehand the α level they will use to decide whether or not to reject the null hypothesis they have designed their study to test. This procedure makes the decision-making process automatic and invulnerable to the investigator's biases or wishful thinking. The investigator does, however, have a choice of significance level, which raises a question about the factors that determine this choice. We begin our discussion of this problem by presenting two examples.

The research division of a drug company has discovered a new drug for the treatment of a specific medical disorder. It is not known whether the drug differs in its therapeutic effectiveness from the standard remedy, but it is known that the two drugs are otherwise comparable in cost, ease of administration, side effects, and so on. An experiment is therefore planned to compare the two drugs' therapeutic effectiveness. The null hypothesis the investigators decide to test is that the two drugs have equal effects ($\mu_1 = \mu_2$); the alternate hypothesis they set up is the unidirectional hypothesis that the new drug is *better*. If the new drug turns out to be significantly better than the old, the company will put it on the market. In considering which α level to use, the researchers ponder the possible kinds of decisions that could result. The H_0 is either true or false, and, after assessing H_0 statistically, they will either accept or reject it. Thus, there are four possibilities, two of them involving ways to come to a *correct* conclusion: the investigator can reject H_0 when it is in fact false and can accept H_0 when it is in fact true. Similarly, there are two ways to come to an *incorrect* conclusion: the investigator can erroneously reject H_0 when it is in fact true (Type I error) and can accept H_0 when it is in fact false (Type II error). These possibilities and the probabilities of each are shown in Table 10.1. Recall from Chapter 9 that the probability of making a Type I error at a given α level is equal to that α, while the probability of a Type II error takes a value somewhat higher than α and is identified as β.

The researchers must weigh the hazards of the two types of errors. If the new drug is better than the old and they accept the false hypothesis that $\mu_1 - \mu_2 = 0$, the drug will not be marketed and those who suffer from the disorder will be deprived of a more potent therapeutic agent. On the other hand, if the new drug is no better than the old and they incorrectly *reject H_0*, the company will waste time and money marketing the new product. Marketing expense, however, is quite minimal, and they decide that a Type I error, concluding that the new drug is more effective when it isn't, would be less costly than a Type II error, concluding that the new drug is equally effective as the old when it

Table 10.1   The Four Possible Outcomes in Evaluating H_0 at a Given α Level and Probabilities of Each

		Correctness of H_0	
		True	False
Conclusion	Accept H_0	Correct $1 - \alpha$	Error (Type II) β
	Reject H_0	Error (Type I) α	Correct $1 - \beta$

is better. Therefore, they decide to use a relatively large value of α, 5 percent or perhaps even 10 percent, to minimize the probability of a Type II error.

Another example: a new method of teaching eighth-grade biology has been devised that would be quite expensive to implement, requiring new books, special laboratory equipment, and retraining of teachers. Before deciding whether or not to introduce the method into a large elementary school, the principal asks that an experiment be conducted in which the exam scores of a group of pupils taught by this method are compared with those of a group taught in the regular way. Because of the expense involved, an α level of 1 percent or even smaller might be used to assess H_0. Setting a more stringent criterion for rejecting the null hypothesis minimizes the probability of a Type I error— that is, lessens the probability that the H_0 will be incorrectly rejected when in fact the new method does not produce better performance than the old. The principal also might insist that the new method produce results not only *better* than the old, but *substantially superior*. She therefore might ask that H_0 be set at some value other than zero.

In both these examples a practical decision is to be based on the conclusion drawn from the experimental data. The α level chosen for assessing the null hypothesis is adjusted up or down, depending on the investigator's judgment about the relative risks of a Type I or Type II error. In purely scientific research the acceptance or rejection of the null hypothesis usually does not have consequences as immediate and as obvious as those in our examples. Nonetheless, errors can be costly. Promising ideas can be abandoned or investigators' time and energy wasted in pursuit of false trails as a consequence of an incorrect decision about H_0. In general, however, scientists prefer to be conservative about rejecting hypotheses and consider a rule that would permit the occurrence of a Type I error more than 5 percent of the time to be unacceptable. Therefore, the highest α value conventionally used is .05. On the other hand, setting α lower than .01 results too often in failure to detect a false H_0 (Type II error) to be acceptable. One therefore finds the 1% and 5% levels used almost exclusively in research reports.

Since there usually is little pressure on anyone to take direct action based on an investigator's decision about the correctness of H_0, most investigators report their data as being significantly different from the null hypothesis if it can be rejected at the 5% level or less. If it is possible to reject H_0 at the 1% level, or at some level lower than 5%, this is the significance level that is actually reported.

POWER OF A STATISTICAL TEST

In their concern not to make Type I errors, scientists have specified relatively stringent criteria for rejecting the null hypothesis, the 5% and 1% significance levels. They are also concerned about minimizing Type II errors and hence about the probability that a statistical test will detect false hypotheses, as noted in Chapter 9. This probability, known as the *power* of the test, is determined by the following formula:

$$\text{Power} = 1 - p(\text{Type II error}) \tag{10.9}$$

We have noted that the probability of a Type II error is symbolized by the Greek letter β (beta). Thus we may also write:

$$\text{power} = 1 - \beta$$

What the formula indicates, we repeat, is the probability that H_0 will be rejected when it is in fact false.

We will discuss power for the simple case in which we test $H_0: \mu_1 - \mu_2 = 0$ and know the values of μ and σ. We can illustrate the method of calculating power by the following example involving two populations. For the first population μ is 93 and for the second μ is 89; both populations have a σ of 8.

We test a random sample of 10 cases from each population and set up the null hypothesis that $\mu_1 - \mu_2 = 0$. Note that the true difference is 93 – 89 or 4 and H_0 is false. For convenience, we will assume that our alternative hypothesis, H_1, is the unidirectional hypothesis that $\mu_1 > \mu_2$. The α level we decide to use in evaluating H_0 is .05.

Since we know the population σ, we are able to compute the z score for the difference between our sample $\overline{X}$'s rather than having to obtain t, an estimate of z. The formula we use is this:

$$z = \frac{(\overline{X}_1 - \overline{X}_2) - (\mu_1 - \mu_2)}{\sigma_{\overline{X}_1 - \overline{X}_2}}$$

Looking at the normal curve table, we see that 45% of the area lies between μ and $z = 1.64$ and 5% above this value of z. Thus with $\alpha = .05$, $H_0: \mu_1 - \mu_2 = 0$ and $H_1: \mu_1 > \mu_2$, we will reject the null hypothesis of no difference between μs if we obtain a z of 1.64 or greater. We know that H_0 is false; that is, our null hypothesis states that the difference is zero but the true difference, in our illustrative example, is 4.

The power of our test, which we have defined as the probability of detecting a false null hypothesis, is thus equal to the probability of obtaining a z of 1.64 or greater. We proceed to determine this probability by first determining $\sigma_{\bar{X}_1 - \bar{X}_2}$. Since each σ equals 8:

$$\sigma_{\bar{X}_1 - \bar{X}_2} = \sqrt{\sigma_{\bar{X}_1}^2 + \sigma_{\bar{X}_2}^2} = \sqrt{\frac{8^2}{10} + \frac{8^2}{10}}$$

$$= \sqrt{\frac{128}{10}} = 3.58$$

Next we substitute in the z formula the critical value of z, the value of $(\mu_1 - \mu_2)$ under H_0, and the value of $\sigma_{\bar{X}_1 - \bar{X}_2}$; we then solve for $(\bar{X}_1 - \bar{X}_2)$:

$$1.64 = \frac{(\bar{X}_1 - \bar{X}_2) - 0}{3.58}$$

so that

$$(\bar{X}_1 - \bar{X}_2) = 1.64(3.58) = 5.87$$

We have now determined that if we test two samples of 10 cases each, we will reject H_0 at $\alpha = .05$ whenever the $\bar{X}_1 - \bar{X}_2$ difference is 5.87 or greater. When we reject H_0 we will accept instead our H_1 that $\mu_1 > \mu_2$. This is illustrated in Figure 10.3. Our next step is to determine the probability of getting a difference between $\bar{X}$'s of +5.87 with the *actual* sampling distribution of difference between means with $\mu_1 = 93$ and $\mu_2 = 89$. We first compute z:

$$z = \frac{5.87 - (93 - 89)}{3.58} = \frac{5.87 - 4.00}{3.58} = .52$$

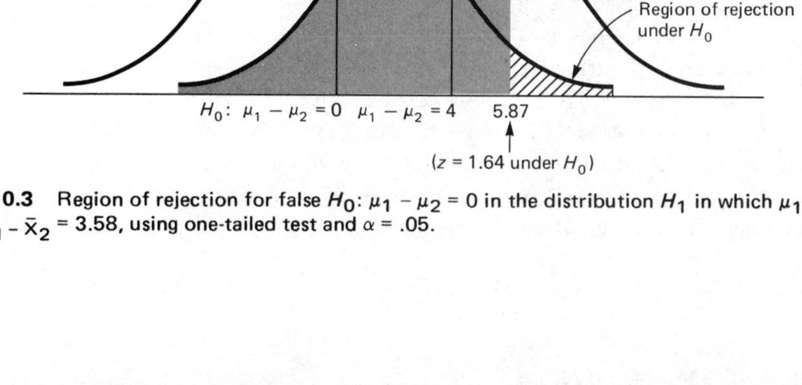

Figure 10.3 Region of rejection for false H_0: $\mu_1 - \mu_2 = 0$ in the distribution H_1 in which $\mu_1 - \mu_2 = 4$ and $\sigma_{\bar{X}_1 - \bar{X}_2} = 3.58$, using one-tailed test and $\alpha = .05$.

Looking at Table B, we find 19.85 percent of the area lies between μ and $z = .52$, and 30.15 percent beyond it. Thus, the probability of rejecting the false H_0 is .3015. This probability is equal to $1 - \beta$, *the power of the test.* We can also see that β, the probability of making a Type II error, is $1 - .3015$ or .6985. We add that a power of .3015 is usually considered too low a value to have in a well-designed experiment. Therefore, an experimenter faced with this set of calculations would seek a new research procedure with a higher value of $1 - \beta$ (power).

Our example has illustrated the calculation of the power of a test when H_0 concerns the difference between μ's. The same general logic is used in determining the power of a test of other types of hypotheses. For example, in testing a hypothesis about μ we would use the z formula $(\overline{X} - \mu)/\sigma_{\overline{X}}$ and follow the same steps outlined above.

Factors Affecting Power of a Test

If you examine the z formula and the steps for determining the power of a test, you will see that the values of the mean and standard error of the sampling distribution (in our example, $\mu_1 - \mu_2$ and $\sigma_{\overline{X}_1 - \overline{X}_2}$, respectively), as well as the value of z itself, all affect the probability figure we obtain. We will consider each of these terms and their implications for power one at a time. Again, we will use the two-population case as an illustration.

In our example the mean of the sampling distribution specified by the false H_0 is zero. The power of the test to detect this false hypothesis increases as the discrepancy between this mean and the true mean of the sampling difference, determined here by $\mu_1 - \mu_2$, increases. We can see this by examining Figure 10.3. If the $\mu_1 - \mu_2$ difference is small, close to the value specified by H_0, the true sampling distribution will show considerable overlap with the distribution under H_0. The value of β will therefore be high and $1 - \beta$, which defines the power of the test, correspondingly low. As $\mu_1 - \mu_2$ grows larger, the true sampling distribution is displaced further and further along the baseline and overlaps less and less with the distribution under H_0. Area β thus approaches a zero value and $1 - \beta$ approaches one. As the difference between H_0 and the true state of affairs increases, then, the probability of rejecting the false H_0 approaches unity.

The second factor affecting power is the standard error of the sampling distribution, in our example $\sigma_{\overline{X}_1 - \overline{X}_2}$, which is equal to:

$$\sqrt{\sigma_{\overline{X}_1}{}^2 + \sigma_{\overline{X}_2}{}^2} \quad \text{or} \quad \sqrt{\frac{\sigma_1{}^2}{N_1} + \frac{\sigma_2{}^2}{N_2}}$$

The basic components of the standard error formula, we can see, are N, the number of cases in each sample, and σ, the standard deviation in each population. Consider first the implications of variations in the N's. As an N increases, the value of the corresponding $\sigma_{\overline{X}}{}^2$ and hence of $\sigma_{\overline{X}_1 - \overline{X}_2}$ will *decrease*. Picture what would happen to the sampling distributions in Figure 10.3 as N increased from 10 per sample to larger numbers. The means of the distributions would remain the same, but with less variability (a smaller $\sigma_{\overline{X}_1 - \overline{X}_2}$), the distributions would each become taller and thinner. This in turn would reduce the overlap of the two sampling distributions. Reduction in overlap, we have already seen, increases $1 - \beta$, the power of the test. Large N's, then, make it less difficult to reject false hypotheses than small N's.

The second component affecting the standard error of sampling distributions is σ, with the standard error decreasing as σ decreases. In designing experiments, therefore, investigators try to minimize the influence of extraneous factors that may result in increased variability among their subjects and hence reduce the power of their test of significance.

To summarize our discussion so far, the formula for z in the two-population case is:

$$z = \frac{(\overline{X}_1 - \overline{X}_2) - (\mu_1 - \mu_2)}{\sqrt{\dfrac{\sigma_1{}^2}{N_1} + \dfrac{\sigma_2{}^2}{N_2}}}$$

The value of z, and hence the power of the test to detect a false H_0, will increase as (1) the difference between μ's increases, (2) the size of σ's decreases, and (3) sample N's increase.

Another factor affecting power is the α level we have chosen, which in turn determines z. We indicated earlier that the lower we set α, the less the probability of committing a Type I (α) error and the greater the probability of committing a Type II (β) error. Since power is defined by $1 - \beta$, this implies that the lower the α, the smaller the power of the test. We can demonstrate this by considering z. With a one-tailed test, as we saw in our example, the z associated with $\alpha = .05$ is 1.64. The value of $(\overline{X}_1 - \overline{X}_2)$ permitting us to reject H_0 was thus 1.64 $(\sigma_{\overline{X}_1 - \overline{X}_2})$ or $1.64(3.58) = 5.87$. We determined that the probability of getting a difference of this size or greater in the true sampling distribution was .3015. The power of the test was thus .3015. If we had set α at .01, z would have been 2.33 and the critical value of $(\overline{X}_1 - \overline{X}_2)$ would have been $2.33(3.58) = 8.34$. By finding the z-score equivalent of 8.34 in the true distribution $[(8.34 - 4.00)/3.58 = 1.21]$, we determine from Table B that only 11.31 percent of the area falls beyond this point. We can see that the power of the test for $\alpha = .01$ is .1131, less than when we used the higher α level of .05.

The final factor affecting the power of a test is the nature of H_1. At a given α level the value of z is less when H_1 is unidirectional and evaluated by a one-tailed test than when a bidirectional H_1 and two-tailed test are used—for example, 1.64 vs. 1.96 for $\alpha = .05$; 2.33 vs. 2.58 for $\alpha = .01$. We have just seen that, other things being equal, the lower the value of this z, the greater the power of the test. Unidirectional H_1's evaluated by one-tailed tests therefore are more powerful than bidirectional H_1's.

DEFINITIONS OF TERMS AND SYMBOLS

Independent variable.   The condition the investigator manipulates in an experiment.

Dependent variable.   The observations the investigator gathers from each subject in an experiment. The purpose of the experiment is to determine whether the conditions of the experiment (values of the independent variable) differentially affect the dependent variable.

Sampling distribution of differences between means. A theoretical frequency distribution consisting of an infinite number of differences $(\overline{X}_1 - \overline{X}_2)$ between pairs of random sample means, the samples in each pair having been drawn from population 1 and population 2, respectively.

Standard error of the difference between $\overline{X}$'s $(\sigma_{\overline{x}_1 - \overline{x}_2})$. The standard deviation of the sampling distribution of differences between $\overline{X}$'s.

Estimated standard error of the difference between $\overline{X}$'s $(s_{\overline{x}_1 - \overline{x}_2})$. An estimate of $\sigma_{\overline{x}_1 - \overline{x}_2}$ based on the data from a single pair of samples.

Null hypothesis (H_0) about difference between μ's. A statistical hypothesis about the specific value of the difference between μ's that is evaluated by means of a t test. Typically, H_0 is that $\mu_1 - \mu_2 = 0$.

t ratio. An estimated z for $(\overline{X}_1 - \overline{X}_2)$, defined as $[(\overline{X}_1 - \overline{X}_2) - (\mu_1 - \mu_2)]/s_{\overline{x}_1 - \overline{x}_2}$. In this formula, $(\overline{X}_1 - \overline{X}_2)$ is determined from the data of the pair of samples that have been observed, $(\mu_1 - \mu_2)$ is the value specified in H_0, and $s_{\overline{x}_1 - \overline{x}_2}$ is the estimated standard error of the sampling distribution of differences.

Degrees of Freedom (df). For a pair of independent random samples, $df = N_1 + N_2 - 2$.

Assumptions underlying t test. The logic of the t test for evaluating a hypothesis about the difference between μ's described in this chapter assumes that (a) the characteristic that has been measured is normally distributed in each of the populations from which the samples were drawn, (b) the population σ's are equal, and (c) the two samples are each randomly selected and independent of each other. If any of these assumptions is seriously violated, use of the t test may lead to inappropriate conclusions about H_0.

Type I and Type II errors. See Chapter 9.

Power of a statistical test. Power refers to the ability of a statistical test to detect false null hypotheses and is defined as $1 - \beta$, where β is the probability of a Type II error. The power of a test of a hypothesis about $\mu_1 - \mu_2$ (a) increases with the magnitude of the $\mu_1 - \mu_2$ difference, (b) decreases with the size of the population σ's and (c) increases with the size of the sample N's.

PROBLEMS

1. Three studies are described below, each of which compares two groups on some measure. For each study, identify the independent variable and the dependent variable. Then describe verbally the specific null hypothesis that would be evaluated in conducting a t test of the data.
    (a) Students who seek help at their college counseling service because they are anxious and upset are assigned at random to one of two types of treatment pro-

grams, insight therapy and cognitive therapy. Six months after their first therapeutic session, all students are given a test that evaluates their current emotional
status.

(b) Groups of "blue collar" factory workers and "white collar" office workers are
administered a questionnaire measuring job satisfaction.

(c) Two groups of infant monkeys were reared in a cage with their mother, isolated
from other monkeys. Those in one group were also placed in a cage with other
young monkeys for 30 minutes several times a week, whereas those in the other
group were not given this social experience. When they were 3 months old, each
animal was placed in a cage alone and the amount of fear each displayed when a
strange mechanical windup toy was placed in the cage was recorded.

2. Assume that information is available on the heights of all men inducted into the U.S.
Army during World War I and World War II and the following results (in inches) are
obtained for the two populations. WWI: $\mu = 65$, $\sigma = 3$; WWII: $\mu = 68$, $\sigma = 3$.

(a) If a very large number of random samples of 100 men each were drawn from
each population, what would be the value of the standard deviation (standard
error of the mean) of each sampling distribution of $\overline{X}$'s?

(b) Suppose the random samples of 100 had been drawn in independent pairs, one
from the men in WWI and the other from WWII, and the difference between $\overline{X}$'s,
in the order $\overline{X}_{WWII} - \overline{X}_{WWI}$, was determined for each pair. What would be the
value of the mean of the resulting distribution of differences between $\overline{X}$'s?

(c) What is the standard deviation of the sampling distribution of differences called?
What would be its value?

(d) What is the probability of drawing a pair of samples of 100 cases each in which
the $\overline{X}$ for the WWII sample is 4 or more score units higher than the $\overline{X}$ for the
WWI sample? (Drawing a sketch of the sampling distribution of differences and
showing its mean and standard error may help you figure out how to answer this
question.)

Note: The following problems call for the conduct of a t test. In each problem, specify
the H_0 and H_1 you are testing, the α level you select, and df.

3. Assume that you have available data from a single pair of random samples of 70 men
each, drawn from the populations described in problem 2, and that you do not know
the population μ's and σ's. The sample $\overline{X}$'s and s's turn out to be:

$$\text{WWI: } \overline{X} = 65.8, s = 3.1; \text{WWII: } \overline{X} = 67.2, s = 2.9.$$

Your interest in these sample data is in estimating whether or not men inducted into
the army in the two wars differed in height.

(a) Set up the null hypothesis that you will test and a bidirectional alternate hypothesis. Explain your choice of H_0.

(b) If you were to test H_0 against a unidirectional H_1, what specific alternative hypothesis would you select? (Use your general knowledge about factors related
to height to answer this question.)

(c) Test the H_0 (with the bidirectional H_1) you set up in (a).

(d) What do you conclude? State your conclusion in terms of the specific research
question the study was designed to answer.

4. A university counseling psychologist finds that students who complain they are
lonely also say they would like someone to talk to and confide in. He wonders
whether, if given an opportunity to disclose something about themselves, they would
be more likely to do so than others because they have greater need or would be less
likely to disclose because they are inhibited or lack the skills. He performs an experi-

ment in which women students who receive high scores or low scores on a loneliness questionnaire are asked by a trained interviewer to talk about a series of moderately intimate topics. Their taped responses are later rated by judges for degree of intimacy of self-disclosure. The results are shown below, with high scores indicating more self-disclosure.

	Lonely	Nonlonely
$\overline{X}$	43.2	47.7
s	4.1	3.6
N	20	20

After conducting the appropriate statistical test, come to some conclusion about loneliness and self-disclosure.

5. A medical researcher has identified a new drug that animal investigations suggest may be a more effective treatment for a genetic defect apparent at birth that results in mental retardation. She is able to find 15 cases of the disorder, but because the drug is scarce and expensive, can treat only 5 of the infants. With parental consent, 5 infants are selected at random for the experimental treatment and the remainder are given a standard drug treatment. When the children are 2 years old, they are given a test of cognitive development. The results are shown below. (High scores indicate high development.)

	Experimental	Standard
$\overline{X}$	58.4	52.6
s	5.6	5.2
N	5	10

In testing H_0, set up a bidirectional H_1 and use $\alpha = .05$. What can be concluded about the relative effectiveness of the two drugs? In conducting your t test, be sure to use Formula 10.6 for calculating $s_{\overline{X}_1 - \overline{X}_2}$ for unequal N's.

6. The investigator described in problem 5 was later able to repeat the experiment with two treatment groups of 25 cases each. The following results were obtained.

	Experimental	Standard
$\overline{X}$	58.9	53.1
s	5.5	5.3

$$s_{\overline{X}_E - \overline{X}_S} = 1.53$$

(a) Using $\alpha = .05$, what do you conclude about the effectiveness of the experimental drug?

(b) If your calculations are correct, your conclusion is different from your conclusion in problem 5. Explain the discrepancy, referring to the concept of power.

Sampling Distributions: Matched Pairs

CHAPTER 11

Random sampling and randomization are two methods of obtaining comparable groups for experiments. The subject of this chapter is a third technique: selection and use of *matched pairs*.

TYPES OF MATCHED PAIRS

Although most research requires that different individuals be used, there are circumstances in which the same individuals can take part in each experimental condition. For example, in a study of speed of recognition of words, a series of randomly ordered five-letter words, half of them highly familiar and half of them less familiar, might be exposed very briefly to determine whether familiarity influences speed of recognition. Each subject is required to guess what each word is, the independent variable being the number of correct guesses of the two types of words. The result is a *pair of scores* for each individual: number correct for familiar words, and number correct for unfamiliar words.

Whenever possible, it is advantageous to use the same individuals in both conditions. In the first place, the actual mechanics of obtaining the individuals and testing them in the experiment are usually simpler because only a single group of individuals is employed. More important, there is no problem in obtaining comparable groups. The individuals serving in the two conditions are ideally matched; individuals are better matches for themselves than anyone else.

Matched pairs can also consist of different individuals who are similar in characteristics that can be assumed to influence the dependent variable to be measured in the experiment. For example, a group of identical twins or pairs of brothers or sisters could be employed. It is also possible to select an individual and then to find another individual who is biologically unrelated but who is a "twin" on one or more characteristics that might affect the particular behavior to be observed. Age, sex, physical or mental abilities, interests, and occupation can be mentioned as possible bases for matching. After pairs of individuals who are matched in some way have been selected, one member of each pair is chosen at random and assigned to one of the experimental conditions. The other member of each pair is then automatically assigned to the other condition. As in the case in which the same individuals are tested under both conditions, the end result is a set of pairs of scores, each pair representing the performance of an individual in one condition and the performance of the individual's "twin" in the other condition.

We have previously described random samples as being independent. That is, except for the restriction that the same individual cannot serve in both groups, the composition of one random sample does not in any way affect the composition of the second sample. In contrast, matched samples are not independent. Once one member of each pair has been assigned to one experimental group, composition of the second group is fixed.

A major advantage in using matched pairs is that the method has the potential of increasing the power of the t test to detect the falsity of the null hypothesis that $\mu_1 - \mu_2 = 0$—that is, the hypothesis that conditions have no differential effect on the dependent variable. To explain this statement, we return to the topic of the power of a test, discussed in the last chapter, and the formula for t when sample N's are equal:

$$t = \frac{\overline{X}_1 - \overline{X}_2}{s_{\overline{X}_1 - \overline{X}_2}} = \frac{\overline{X}_1 - \overline{X}_2}{\sqrt{\dfrac{s_1^2}{N_1} + \dfrac{s_2^2}{N_2}}}$$

With $(\overline{X}_1 - \overline{X}_2)$ and sample N's held constant, the value of t is inversely related to the values of s_1^2 and s_2^2, our estimates of the population σ^2 values. As these measures of variability become smaller, $s_{\overline{X}_1 - \overline{X}_2}$ becomes smaller.[1] In turn, as the standard error of the difference becomes smaller, t becomes larger and H_0 is more often rejected.

Each s^2 reflects, of course, the amount of variability among the individuals within each group. Where does this variability in subjects' performance come from? In part it comes from fluctuations in external conditions: The room temperature and lighting may not always be exactly the same, the experimenter's behavior may vary slightly from one occasion to another, and so forth. In part it comes from temporary variations in the subjects themselves. A particular individual may not be feeling well, may be preoccupied with some personal matter, or may suffer some momentary lapse of attention to the experimental task. Although experimenters attempt to minimize these random fluctuations in subjects and conditions within practical limits, perfect control is impossible, and some variability among subjects' behavior due to these sources is therefore inevitable.

Another source of variability comes from differences among subjects that are relatively stable over time. Individuals vary in their ability to perform the experimental task and in the case of human subjects, in such characteristics as cooperativeness, their desire to impress the experimenter, their interest in psychology experiments, and their general motivation to perform well. In many experiments, much of the variability among subjects within experimental groups is due to stable individual differences of this sort.

What the existence of these individual differences implies is that if individuals were tested on similar tasks on two occasions, their performance would be likely to be *posi-*

---

[1] Although the formula $s_{\overline{X}_1 - \overline{X}_2} = \sqrt{s_1^2/N_1 + s_2^2/N_2}$ makes clear the contribution of within-group variability (s) to the standard error of the difference between $\overline{X}$'s, it may not be intuitively obvious why the variability of the sampling distribution of the differences between $\overline{X}$'s is related to s. If there were *no* variability between individuals in each population, that is, if $\sigma = 0$, then all scores in a sample would take the same value as $\overline{X}$ and all $\overline{X}$'s the same value as μ. Thus the difference between $\overline{X}$'s of a series of pairs of random samples would always be the same, so that $s_{\overline{X}_1 - \overline{X}_2} = 0$, and each $\overline{X}_1 - \overline{X}_2$ would be equal to $\mu_1 - \mu_2$. As σ's increase in value, the distribution of $\overline{X}$'s from each population becomes more variable due to sampling error, and the differences between pairs of sample $\overline{X}$'s also become more variable for the same reason.

tively correlated. That is, the person who performed best on the first occasion is also likely to perform well on the second occasion, the person who performed worst on the first occasion is also likely to exhibit a poor performance on the second occasion, and so forth. A positive correlation is also expected when the same individual or a pair of previously matched individuals is observed under two experimental conditions. That is, even if conditions have an effect, so that the two $\overline{X}$'s differ, the *relative position* of each member of pairs of scores may be similar; that is, within a pair, an individual's standing in the distribution of scores for the first condition may be similar to the individual's (or his "twin's") standing in the distribution of scores for the second condition.

In discussing the concept of correlation in Chapter 6, we noted that the correlation coefficient provides an indication of the amount of *explained* variability in distributions of two measures; the greater the correlation, the greater the amount of *explained*, as opposed to *unexplained*, variability. By finding the correlation between pairs of scores in an experiment employing matched pairs of individuals or the same individuals in both conditions, we have a device for accounting for some of the variability within groups. As will be explained in the next section, this correlation can be used to reduce $s_{\overline{X}_1 - \overline{X}_2}$, which can be regarded as a measure of unexplained variability in a sampling distribution of differences between $\overline{X}$'s. With a sufficient reduction in $s_{\overline{X}_1 - \overline{X}_2}$, the power of the t test increases.

THE t TEST WITH MATCHED SAMPLES

When matched pairs are used, there will be, as usual, two distributions of scores at the conclusion of the research. Just as with random or randomized groups, the distributions represent performance under the two experimental conditions, and the difference between the $\overline{X}$'s is evaluated by the t test. The formula for t, $(\overline{X}_1 - \overline{X}_2)/s_{\overline{X}_1 - \overline{X}_2}$, remains the same, but the equation for $s_{\overline{X}_1 - \overline{X}_2}$ is expanded to take the correlation between the pairs of scores into account. This formula is:

$$s_{\overline{X}_1 - \overline{X}_2} = \sqrt{s_{\overline{X}_1}^2 + s_{\overline{X}_2}^2 - 2r_{12}s_{\overline{X}_1}s_{\overline{X}_2}} \qquad (11.1)$$

where $s_{\overline{X}_1}$ = standard error of the mean of one group

$s_{\overline{X}_2}$ = standard error of the mean of the other group

r_{12} (read r sub one two) = the Pearson product-moment correlation between the scores of two groups—that is, between the two distributions of measures being tested for the significance of the difference.

Note that the initial part of the formula, up to the minus sign, is the same as the formula for random samples with equal N's and consists of the sum of the two $s_{\overline{X}}^2$'s. For matched samples, the term $(2r_{12}s_{\overline{X}_1 - \overline{X}_2})$ is subtracted from this sum.

Let us consider the correlation coefficient r_{12} in that term. *It is the correlation between the two distributions, the $\overline{X}$'s of which are to be tested for significance.* Recall that to compute a Pearson r between two distributions, we must be able to pair a score from one distribution with a score from the other. In any experiment using matched groups, each pair of scores comes from the already paired subjects. Thus, all the terms under the radical sign in the $s_{\bar{X}_1 - \bar{X}_2}$ formula, the standard errors and the correlation, are obtained from the two distributions of scores resulting from the experiment. Either of these distributions may be identified by the subscript 1 and the other by the subscript 2.

What happens to the $s_{\bar{X}_1 - \bar{X}_2}$ for matched groups as r_{12} takes various values? When r_{12} is zero, the last term under the radical (everything after the minus sign) becomes zero, and we have the $s_{\bar{X}_1 - \bar{X}_2}$ formula for random groups. If there is no correlation between the groups, they are, in that respect, the same as independent random groups. If r_{12} is some positive value, the $s_{\bar{X}_1 - \bar{X}_2}$ will be *smaller* than when r_{12} is zero, and the higher the r_{12} the greater the reduction in the $s_{\bar{X}_1 - \bar{X}_2}$. Since a reduction in $s_{\bar{X}_1 - \bar{X}_2}$ leads to an increased t value with a given difference between two $\bar{X}$'s, we can see that this difference in $\bar{X}$'s is more likely to result in a significant t when divided by the $s_{\bar{X}_1 - \bar{X}_2}$ obtained from matched samples than from random samples. We also can see why there would be little point in using matched samples in an experiment unless the members of a pair originally were fairly similar. Unless the r_{12} is at least some medium to high value, there will be little reduction in $s_{\bar{X}_1 - \bar{X}_2}$.

Having computed the $s_{\bar{X}_1 - \bar{X}_2}$ for matched groups, we proceed to test the null hypothesis that the μ values are equal by obtaining t by the usual formula: $(\bar{X}_1 - \bar{X}_2)/s_{\bar{X}_1 - \bar{X}_2}$. In assessing the t value, the df that we use is N - 1, the number of *pairs* of measures minus one. Notice that df is less than if independent groups had been used, in which case df would have been $N_1 + N_2 - 2$. This means that for correlated samples a larger t will be required for significance at a given α level than if a t for independent samples had been calculated instead. When r is large enough (in a positive direction), the reduction in $s_{\bar{X}_1 - \bar{X}_2}$ that it brings about more than compensates for this problem, making it easier to find a significant difference with matched pairs when the null hypothesis is false.

Computing $s_{\bar{X}_1 - \bar{X}_2}$ for matched pairs requires a good deal of calculation. Fortunately, there are two shorter methods for testing the null hypothesis about the difference between μ values for matched pairs, both of which lead to the same conclusion about the null hypothesis as the method just described. The first is the *direct-difference method* for computing t.

Direct-Difference Method for Computing *t*

We start by describing an experiment in which the same subjects were used in both conditions of the study. The investigator was concerned with the influence that children's physical attractiveness has on adults' judgment of their behavior.[2] She gave 20 adults a series of descriptions of minor offenses, each supposedly committed by a real child. The subjects were asked to indicate on an objective rating scale how severe the punishment for each misbehavior ought to have been. A photograph of the purported miscreant accom-

---

[2]This hypothetical experiment is similar to an actual investigation performed by K. K. Dion: Physical attractiveness and evaluations of children's transgressions, *Journal of Personality and Social Psychology*, 1972, *24*, 207–213.

panied each description, half of the photographs picturing children who had previously been judged to be physically attractive and the other half picturing children judged to be relatively unattractive. All 20 subjects therefore served in both conditions of the experiment. The raw data are presented in Table 11.1. The table shows the average severity of punishment rating made by each subject for the attractive and unattractive children, high scores indicating more severe punishment.

To perform the *t* test for matched groups by the direct-difference method, we first obtain for each subject the *difference* between the pair of scores. This has been done in Table 11.1 in the column labeled *Difference*. In obtaining the difference score, it does not matter which raw score is subtracted from which as long as it is done in the same way for every subject. In Table 11.1 the attractive children's severity score is subtracted from the unattractive children's severity score; thus the difference score is negative whenever the attractive children's score is larger. The difference scores, for which we shall use the symbol D, can be considered and treated in the same way as a distribution of raw scores.

Once the distribution of direct differences, each difference having its correct algebraic sign, has been obtained, *all* further computations are based on it. First, we compute the algebraic mean of the differences ($\overline{D}$). This is done by summing all the positive values, summing all the negative values, adding the sum of the negative values to the sum of the positive values to get ΣD, and then dividing ΣD by N to get $\overline{D}$. In any matched-groups experiment N is always the number of *pairs* of raw scores, or the number of direct differences; thus, in this experiment N is 20.

The algebraic mean of the differences ($\overline{D}$) is *always* the same as the difference between the $\overline{X}$'s of the two raw-score distributions. Accordingly, in Table 11.1 $\overline{D}$ is 4.65 and identical with the difference between the two raw-score $\overline{X}$'s (31.55 and 26.90). The $\overline{D}$ is therefore the *numerator* in the *t* ratio; it is not necessary to compute the $\overline{X}$'s of the raw-score distributions, except as a check on accuracy of computation.

The $s_{\overline{X}_1 - \overline{X}_2}$ that is needed is obtained by computing *the standard error of the mean difference* ($s_{\overline{D}}$). Except for the fact that we are using difference scores, we obtain the $s_{\overline{D}}$ in the same way as any standard error of the mean.

First, we compute s^2 for the distribution of difference scores, which we symbolize as $s_D{}^2$. The most convenient way to compute $s_D{}^2$ is to use the raw score formula (see Formula 5.8 in Chapter 5). Rewriting this formula to indicate that we are now concerned with a distribution of difference scores and letting D stand for difference, we thus have:

$$s_D{}^2 = \frac{\sum D^2 - \left(\sum D\right)^2 / N}{N - 1} \tag{11.2}$$

Then we compute $s_{\overline{D}}$, the standard error of the mean difference, by the formula:

$$s_{\overline{D}} = \sqrt{\frac{s_D{}^2}{N}} \tag{11.3}$$

The standard error of the mean difference ($s_{\overline{D}}$) *is identical with the standard error of the difference between means that would be obtained by use of the long formula involving the correlation.*

Table 11.1 Severity of Punishment Ratings of Physically Attractive and Unattractive Children

Offense	Unattractive Offender	Attractive Offender	Difference (U − A)	(Difference)^2
1	15	17	−2	4
2	49	36	13	169
3	21	21	0	0
4	17	14	3	9
5	21	15	6	36
6	25	25	0	0
7	20	22	−2	4
8	56	38	18	324
9	16	19	−3	9
10	31	31	0	0
11	28	33	−5	25
12	44	39	5	25
13	35	29	6	36
14	48	41	7	49
15	32	31	1	1
16	37	27	10	100
17	45	21	24	576
18	29	27	2	4
19	28	32	−4	16
20	34	20	14	196

$$\sum D(+) = \overline{109} \qquad \sum D^2 = \overline{1583}$$
$$\sum D(-) = \overline{-16}$$
$$\sum D = \quad 93$$

$$\overline{X}_U = 31.55$$
$$\overline{X}_A = 26.90$$

$$\overline{D} = \frac{\sum D}{N} = \frac{93}{20} = 4.65$$

$$s_D{}^2 = \frac{\sum D^2 - \left(\sum D\right)^2 \big/ N}{N-1} \qquad s_{\overline{D}} = \sqrt{\frac{s_D{}^2}{N}}$$

$$= \frac{1583 - (93)^2/20}{20 - 1} \qquad\qquad = \sqrt{\frac{60.56}{20}} = \sqrt{3.03}$$

$$= \frac{1150.55}{19} = 60.56 \qquad\qquad = 1.74$$

$$H_0 : \ \mu_A - \mu_U = 0$$

$$H_1 : \ \mu_A - \mu_U \neq 0$$

$$t = \frac{\overline{D}}{s_{\overline{D}}} = \frac{4.65}{1.74} = 2.67$$

with $df = N - 1 = 19$, $t_{.05} = 2.09$, and $t_{.01} = 2.86$

Since $\overline{D}$ is equal to $(\overline{X}_1 - \overline{X}_2)$ and $s_{\overline{D}}$ is equal to $s_{\overline{X}_1 - \overline{X}_2}$, it follows that:

$$t = \frac{\overline{D}}{s_{\overline{D}}} \qquad\qquad (11.4)$$

The form of the numerator of this formula indicates, you will notice, that we are testing the usual null hypothesis that the two μ's are equal in value; therefore, we do not have to include the expression $\mu_1 - \mu_2$.

For the data in Table 11.1, $s_{\overline{D}} = 1.74$. Our t ratio is therefore $4.65/1.74$, or 2.67. In any experiment using matched groups the number of degrees of freedom (df) for evaluating the t is N - 1, where again N is the number of pairs of measures. In our example there are twenty pairs of measures, so df is 19. With 19 df our t of 2.67 is significant at the 5% level. We conclude (with no great pleasure) that children who are physically attractive are judged less severely for their misbehavior than their less attractive peers.

The direct-difference method automatically takes into account the correlation that exists between the raw-score distributions (r_{12}), regardless of the size or algebraic sign of the correlation. Therefore, as we have said, the t value obtained by this method will be identical with the t value obtained by use of the long formula for the $s_{\overline{X}_1 - \overline{X}_2}$. We rarely use the long formula, which would probably save time only if we wanted to know the value of the correlation and thus already had computed r_{12}.

Sandler's A Statistic

An even simpler method of testing the hypothesis that $\mu_1 - \mu_2 = 0$ for matched pairs is provided by Sandler's A statistic, named after its discoverer, Joseph Sandler. The A statistic is based on the same difference or D scores and squares of difference scores as are obtained in the direct-difference method just described. The formula for A is as follows:

$$A = \frac{\sum D^2}{\left(\sum D\right)^2} \qquad\qquad (11.5)$$

To interpret the A statistic, we must refer to Table H in Appendix II. We use Table H to find the critical value of A for $df = N - 1$ associated with our chosen α level. In assessing our computed A, we reverse the procedure used in assessing t. That is, if our computed A is equal to *or less than* the tabled value at the given α level, H_0 is *rejected* at that level.

An exact equivalence can be shown between the critical value of t at any given probability level and the critical value of A at that level. The conclusion that is reached about $H_0: \mu_1 - \mu_2 = 0$ is therefore the same for t and A.

The data from our previous example can be used to illustrate the application of the A statistic. Filling in the values of $\sum D$ and $\sum D^2$ from Table 11.1, we find:

$$A = \frac{\sum D^2}{\left(\sum D\right)^2} = \frac{1583}{(93)^2} = .183$$

Reference to Table H shows that with $df = 19$, the critical values of A for a two-tailed test are 0.267 at $\alpha = .05$ and 0.166 for $\alpha = .01$. Our calculated A of 0.183 is less than the value of A at $\alpha = .05$ but is greater than the value of A at $\alpha = .01$. Since the *smaller* the A the more significant it is, we are able to reject H_0 at $\alpha = .05$ but not at $\alpha = .01$. This conclusion is the same as the one we reached earlier for the direct difference t.

TESTING THE SIGNIFICANCE OF r and r_S

In conducting experiments using the matched pairs technique, our interest is in comparing $\overline{X}$'s to determine whether the experimental conditions differentially influence the performance measure. Matched pairs are also used to determine the degree of covariation or *correlation* between two variables. In order to compute a Pearson r, for example, we must first obtain two measures from each member of a single group of individuals or, in a group of previously paired individuals, a measure from each member of the pair.

In computing a correlation, our interest typically is in the relationship between the variables in the population as a whole rather than in the particular sample of individuals we happen to measure. We therefore set up a null hypothesis about the population r and use our sample r to evaluate this hypothesis. The null hypothesis that is usually tested is that the population correlation is zero.

Testing the null hypothesis that the population r is zero is quite simple. At any given α level, the value of r required for significance turns out to be the same for all samples with the same df. The df is given by N - 2, one less than you might have expected; N, remember, is the number of *pairs* of scores. Table D in Appendix II lists the critical values of r for the 5% and 1% levels. Whenever our obtained r is equal to or greater *in absolute size* than the tabled value, we can conclude it is significant at the α level we have specified. That is, we reject the hypothesis that the population r is zero and conclude that there is some degree of relationship between the variables.

As an illustration, suppose that 152 ten-year-old children were given a test of verbal ability and a test of motor coordination and r was calculated to be .17. Table D shows that for $df = 150$, the critical values of r are .159 and .208 at the 5% and 1% significance levels. We reject at $\alpha = .05$ the null hypothesis that the population r is zero and conclude instead that children who are high in verbal ability tend to be somewhat more coordinated than those of lesser verbal ability.

Two assumptions are required for testing the null hypothesis about r. First, the sample of pairs must have been obtained by random sampling from the population concerned. Second, the population of X and Y scores must have a distribution that we will characterize by saying that X and Y must each be normally distributed, and that the relationship between X and Y must be linear. This is by no means a complete statement of the second assumption, but it suggests what to consider before testing the null hypothesis about the population r. If the sample is randomly drawn, if X and Y are each normally distributed, and if the correlation between X and Y is linear, one may reasonably decide to perform this significance test.

The null hypothesis that the population value is zero can also be tested for the Spearman r (r_S) calculated from rank-order data, as discussed in Chapter 6. Table E in Appendix II presents the critical values of r_S at the 5% and 1% significance levels. As in the case of r, the required values for significance depend upon the number of pairs, N. In Table E, we use N directly and do not have to find *df*. To illustrate the use of the table, suppose we have calculated an r_S of .514 between two sets of ranks obtained from a sample of twelve individuals. Can we conclude that r_S in the population from which the sample was drawn is greater than zero? When we refer to Table E, we see that with N = 12, an r_S of .591 is required for significance at the 5% level. Since our obtained r_S is below this value, we accept the hypothesis of no relationship between the two variables at the 5% level. Notice that in this case an r_S that seems quite large (.514) is not significantly different from zero at the 5% level because it is based upon a relatively small N.

DEFINITIONS OF TERMS AND SYMBOLS

Matched pairs. A method of obtaining comparable groups in a two-group experiment. The same individual is tested twice, once under each experimental condition, or a series of pairs of individuals who are similar in one or more characteristics is selected, one member of each pair being assigned at random to one experimental condition and the second member then automatically assigned to the second condition. The result in both cases is a set of pairs of scores.

Direct-difference method. A method for computing t to test H_0: $\mu_1 - \mu_2 = 0$ that is used with matched pairs and is based on an analysis of the difference (D) between each pair of scores $(X_1 - X_2)$.

Mean difference $(\overline{D})$. The mean of the sample of difference (D) scores, obtained by summing the D's and dividing by N, the number of paired scores.

Estimated standard error of mean difference $(s_{\overline{D}})$. The estimated value of the standard deviation of the sampling distribution of mean difference scores $(\overline{D})$ that would otherwise be obtained from an infinite number of samples consisting of matched pairs. The estimated standard error, $s_{\overline{D}}$, is determined from the data of a single sample of D scores by the formula $s_D/\sqrt{N}$.

Degrees of freedom (df) for matched pairs. Degrees of freedom (df) used in evaluating t for matched pairs is N − 1, where N is the number of pairs.

Sandler's A statistic. A simple method of testing the hypothesis that $\mu_1 - \mu_2 = 0$ based on the difference (D) scores from a sample of matched pairs. The A statistic, defined as $\Sigma D^2/(\Sigma D)^2$, is interpreted by means of a special A table (shown in Table H). At any given α level, there is an exact equivalence between the critical value of A and the critical value of t, leading to the same conclusion about H_0.

Null hypothesis about r. A statistical hypothesis about the population value of Pearson r. The null hypothesis that is typically tested is that the population $r = 0$. This hypothesis can be evaluated at a given α level by consulting the critical values of the sample r associated with the sample df, listed in Table D. The df is N − 2, where N is the number of pairs.

Null hypothesis about Spearman r (r_S).   A statistical hypothesis about the population value of the Spearman rank-order correlation coefficient, r_S, typically that the population value $r_S = 0$.   The null hypothesis that $r_S = 0$ can be evaluated at a given α level by consulting the critical values of the sample r_S associated with N, the number of pairs in the sample.   These values are listed in Table E.

PROBLEMS

1. In an experiment designed to test whether schizophrenics are more likely to exhibit a deficit in information processing than nonschizophrenics, 15 hospitalized schizophrenics are first selected.   Then each of these patients is matched with a nonpsychiatric patient with a medical problem on the basis of age, education, and socioeconomic background.   Shown below are the $\overline{X}$ and standard error of the mean ($s_{\overline{X}}$) for each group, derived from their performance on an information-processing task.   Also shown is the correlation between the scores for the matched pairs of subjects.

	Schizophrenic	Medical
$\overline{X}$	84.2	91.1
$s_{\overline{X}}$	2.3	2.6
N = 15	$r = .45$	

   (a) What does the r indicate?
   (b) Calculate t, using the *random groups* method for determining $s_{\overline{X}_1 - \overline{X}_2}$ (see Chapter 10).   What do you conclude about the performance of the two groups of patients, using $\alpha = .05$?
   (c) Recalculate t, using Formula 11.1 to determine $s_{\overline{X}_1 - \overline{X}_2}$.   Do you come to a different conclusion from the one in (b)?   Why?

2. In a study of taste aversion, 10 laboratory rats were fed twice a day, their diet alternating between a sugar-flavored and a chocolate-flavored mash.   As determined by amount eaten, the rats found the two mixtures equally palatable.   After several weeks on these diets, the rats were injected with a chemical immediately after consuming the chocolate-flavored mash.   The chemical was one that had the effect of making the animals temporarily ill.   The amount of each type of food each animal ate during the next feeding cycle is shown below.

Subject	Chocolate	Sugar
1	10	9
2	4	7
3	14	18
4	5	5
5	11	19
6	9	14
7	16	22
8	11	16
9	12	9
10	7	6

(a) Calculate t by the direct difference method. Using $\alpha = .05$, what do you conclude about the effect of the chemically induced illness on the palatability of the chocolate-flavored mash?

(b) Compare the two groups by means of Sandler's A statistic. What do you conclude? [Your conclusion should be the same as in (b), above.]

3. Several sample r's and r_S's are listed below, along with N. Determine whether each is significant at the specified α level.

(a) Before determining significance, state what H_0 you are testing about all these correlations.

(b) $r = .35$, N = 20. Use $\alpha = .05$.

(c) $r = .56$, N = 30. Use $\alpha = .01$.

(d) $r = .15$, N = 205. Use $\alpha = .05$.

(e) $r_S = .42$, N = 16. Use $\alpha = .01$.

(f) $r_S = .48$, N = 24. Use $\alpha = .01$.

One-Way Analysis of Variance

CHAPTER 12

In previous chapters we have explained techniques for analyzing data from two different experimental conditions in order to test hypotheses about the effects of those conditions. We saw how to test the significance of differences between $\overline{X}$'s for two independent groups treated differently and also how to test the significance of differences between $\overline{X}$'s for the same group or matched pairs of subjects tested under two different conditions. As yet, however, we have not considered the statistical analysis of findings from experiments in which the effects of three or more conditions are compared. This chapter will describe a technique called analysis of variance, one of the most widely used methods for making such analyses.

THE PURPOSE OF ANALYSIS OF VARIANCE

First of all, we ask why an experiment should ever include more than two conditions. Two answers may be given: (a) We may be interested in studying more than two conditions at a time, and (b) the data obtained with two conditions may give quite a different answer to our experimental problem than would comparable data from three or more conditions. Referring to our first point, suppose that an investigator is interested in the relative effectiveness of various methods of teaching French. We can see that there might be three, four, or even more teaching methods the investigator would like to compare rather than only two. However (and this is our second point), if only two teaching methods were studied and no differences were found between them, it might be concluded that methods never differentially affect students' performance, when in fact some other method that could have been included in an experiment with more than two conditions could have affected performance appreciably.

Let us consider another experiment in which more than two experimental conditions would be desirable. If we wished to determine the course of forgetting with the passage of time, we could train each of two groups of people to recite a list of twelve nonsense syllables and test one group for retention one hour after training and another group 24 hours after training. This would give us two points on the recall curve (the curve showing how much is remembered at various lengths of time after training), but it certainly would not give us enough information to permit us to plot the intermediate points on the curve or to estimate the amount of recall after, say, 48 hours. Here too, we need more than two groups in order to obtain an adequate picture of the relationship between our experimental variable and the behavior being observed.

Once it is agreed that experiments involving more than two groups or conditions are sometimes desirable, we face the necessity of testing the significance of the differences among the $\overline{X}$'s obtained with the several conditions. Our first thought would surely be to compute t ratios to assess for each pair of groups the null hypothesis that the populations from which the groups were drawn had the same μ. We would thus test the significance of the difference between $\overline{X}$'s for all possible pairs of conditions. However, the use of separate t tests for analyzing the results of an experiment based upon more than two groups would lead to results that could not easily be interpreted. If we obtain one significant t value out of six, what does this mean? Had we performed a single t test, we would have had .05 probability of obtaining a *t* significant at the 5% level by random sampling alone, but with six *t* tests our chances of having at least one of them significant because of random sampling is much larger. To carry this argument to its extreme, if a million *t* tests were involved in an experiment, a single significant *t* value would certainly be no indication that the conditions being varied had an effect.

We would not be handicapped by the fact that increasing the number of *t* ratios to be computed for an experiment increases the probability that a significant *t* will be found by chance if we could state how this probability changes with the number of *t* values to be computed. Unfortunately, this cannot readily be done, and so our use of *t* leads us to results we cannot evaluate easily.[1] Consequently, a single test for the significance of the differences among all the $\overline{X}$'s of the experiment seems necessary as a means of permitting a more correct determination of the significance of differences obtained in the experiment. This chapter will introduce you to one such test, *analysis of variance*. This technique, which is often identified by the acronym **ANOVA**, is appropriate for use with two or more groups. However, it is most frequently employed with three or more groups, since the *t* test is available for two-group experiments.

First, we will describe the principles of analysis of variance (ANOVA) as they apply to experiments employing two or more conditions, each representing the manipulation of a single type of independent variable; for example, different kinds of diet or amounts of time between memorization and a test of recall. Since only one basic variable is being studied, this method is identified as *one-way analysis of variance*. (In the next chapter, *two-way analysis of variance*, suitable for analyzing data of experiments in which two types of independent variables are manipulated, will be discussed). After presenting the principles of one-way analysis of variance (ANOVA), we will demonstrate how to apply the method to the data of investigations employing two or more conditions.

THE ABC's OF ANALYSIS OF VARIANCE

This section will describe the analysis of variance technique in very general terms. The basic underlying fact is that the total variability of a set of scores from several groups may be divided into two or more categories. In the type of experimental design we are considering, there are two categories of variability: the variability of subjects within each group and the variability between the different groups. To visualize these two kinds of variability, imagine that we have four experimental conditions, each of which is applied to a different group. Suppose each condition yields widely different scores, so that members of group 1 all have smaller scores than members of group 2, who in turn have smaller scores than members of group 3, and so on. This gives us the situation graphed in Figure 12.1. Notice that $\overline{X}$ values are given for each group separately and also for the set of four groups combined, $\overline{X}$ in the latter case being called $\overline{X}_{tot}$ ($\overline{X}$ of total). Now we can identify

---

[1]There is a conservative procedure for evaluating the significance of several *t* tests performed at once. This procedure, called Dunn's test, consists of changing the usual .05 (or .01) significance level to a more stringent level that depends upon the number of *t* tests being performed. See G. Keppel, *Design and analysis* (Englewood Cliffs, N.J.: Prentice-Hall, 1973), pp. 147–149.

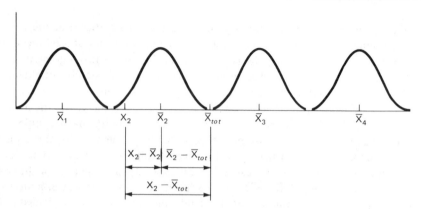

Figure 12.1 An example of four groups of scores illustrating variability within groups and variability between groups.

the variability observed in the experiment: the variability of the scores from $\overline{X}_{tot}$ is called the *total variability*, the variability of the scores from the $\overline{X}$ of their group is called the *variability within groups*, and the variability of the group $\overline{X}$'s from $\overline{X}_{tot}$ is called the *variability between groups*.

Consider, for example, the people who have the score marked X_2. The deviation $(X_2 - \overline{X}_{tot})$ contributes to the total variability, but that deviation is composed of the two quantities $(X_2 - \overline{X}_2)$ and $(\overline{X}_2 - \overline{X}_{tot})$, as Figure 12.1 makes clear. Since $(X_2 - \overline{X}_2)$ is the difference between a score and its group $\overline{X}$, and $(\overline{X}_2 - \overline{X}_{tot})$ is the difference between the group $\overline{X}$ and $\overline{X}_{tot}$, the contribution of X_2 to the total variability is made up of one part that is variability within a group and another that is variability between groups. Since any score's deviation from $\overline{X}_{tot}$ may be divided algebraically into these two parts, it is possible to express the overall variability observed in an experiment as the sum of these two types of variability—the sum of the variability between groups and the sum of the variability within groups.

Once we have assimilated the idea that one kind of variability in an experiment is between groups and another kind is within groups, we can let our previous experience with statistical tests suggest how this idea can be employed in analyzing differences in $\overline{X}$ values obtained from several groups. Thinking back to our study of t and our test of the null hypothesis that $\mu_1 - \mu_2 = 0$, we remember that the numerator of the t ratio is equal to $\overline{X}_1 - \overline{X}_2$ and that the denominator is $s_{\overline{X}_1 - \overline{X}_2}$. Since $\overline{X}_1 - \overline{X}_2$ is a measurement of the difference between groups, it could be called an expression of the variability between groups. Similarly, $s_{\overline{X}_1 - \overline{X}_2}$ could be considered a measurement of the variability within groups, making t the ratio of variability between groups to variability within groups. In that case we could say that a significant t occurs only when the ratio of between-groups variability to within-groups variability is as large as the significant t values presented in Table C. Applying the same reasoning to experiments involving more than two groups, we first assume the null hypothesis that the μ values of the groups are all the same $(\mu_1 = \mu_2 = \mu_3 = \cdots = \mu_k)$, so that the overall difference is zero. We then obtain a ratio of these two types of variability and, provided this ratio is sufficiently large, we call the differences among $\overline{X}$ values significant. This ratio, we will discover shortly, is identi-

ANOVA
is

fied by the letter F. The analysis of variance technique is simply a method that provides an objective criterion for deciding whether the variability between groups is large enough in comparison with the variability within groups to justify the inference that the means of the populations from which the different groups were drawn are not all the same.

The relationship between the t and the F ratio is more formal than the parallel we have just described. In fact, when only two groups are being compared, F has the value of t^2. Further, when one is significant (or nonsignificant) at a given α level, so is the other.

THE CONCEPT OF SUMS OF SQUARES

Basic Ideas

The cornerstone in the analysis of variance technique is the *sum of squares*, an old concept masquerading under a new name. *The sum of squares is simply Σx^2*, *the sum of squared deviations from the mean*, an expression we used earlier in defining the variance, or the standard deviation squared. (We remember that the variance equals $\Sigma x^2/N$, where x is the deviation of the score X from the mean.)

When the value of x is large [that is, when a score (X) is far from $\overline{X}$] x^2 is also large. Thus a large Σx^2 (sum of squares) occurs when the scores tend to be widely dispersed about $\overline{X}$. But this is also what we mean by great variability. Consequently, a large value for the sum of the squares indicates a large amount of variability; small sums of squares values indicate small amounts of variability.

Types of Sums of Squares Values

We have already stated that the total variability in an experiment may be divided into two parts: the variability of subjects within groups, and the variability between different groups. Now we wish to find ways of computing sums of squares that will correspond to the total variability and to its two parts. These three sums of squares, which we shall call, respectively, total sum of squares (SS_{tot}), within-groups sum of squares (SS_{wg}), and between-groups sum of squares (SS_{bg}), will be discussed presently. First, however, we must list some new symbols that will be used in the remainder of this chapter:

$\overline{X}_{tot}$ = the mean of all scores in the experiment (also called the grand mean or the overall mean)

$\overline{X}_g$ = a general expression for the mean of the scores in any group (g)

N_{tot} = the total number of scores in the experiment

N_g = a general expression for the number of scores in any group

k = the number of groups

$\sum_{tot} X$ = the sum of all the scores in the experiment

$\sum_{tot} X^2$ = the sum of all the squared scores in the experiment

We also need to distinguish between the scores of the different groups. A score from group 1 is indicated by X_1, a score from group 2 by X_2, a score from group 3 by X_3, and so forth. This numerical subscript is similarly used to identify the mean and N of each group. For example, $\overline{X}_4$ is the mean of group 4 and N_4 is the number of cases in that group.

The Definition of SS_{tot}. The total sum of squares (SS_{tot}) is just what our previous statement would suggest: *the sum of the squared deviations of every score from the grand mean* $(\overline{X}_{tot})$ *of all the scores in the experiment.* This definition may also be expressed in an equation:

$$SS_{tot} = \sum (X - \overline{X}_{tot})^2 \tag{12.1}$$

Because SS_{tot} is to represent all the variability recorded in an experiment, we must find $(X - \overline{X}_{tot})^2$ for each X in the entire experiment and then sum all of these $(X - \overline{X}_{tot})^2$ quantities. Accordingly, the number of squared deviations to be added will be equal to the total number of subjects in the experiment (N_{tot}). To repeat, we subtract $\overline{X}_{tot}$ from each person's score (X), square this deviation, and then sum these squared deviations.

The Definition of SS_{wg}. The sum of squares within groups is most easily understood as a combination of sums of squares within separate groups. For example, we define the sum of squares within group 1 (SS_{wg1}) by the following equation:

$$SS_{wg1} = \sum (X_1 - \overline{X}_1)^2$$

Note that X_1 is a score in group 1, and that we must find $(X_1 - \overline{X}_1)^2$ for every X_1 in group 1. The sum of these squared deviations is an expression of the variability within group 1. You can see that it serves the same purpose for group 1 that SS_{tot} does for the entire experiment.

We define the sum of squares within group 2 (SS_{wg2}) by writing the preceding equation in terms of X_2 and $\overline{X}_2$ rather than X_1 and $\overline{X}_1$. A similar change is made for group 3 and for any other group in the experiment. Once the sum of squares within each of the groups is known, we add them, because we want only one SS_{wg} expression. Accordingly, we define SS_{wg} for the entire experiment as the sum of the SS_{wg} values for the separate groups. Writing this definition in terms of the quantities that make up the individual SS_{wg} values, we have for the case of three groups:

$$SS_{wg} = SS_{wg1} + SS_{wg2} + SS_{wg3}$$

With a different number of groups (k), the number of terms on the right-hand side of the equation would equal the new value of k. For example, with five groups (when $k = 5$), SS_{wg} would be defined as:

$$SS_{wg} = SS_{wg1} + SS_{wg2} + SS_{wg3} + SS_{wg4} + SS_{wg5}$$

A general expression for SS_{wg} can be written as:

$$SS_{wg} = SS_{wg1} + SS_{wg2} + \cdots + SS_{wgk} \qquad (12.2)$$

or

$$SS_{wg} = \sum(X_1 - \overline{X}_1)^2 + \sum(X_2 - \overline{X}_2)^2 + \cdots + \sum(X_k - \overline{X}_k)^2$$

where the subscript k indicates the last in a set of k groups.

The Definition of SS_{bg}. From the above we would expect that SS_{bg} would be based upon the squared deviation of each $\overline{X}_g$ from $\overline{X}_{tot}$, just as other SS's are based upon deviations of scores from $\overline{X}$'s. This is exactly what is done. First we find $(\overline{X}_g - \overline{X}_{tot})^2$ for each group. Then we weight each $(\overline{X}_g - \overline{X}_{tot})^2$ quantity with the number of cases in its group (N_g) and add the weighted quantities $N_g(\overline{X}_g - \overline{X}_{tot})^2$ to find SS_{bg}. In general, then, we can say the following:

$$SS_{bg} = \sum [N_g(\overline{X}_g - \overline{X}_{tot})^2] \qquad (12.3)$$

Taking a specific number of groups ($k = 3$), we can write this equation in a new form:

$$SS_{bg} = N_1(\overline{X}_1 - \overline{X}_{tot})^2 + N_2(\overline{X}_2 - \overline{X}_{tot})^2 + N_3(\overline{X}_3 - \overline{X}_{tot})^2$$

since N_g is N_1 for the first group, N_2 for the second group, and N_3 for the third group and $\overline{X}_g$ has comparable values.

Relationship between SS_{tot} and SS_{wg}

We have said that SS_{wg} typically accounts for part, but not all, of SS_{tot}. For instance, if we compared the mathematical aptitude scores of physics and English majors, the variability of scores within the two groups would account for part of the total variability (SS_{tot}). This is not the sole source of variability, however; we would expect the physics majors to exhibit greater mathematical aptitude than the English majors, and this difference between groups would be reflected in an SS_{bg} value, which would also be part of SS_{tot}.

In some other experiments there might be no difference, or almost none, between the groups. For example, three teaching methods—a lecture method, a discussion method, and a project method—used with three different groups of students of freshman history might result in identical $\overline{X}$'s on their exams. Then each $\overline{X}_g$ would have the same value as $\overline{X}_{tot}$, and there would be no SS_{bg}. In such a case all of SS_{tot} would be accounted for by SS_{wg}. In the next three paragraphs this fact, that SS_{wg} is *part* of SS_{tot} when there are differences in the $\overline{X}$'s of the groups, and that SS_{wg} is *all* of SS_{tot} when there are no differences in the $\overline{X}$'s of the groups, is elaborated for the case of three groups.

If all groups have the same mean, then $\overline{X}_1 = \overline{X}_2 = \overline{X}_3$. Furthermore, each $\overline{X}_g$ (group $\overline{X}$) is equal to $\overline{X}_{tot}$. Consequently, we may substitute $(X - \overline{X}_{tot})$ for each $(X - \overline{X}_g)$ value in our definition of SS_{wg}. But $(X - \overline{X}_{tot})$ is the basic component in the compu-

Figure 12.2 Demonstration that SS_{tot} is equal to SS_{wg} when all group $\overline{X}$'s are equal to $\overline{X}_{tot}$.

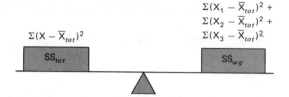

$$\Sigma(X - \overline{X}_{tot})^2$$

$$\Sigma(X_1 - \overline{X}_{tot})^2 +$$
$$\Sigma(X_2 - \overline{X}_{tot})^2 +$$
$$\Sigma(X_3 - \overline{X}_{tot})^2$$

tation of SS_{tot}. Therefore, SS_{tot} will now equal SS_{wg}, as the set of balances in Figure 12.2 shows.

The reason that SS_{wg} balances SS_{tot} is suggested by the objects in the scale. Because every X must come from some group, it is either an X_1, an X_2, or an X_3 in this situation. Therefore, the quantity on the right of the balance is exactly the same as that on the left; on each side of the balance we are subtracting $\overline{X}_{tot}$ from each score separately, squaring the differences and summing them. Accordingly, $SS_{tot} = SS_{wg}$ when all $\overline{X}_g$ values are equal, as in the case of the three teaching methods that proved equally effective.

Of course, the situation above is an exceptionally rare one. Even when three experimental conditions have identical effects, the $\overline{X}_g$ values usually differ slightly from one another. In this case the $\overline{X}_g$ values are not equal to $\overline{X}_{tot}$, SS_{tot} is greater than SS_{wg}, and the scale no longer balances, as illustrated in Figure 12.3.

We can put the scale in balance again if we add an appropriate amount to the right-hand side. This amount will always be the value of SS_{bg}, as we have indicated in Figure 12.4. However, SS_{bg} is not simply a convenient quantity dreamed up to account for the difference between SS_{tot} and SS_{wg}. It is a meaningful measurement of the difference between group $\overline{X}$s and will prove extremely useful to us.

We now know the definitions of SS_{tot} and its two components, SS_{bg} and SS_{wg}. We also understand how the differences among $\overline{X}_g$ values affect these SS values. Before we can apply our knowledge in a practical way, however, we must know how to compute the values for SS_{tot}, SS_{bg}, and SS_{wg}.

Computational Formulas for Finding Sums of Squares

Application of the basic formulas we have just discussed for finding sums of squares is a long, laborious process that involves determining the deviation of each score from the overall or grand mean, the deviation of each score from its own mean, squaring each of these sets of deviations, and so on. We can simplify our work in calculating SS values by

Figure 12.3 Demonstration that SS_{tot} is greater than SS_{wg} when the group $\overline{X}$'s are not equal to $\overline{X}_{tot}$.

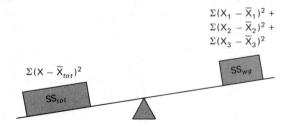

$$\Sigma(X - \overline{X}_{tot})^2$$

$$\Sigma(X_1 - \overline{X}_1)^2 +$$
$$\Sigma(X_2 - \overline{X}_2)^2 +$$
$$\Sigma(X_3 - \overline{X}_3)^2$$

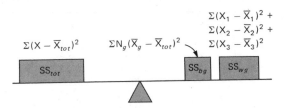

Figure 12.4 Demonstration that $SS_{tot} = SS_{wg} + SS_{bg}$.

using special computational formulas that lead to precisely the same answers as do the basic formulas but are usually faster. Although these new equations can be derived algebraically from the basic definitions, we will present them without proof. Following this presentation, we will demonstrate their application to a set of data.

First, SS_{tot} is found by the computational formula:

$$SS_{tot} = \left(\sum_{tot} X^2\right) - \frac{\left(\sum_{tot} X\right)^2}{N_{tot}} \tag{12.4}$$

The formula states that SS_{tot} is computed by first squaring each X and adding all the X^2 values together, giving us $\sum_{tot} X^2$. Next the sum of all the X values is found; this *sum* is squared and then divided by the total N, giving us $\left(\sum_{tot} X\right)^2 / N$. Finally, this latter result is subtracted from $\sum_{tot} X^2$ to give SS_{tot}.

Next, SS_{bg} is found by the formula:

$$SS_{bg} = \sum_g \left[\frac{\left(\sum X_g\right)^2}{N_g}\right] - \frac{\left(\sum_{tot} X\right)^2}{N_{tot}} \tag{12.5}$$

Thus, we determine SS_{bg} by first finding for each group its $\sum X$, squaring it, and dividing the result by the corresponding N. The resulting quantities from all the groups are added together, as indicated by the symbol $\sum_g$, and the quantity obtained by dividing $\left(\sum_{tot} X\right)^2$ by N_{tot} subtracted from this sum. (You will note that the latter quantity was also used in determining SS_{tot} so it needs to be calculated only once.) The computational formula for SS_{bg} could also be written for k groups as follows:

$$SS_{bg} = \left[\frac{\left(\sum X_1\right)^2}{N_1} + \frac{\left(\sum X_2\right)^2}{N_2} + \cdots + \frac{\left(\sum X_k\right)^2}{N_k}\right] - \frac{\left(\sum_{tot} X\right)^2}{N_{tot}}$$

Table 12.1 Application of the Computational Formulas for SS_{tot}, SS_{bg}, and SS_{wg} to the Data from the Hypothetical Attitude Experiment

X_1	$X_1{}^2$	X_2	$X_2{}^2$	X_3	$X_3{}^2$
68	4624	78	6084	94	8836
63	3969	69	4761	82	6724
58	3364	58	3364	73	5329
51	2601	57	3249	67	4489
41	1681	53	2809	66	4356
40	1600	52	2704	62	3844
34	1156	48	2304	60	3600
27	729	46	2116	54	2916
20	400	42	1764	50	2500
18	324	27	729	32	1024
420	20448	530	29884	640	43618

$$\sum_{tot} X = \sum X_1 + \sum X_2 + \sum X_3 = 420 + 530 + 640 = 1590$$

$$\sum_{tot} X^2 = \sum X_1{}^2 + \sum X_2{}^2 + \sum X_3{}^2 = 20448 + 29884 + 43618 = 93950$$

$$N_{tot} = N_1 + N_2 + N_3 = 10 + 10 + 10 = 30$$

$$SS_{tot} = \left(\sum_{tot} X^2\right) - \frac{\left(\sum_{tot} X\right)^2}{N_{tot}}$$

$$= 93950 - \frac{(1590)^2}{30}$$

$$= 93950 - 84270$$

$$= 9680$$

$$SS_{bg} = \sum_{g}\left[\frac{\left(\sum X_g\right)^2}{N_g}\right] - \frac{\left(\sum_{tot} X\right)^2}{N_{tot}}$$

$$= \frac{\left(\sum X_1\right)^2}{N_1} + \frac{\left(\sum X_2\right)^2}{N_2} +$$

$$\frac{\left(\sum X_3\right)^2}{N_3} - \frac{\left(\sum_{tot} X\right)^2}{N_{tot}}$$

$$= \frac{(420)^2}{10} + \frac{(530)^2}{10} +$$

$$\frac{(640)^2}{10} - \frac{(1590)^2}{30}$$

$$= \frac{176400}{10} + \frac{280900}{10} +$$

$$\frac{409600}{10} - 84270$$

$$= \frac{866900}{10} - 84270$$

$$= 2420$$

$$SS_{wg} = \sum_{g}\left[\left(\sum X_g{}^2\right) - \frac{\left(\sum X_g\right)^2}{N_g}\right]$$

$$= 20448 - \frac{(420)^2}{10} +$$

$$29884 - \frac{(530)^2}{10} +$$

$$43618 - \frac{(640)^2}{10}$$

$$= 2808 + 1794 + 2658$$

$$= 7260$$

OR

$$SS_{wg} = SS_{tot} - SS_{bg}$$

$$= 9680 - 2420$$

$$= 7260$$

Finally, since we know that SS_{tot} is composed of SS_{bg} and SS_{wg}, we can find SS_{wg} by subtraction:

$$SS_{wg} = SS_{tot} - SS_{bg} \qquad \text{(12.6)}$$

We could also compute SS_{wg} directly, instead of by subtraction, by calculating SS_{wg} for each group separately and then adding these values together to obtain the overall SS_{wg}. These operations are expressed by the following formula:

$$SS_{wg} = \sum_g \left[\left(\sum X_g^2 \right) - \frac{\left(\sum X_g \right)^2}{N_g} \right] \qquad \text{(12.7)}$$

With the computational formulas for SS_{tot}, SS_{bg}, and SS_{wg} before us, we can illustrate their use with a specific set of scores. Table 12.1 presents the results of a hypothetical experiment on the effects of courses in music appreciation on attitudes toward classical music. Group 1 members took no courses in music appreciation, group 2 took one course, and group 3 took two courses before having their attitudes measured. Our first step, as shown in Table 12.1, is to find the $\sum X$ and $\sum X^2$ for each group and then to sum each of these quantities across groups to obtain $\sum_{tot} X$ and $\sum_{tot} X^2$. We then compute SS_{tot}, SS_{bg}, and SS_{wg} using our computational formulas. You will notice that we have obtained the value of SS_{wg} twice, first by subtracting SS_{bg} from SS_{tot} and then by computing it directly. The fact that the two procedures yield the same value for SS_{bg} not only shows that our calculations were accurate, but also demonstrates our earlier statement that $SS_{bg} + SS_{wg}$ always equals SS_{tot}.

One more fact about the computation of SS values is important to us. If we look carefully at the basic formulas for SS values, we see that no SS can ever have a negative value. This is not obvious from the computational formulas, but must also be true for them. Consequently, a negative SS value is always a sign that a mistake in arithmetic has been made and that recalculations are necessary.

THE CONCEPT OF VARIANCE OR MEAN SQUARE

The use of sums of squares as measurements of variability has only one disadvantage: The size of SS depends very much on the number of measurements on which it is based. Other things being equal, a sample of 100 measurements from a given population will yield about twice as large an SS as a sample of 50 from that population, because there will be twice as many squared deviations involved in the former. For this reason, a large SS may result either from a large variability of the characteristic being measured or from a large N in the sample. Consequently, SS is an unsatisfactory measurement of variability and can only be of limited usefulness, unless it leads to some new statistic that does not depend so greatly upon N.

Fortunately, such a statistic does exist: the *variance*. Recall from Chapter 5 that the variance (or standard deviation squared) is defined as $\Sigma x^2/N$, the mean of the squared deviations from the mean. In the present context we will call this quantity the *mean square* rather than the variance, because mean square (MS) has come to be used by statisticians in place of variance in discussing the analysis of variance technique.

Recall also from Chapter 5 that the formula for s^2, the unbiased estimate of the population variance (or mean square) obtained from the data of a single sample, is $\Sigma x^2/(N-1)$, where $N-1$ is the number of degrees of freedom associated with the sample. Since we have agreed in this chapter to call the sum of the squared deviations (Σx^2) sum of squares or SS, we can now present a general expression for the unbiased estimate of the population variance or mean square: $MS = SS/df$.

As we said when we discussed the ABC's of analysis of variance, we want to judge whether our group $\overline{X}$'s differ significantly by comparing a measure of variability between groups, which you can now surmise to be a mean square between groups, with a measure of variability within groups, which will be the mean square within groups. We then discover how large a value of the ratio of these two variabilities is required to indicate that the group $\overline{X}$'s differ more than would be expected if the null hypothesis of equal μ's were true. With this in mind, it appears that what we want now is to find one MS (mean-square) value to represent variability between groups and another MS value to represent variability within groups. Once these values (MS_{bg} and MS_{wg}, respectively) are known for any experiment, the size of their ratio can be used as the basis for drawing statistical conclusions about the data being analyzed.

To find each MS value, we simply divide the appropriate SS by its df. Thus MS_{bg} is based upon SS_{bg} and df_{bg}, the number of degrees of freedom between groups:

$$MS_{bg} = \frac{SS_{bg}}{df_{bg}} \tag{12.8}$$

where $df_{bg} = k - 1$. We note that df_{bg} has a meaning similar to that of df in the t test. However, it is based upon k, the number of groups, rather than upon the number of subjects, as with the t test.

The second MS value, MS_{wg}, is the ratio of SS_{wg} to df_{wg}:

$$MS_{wg} = \frac{SS_{wg}}{df_{wg}} \tag{12.9}$$

where $df_{wg} = N_{tot} - k$.

We may easily remember the value of df_{wg} by saying that the df within a single group is one less than the number of scores in the group (that is, $N_g - 1$), thus making the df within all k groups equal to $(\Sigma N_g) - k$ or $N_{tot} - k$, since 1 df is lost in each of the k groups.

We now illustrate the determination of MS_{bg} and MS_{wg} with a computational example. By referring to Table 12.1, we find that $SS_{bg} = 2420$ and $SS_{wg} = 7260$ in our hypothetical attitude experiment. Since $N_{tot} = 30$ and $k = 3$ for this experiment, $df_{bg} = 3 - 1 = 2$ and $df_{wg} = 30 - 3 = 27$. Notice that $df_{bg} + df_{wg} = 2 + 27 = 29$, which is one less than N_{tot}. From this example it can be seen that total degrees of freedom, df_{tot}, is given by $df_{tot} = df_{bg} + df_{wg} = N_{tot} - 1$.

First we find MS_{bg}:

$$MS_{bg} = \frac{2420}{2} = 1210$$

Then we compute MS_{wg}:

$$MS_{wg} = \frac{7260}{27} = 268.9$$

MS_{bg} will not be greatly larger than MS_{wg} in an experiment unless the group $\overline{X}$'s differ appreciably. We have just found an MS_{bg} of 1210, compared with an MS_{wg} of 268.9, which leads us to suspect that the variability between group $\overline{X}$ values is too great to be attributed to random differences between samples from populations with identical means. To make a more precise comparison of these MS values, we now form what is called the F ratio and evaluate it.

THE F RATIO

The F ratio, which is simply a numerical expression of the relative size of MS_{bg} and MS_{wg}, is defined by the equation below:

$$F = \frac{MS_{bg}}{MS_{wg}} \tag{12.10}$$

When an F ratio has been found in an experiment, a decision regarding the significance of the differences in group $\overline{X}$'s is made by comparing that ratio with the F values to be expected if the null hypothesis that the population μ values are equal were true. The information necessary for this comparison is presented in Table F in Appendix II. (Note the happy accident that the values of F appear in Table F.) Table F shows the F values required at the 5% and 1% levels of significance. That is, this table presents the F values that are so large that they would be exceeded only 5% or 1% of the time by random sampling if the null hypothesis were true. If MS_{bg} is enough greater than MS_{wg}, so that the resulting F ratio exceeds the tabled F at the 5% (or 1%) level, then we realize it would be very unlikely that such an F would occur unless the null hypothesis were false. Therefore, we would reject the null hypothesis that the population μ values are equal and say that our obtained F is significant at the 5% (or 1%) level.

One step in this procedure remains to be explored: We must know where to look in Table F for the significant values of the F ratio with which to compare the F obtained in a particular experiment. This step is much like the procedure for finding a significant t value in Table C. There we had to look for a t associated with a df value. Now we must find an F associated with two *df* values, because the degree of significance of any obtained F depends upon two factors—the number of groups and the number of subjects within the groups. Accordingly, we employ the values of df_{bg} and df_{wg} in using Table F.

Knowing that $df_{bg} = 2$ and $df_{wg} = 27$ in our attitude experiment, we can find a cell in Table F that is appropriate for that experiment. Since the *df* for the MS in the numerator of the F ratio is df_{bg} and the *df* for the MS in the denominator is df_{wg}, we see from the table that we must go to the *column* labeled *df* = 2 and proceed down to the *row* labeled *df* = 27. There we find two F values, one for the 5% level of significance and the other for the 1% level. The F ratio at the 5% level of significance is 3.35 and at the 1% level, 5.49. An F obtained in an experiment with three groups, each of 10 subjects, must be compared with these values to determine whether or not it is significant.

To illustrate the use of the F ratio, we again employ our data from the attitude experiment. Since $MS_{bg} = 1210$ and $MS_{wg} = 268.9$, we find F with the following equation:

$$F = \frac{1210}{268.9} = 4.50$$

This F ratio of 4.50 is significant at the 5% level because it is greater than 3.35, the F value tabled at the 5% level for 2 and 27 *df*. However, it is not significant at the 1% level because it is less than 5.49, the value required for significance at that level. Thus, in this situation we would reject the null hypothesis if we were using a 5% level of significance but accept it if we were using a 1% level. We may notice that, if df_{wg} had been between 30 and 40 rather than at 27 as it was, Table F would not have indicated the exact F values required for significance at the 5% and 1% levels. In that case we would have used the tabled values for 2 and 30 *df*, because they are higher than those for 2 and 40 *df*, thereby behaving with extra caution when we conclude that an F is significant at the 5% or 1% level.

Assumptions Underlying the Analysis of Variance

Like every other statistical test we have studied, the analysis of variance involves certain assumptions that had to be made in order to derive the table of significant values for the test. Table F is known to present the F values required for significance at the 5% and 1% levels when the null hypothesis is assumed to be true and three assumptions are satisfied: (a) each of the *k* populations from which the groups in the experiment were drawn is normally distributed; (b) the σ^2 values for the *k* populations are equal; (c) the subjects of the experiment have been randomly and independently drawn from their respective populations.

If we know that these assumptions have been met, we can accept conclusions based upon the use of Table F at their face value. If these assumptions are not satisfied, the mathematical demonstrations that the values presented in Table F are the exact values required at the 5% and 1% levels can no longer be made. However, the practical usefulness of the analysis of variance procedure may be nearly as great when one or two of these assumptions are fulfilled as when all are satisfied.[2] If one of these assumptions appears not to be met, the experimenter may prefer to perform the analysis of variance and interpret it conservatively—for example, require that F values reach the tabled values for a higher level for significance such as 1% when the experimenter would otherwise have required only a 5% level for concluding that F is significant.[3]

To decide whether the assumptions of normality and equal σ^2's have been met is not easy. Since the groups involved in this type of experiment are usually small, there are seldom enough scores to indicate the shape of the distribution for each population. If the sample distribution is very far from normal, we may suspect that the distribution of the population is not normal. Similarly, very great differences among the s^2's for the different groups suggest that the population σ^2's may be unequal. Statistical tests of normality and equality of σ^2's could be presented here; we omit them because of the space they would require and the impracticability of applying a test of normality except when N is large.

A Step-by-Step Computational Procedure

We now can list in sequence all the steps required to perform a one-way ANOVA. Then we will illustrate each step by analyzing the results of an experiment.

Basic data

1. Find for each separate group ΣX, ΣX^2, and N.
2. By summing the appropriate figures for each of the groups, find $\sum_{tot} X, \sum_{tot} X^2$, and N_{tot}.

Calculational steps

Find the SS's by the formulas:

3. $SS_{tot} = \left(\sum_{tot} X^2 \right) - \dfrac{\left(\sum_{tot} X \right)^2}{N_{tot}}$

4. $SS_{bg} = \sum_{g} \left[\dfrac{\left(\sum X_g \right)^2}{N_g} \right] - \dfrac{\left(\sum_{tot} X \right)^2}{N_{tot}}$

5. $SS_{wg} = SS_{tot} - SS_{bg}$

---

[2]The ability of a statistical test to yield approximately correct results when certain assumptions are not met is known as its robustness. See H. Scheffé, *The analysis of variance* (New York: Wiley, 1959), chapter 10.

[3]If there is a great discrepancy between the assumptions of analysis of variance and the characteristics of the obtained data, it may be advisable to use a so-called nonparametric test such as the Kruskal-Wallis test, presented in Chapter 15. Nonparametric tests require no assumptions to be made about population parameters.

or

$$SS_{wg} = \sum_g \left[\left(\sum X_g^2 \right) - \frac{\left(\sum X_g \right)^2}{N_g} \right]$$

Find df's by the formulas:

6. $df_{bg} = k - 1$

7. $df_{wg} = N_{tot} - k$

Find MS's by the formulas:

8. $MS_{bg} = \dfrac{SS_{bg}}{df_{bg}}$

9. $MS_{wg} = \dfrac{SS_{wg}}{df_{wg}}$

Find and determine the significance of F:

10. $F = \dfrac{MS_{bg}}{MS_{wg}}$

11. Entering Table F with the df_{bg} and df_{wg} found in steps 6 and 7, find the Fs required for significance at the 5% and the 1% levels.

12. If the obtained F exceeds the required F at the 5% level (or the 1% level) reject the null hypothesis at that level; otherwise, accept it.

We can illustrate these steps by referring to the results of a hypothetical experiment in which students were each asked to read a prepared speech to an audience of one, three, five, or ten people, and judged on the effectiveness of their presentation. Let us perform an ANOVA to find out whether the four audience sizes resulted in significantly different ratings of effectiveness. Table 12.2 presents data for the four groups of students together with the X^2 value for each X.

Basic data

Step 1 is to find $\sum X$, $\sum X^2$, and N for each group, and step 2 to add these quantities together to get $\sum_{tot} X$, $\sum_{tot} X^2$, and N_{tot}. These values are presented in Table 12.2.

Calculational steps

The SS's are first calculated as follows:

3. $SS_{tot} = \left(\sum_{tot} X^2 \right) - \dfrac{\left(\sum_{tot} X \right)^2}{N_{tot}}$

$$= 10592 - \frac{(416)^2}{21} = 10592 - 8240.76$$

$$= 2351.24$$

Table 12.2 Preliminary Steps in the Analysis of Variance of Effectiveness of Speech Presentation (X) by Students with an Audience of One, Three, Five, or Ten Persons

Group I 1 person		Group II 3 persons		Group III 5 persons		Group IV 10 persons	
X_1	$X_1{}^2$	X_2	$X_2{}^2$	X_3	$X_3{}^2$	X_4	$X_4{}^2$
24	576	16	256	8	64	28	784
24	576	8	64	36	1296	8	64
16	256	28	784	16	256	20	400
12	144	24	576	20	400	8	64
32	1024	48	2304	24	576	8	64
		8	64				
$\sum X_1 = 108$	$\sum X_1{}^2 = 2576$	$\sum X_2 = 132$	$\sum X_2{}^2 = 4048$	$\sum X_3 = 104$	$\sum X_3{}^2 = 2592$	$\sum X_4 = 72$	$\sum X_4{}^2 = 1376$

$$\sum_{tot} X = \sum X_1 + \sum X_2 + \sum X_3 + \sum X_4 = 108 + 132 + 104 + 72 = 416$$

$$\sum_{tot} X^2 = \sum X_1{}^2 + \sum X_2{}^2 + \sum X_3{}^2 + \sum X_4{}^2 = 2576 + 4048 + 2592 + 1376 = 10592$$

$$N_{tot} = N_1 + N_2 + N_3 + N_4 = 5 + 6 + 5 + 5 = 21$$

4. $SS_{bg} = \sum_g \left[\dfrac{\left(\sum X_g\right)^2}{N_g} \right] - \dfrac{\left(\sum\limits_{tot} X\right)^2}{N_{tot}}$

$= \left[\dfrac{(108)^2}{5} + \dfrac{(132)^2}{6} + \dfrac{(104)^2}{5} + \dfrac{(72)^2}{5} \right] - \dfrac{(416)^2}{21}$

$= 8436.80 - 8240.76$

$= 196.04$

5. $SS_{wg} = SS_{tot} - SS_{bg}$

$= 2351.24 - 196.04$

$= 2155.20$

Then *df*'s and MS's are found as follows:

6. $df_{bg} = k - 1 = 4 - 1 = 3$

7. $df_{wg} = N_{tot} - k = 21 - 4 = 17$

8. $MS_{bg} = \dfrac{SS_{bg}}{df_{bg}} = \dfrac{196.04}{3}$

$= 65.35$

9. $MS_{wg} = \dfrac{SS_{wg}}{df_{wg}} = \dfrac{2155.20}{17}$

$= 126.78$

Finally, the F is obtained and its significance determined:

10. $F = \dfrac{MS_{bg}}{MS_{wg}} = \dfrac{65.35}{126.78} = .52$

11. Looking in Table F for the case where $df_{bg} = 3$ and $df_{wg} = 17$, we find that F must be 3.20 or greater to be significant at the 5% level, and 5.18 or greater to be significant at the 1% level.

12. Our F of .52 is not significant, since it is much less than the F of 3.20 required for significance at the 5% level. In fact, as Table F shows, no F below 1 is significant regardless of the *df* values. Therefore, we cannot reject the null hypothesis that the groups come from populations with the same μ.

Table 12.3 summarizes the results of our ANOVA, listing the SS's, *df*'s, MS's, and the F we have calculated as well as the *p* value of F. This table is an example of the standard method used to summarize an analysis of variance.

Table 12.3 Summary of the Analysis of Variance of the Experiment on Effects of Audience Size on Speech Effectiveness

Source of variation	SS	df	MS	F	p
Between groups	196.04	3	65.35	.52	> .05
Within groups	2155.20	17	126.78		
Total	2351.24	20			

A MULTIPLE-COMPARISON TEST FOLLOWING ANALYSIS OF VARIANCE

When we have performed a one-way ANOVA and found a significant F between the groups, we are able to reject the hypothesis that all the groups were drawn from populations with the same mean. But when more than two groups are being compared, a significant F does not imply that each sample $\overline{X}$ necessarily differs significantly from every other sample $\overline{X}$; quite often we are interested in pinpointing more precisely where the differences lie. Our experimental interests, for example, may lead us to inquire about the significance of the differences between specific pairs of means. Or in other instances we may observe that the data pattern themselves in interesting ways. We may find, for example, that the means appear to fall into subsets or clusters, and we may want to determine whether the clusters differ significantly from each other, or we may find that one of the means is markedly different from all the rest. What is required, then, is some technique that will allow us to follow up the analysis of variance with further statistical comparisons.

Although *t* might seem an appropriate test for this purpose, several statistical difficulties prohibit its use, as we noted earlier. But a number of other tests for making multiple comparisons following analysis of variance have been suggested by statisticians. Here we present one such multiple-comparison test, a method devised by Tukey[4], which we will call the *honestly significant difference* test or *hsd*.

The Tukey *hsd* usually is performed only if the F obtained in the analysis of variance is significant, but it is theoretically permissible to perform whatever the significance of F. The test has the advantage of allowing us to compare any pair of means, or any pair of subsets of means, that we wish to select and of placing no limits on the number of comparisons that may be made with any set of data. We will illustrate its use only for the simple case in which we wish to compare individual pairs of means and in which the N's in each group are equal. Procedures are available that permit comparisons of subsets of means (e.g., $\overline{X}$'s of groups 1 and 2 vs. $\overline{X}$'s of groups 3, 4, and 5) and of groups of unequal size, but they are more complex.

The test involves computing a quantity that we will identify as *hsd*. The difference between any pair of means is significant at a given α level if it equals or exceeds the absolute value of *hsd* we have computed.

---

[4]J. W. Tukey, The problem of multiple comparisons (mimeographed, Princeton, N.J., 1953). The computational procedures are adapted from R. P. Runyon and A. Haber, *Fundamentals of Behavioral statistics*, 2nd ed. (Reading, Mass.: Addison-Wesley, 1971).

The formula for *hsd* is as follows:

$$hsd = q_\alpha \sqrt{\frac{MS_{wg}}{N_g}} \qquad\qquad (12.11)$$

In this equation, MS_{wg} is the within-groups mean square calculated in the analysis of variance, and N_g is the number of cases in each group. The value of q_α, also called the Studentized range statistic, is found from the appropriate entry in Table G in Appendix II. We enter Table G at the specified α level for the *df* associated with MS_{wg} and k, the total number of groups in the experiment. Suppose, for example, that we wished to compare one or more pairs of means, using $\alpha = .05$, from a 4-group experiment in which there were 5 subjects per group. Looking at Table G for $k = 4$ and $df_{wg} = 20 - 4 = 16$, we find that $q_{.05} = 4.05$.

We will illustrate the application of the test by using the data from the hypothetical attitude experiment reported in Table 12.1 that involved three groups of 10 cases each. We repeat below some of the results obtained in our analysis of these data that we will need to determine *hsd*.

$$\overline{X}_1 = 42 \qquad\qquad \overline{X}_2 = 53 \qquad \overline{X}_3 = 64$$

$$MS_{wg} = 268.9 \qquad df_{wg} = 27 \qquad k = 3$$

where k = the total number of groups and $df_{wg} = N_{tot} - k$.

The first step, assuming we wish to compare all possible pairs of means, is to find the *absolute difference* between each pair of $\overline{X}$'s (for example, the absolute difference between $\overline{X}_1$ and $\overline{X}_2$ is $53 - 42 = 11$). In Table 12.4 we show these differences for our experiment.

The next step is to compute *hsd*. Suppose that we set $\alpha = .05$. In order to find $q_{.05}$ we consult Table G to find the value associated with this alpha level for $df_{wg} = 27$ and $k = 3$. Unfortunately, by looking down the *df* column at the left, we discover that just as in the case of the F table, Table G is incomplete and has no entry for $df = 27$. Again, we choose to be conservative and use the next lowest *df*, which turns out to be 24. Going across this row to the $k = 3$ column, we find that the tabled value for q at $\alpha = .05$ is

Table 12.4   Absolute Differences between Pairs of Means in Hypothetical Attitude Experiment

	$\overline{X}_1$	$\overline{X}_2$
$\overline{X}_2$	11	—
$\overline{X}_3$	22	11

3.53. We now enter this value, along with the values of MS_{wg} and N_g, into the *hsd* equation:

$$hsd = q_{.05} \sqrt{\frac{MS_{wg}}{N_g}} = 3.53 \sqrt{\frac{268.9}{10}} = 3.53(5.19) = 18.32$$

We have now computed the critical value that the difference between a pair of means must equal or exceed in order to be significant at the 5% level. By inspecting Table 12.4 we see that the difference between $\overline{X}_1$ and $\overline{X}_3$ is the only one that exceeds 18.32. The mean differences for the remaining pairs are not significant.

We said above that the Tukey test places no limits on the number of comparisons that may be made with any set of data. By this we mean that the probability of finding a significant difference by chance at a given α level does not increase with each additional comparison, as it does in the case of *t*. When all population means are equal and a series of Tukey multiple-comparison tests such as the three *hsd* tests just reported are made at a given α level (let us say, .05), the probability is .05 that a Type I error will occur *anywhere* in the set of comparisons that are to be performed. Since the 5% significance level applies to the *total* experiment, and not just to individual comparisons, we can perform as many comparisons as we choose and remain confident that we are not increasing our chances of a Type I error.

In concluding this chapter, we should again mention that we have presented only the most elementary form of analysis of variance. This one-way ANOVA, as it is called, is the type used for testing the null hypothesis that the population means are equal in experiments involving several groups but only one basic variable—for example, the number of courses in music appreciation, number of hours of food deprivation, methods of teaching French, and so on. In the next chapter we will discuss an extension of the one-way analysis of variance technique, one that is used to analyze the data from a more complicated type of experiment in which the effects of not one but two variables are simultaneously studied.

DEFINITIONS OF TERMS AND SYMBOLS

One-way analysis of variance (ANOVA).   A statistical technique for testing the null hypothesis that two or more samples come from populations with the same μ's.

Total variability.   In a study involving two or more groups, the variability of the scores from the overall mean of the total set of scores ($\overline{X}_{tot}$).

Variability within groups.   The variability of scores from the mean of the particular group to which the scores belong.

Variability between groups. The variability of the group $\overline{X}$'s from the overall mean $(\overline{X}_{tot})$.

Total sum of squares (SS_{tot}). The sum of the squared deviations of the scores from the overall mean $(\overline{X}_{tot})$. The total sum of squares can be broken down into two components: the sum of squares within groups and the sum of squares between groups.

Sum of squares within groups (SS_{wg}). The sum of the squared deviations of the individual scores from the mean of the group to which they belong.

Sum of squares between groups (SS_{bg}). The sum of the squared deviations of the group $\overline{X}$'s from $\overline{X}_{tot}$.

Mean square (MS). A mean square (MS) is an unbiased estimate of the population variance, obtained by dividing a sum of squares by its corresponding degrees of freedom.

Mean square within groups (MS_{wg}). The within-groups MS or variance is obtained by dividing SS_{wg} by df_{wg} where df_{wg} equals the total number of scores minus the number of groups ($N_{tot} - k$).

Mean square between groups (MS_{bg}). The between-groups MS or variance is obtained by dividing SS_{bg} by df_{bg} where df_{bg} equals number of groups minus one ($k - 1$).

F ratio. The F ratio in a one-way analysis of variance is defined as MS_{bg}/MS_{wg}. The value of F is compared with the critical value of F for df_{bg} and df_{wg} at $\alpha = .05$ or $.01$ to evaluate the null hypothesis that population means are equal.

Assumptions of analysis of variance. The analysis of variance for independent groups assumes that (a) each of the populations from which the groups were drawn is normally distributed, (b) the variances (σ^2 values) of the populations are equal, and (c) the members of each group have been randomly and independently selected from their respective populations. If any of these assumptions is seriously violated, the analysis of variance technique should not be applied.

Tukey honestly significant difference (hsd) test. The *hsd* test is an example of a multiple-comparison test used following analysis of variance to determine the significance of the difference between pairs of $\overline{X}$'s or pairs of subsets of $\overline{X}$'s. The value of the *hsd* statistic at a given α level gives the minimal difference between pairs of $\overline{X}$'s that is significant at that α level.

PROBLEMS

In all problems requiring computation of an analysis of variance (ANOVA) show all your computational steps and place the results in a summary table, patterned after Table 12.3 in the text. For ease of computation, round off SS's to the nearest whole number, MS's to the first decimal place, and F's to the second decimal place.

1. In a study on the effects of object orientation on identification of shape, three groups of five subjects each were shown a series of pairs of common objects in a special viewing apparatus. In each pair, one of the objects was shown in a normal, up-

right position. For one group of subjects (group 1), the second member of each pair was tilted 45 degrees from upright. For group 2, it was tilted 90 degrees, and for the group 3 it was upside down (180 degrees from upright). For each pair, subjects pressed a button as soon as they determined whether the tilted figure was on the right or left, the time being recorded from exposure of the stimulus pair to the button press. The total reaction time (in seconds) over the series of pairs was determined for each subject. The sums and sums of squares for each group are shown below.

$$\sum X_1 = 80 \qquad \sum X_2 = 64 \qquad \sum X_3 = 51 \qquad N_g = 5$$
$$\sum X_1^2 = 1300 \qquad \sum X_2^2 = 852 \qquad \sum X_3^2 = 535$$

(a) Determine the $\overline{X}$ for each group and then graph the $\overline{X}$'s.
(b) State the null hypothesis to be tested by means of a one-way analysis of variance. Be specific, describing H_0 for this specific experiment.
(c) Conduct the appropriate ANOVA.
(d) After considering the outcome of the ANOVA and the $\overline{X}$'s graphed in (a), describe your conclusions about the effect of object orientation.

2. An educational psychologist teaches a large introductory course in which students also participate in smaller discussion sections. He decides to experiment with different ways to conduct the discussion sections and assigns the 120 students at random to four equal sections, each taught by a different method. At the end of the course, the students' final examination scores are analyzed. The results of the ANOVA are summarized below. Also shown are the means for each of the four groups of 30.

Source	df	MS	F
Between groups	3	94.73	3.47*
Within groups	116	27.30	
Total	119		

*$p < .05$.

$$\overline{X}_1 = 81.5 \qquad \overline{X}_2 = 80.3 \qquad \overline{X}_3 = 88.3 \qquad \overline{X}_4 = 85.9$$

Perform a follow-up analysis of the means of the individual method groups with $\alpha = .05$. What do you conclude?

3. Three experiments were conducted, each involving three or more groups. For each experiment, the mean, median, and range of scores are shown below for each of the experimental groups. From the information given, does it appear that one or more assumptions underlying analysis of variance is violated? If your answer is yes, specify the assumption(s).

(a) Group	$\overline{X}$	Mdn	Range		(b) Group	$\overline{X}$	Mdn	Range
1	62	67	53–72		1	104	106	85–126
2	72	74	60–83		2	98	97	79–116
3	81	89	73–92		3	108	109	88–127
					4	115	113	95–134

(c) Group	$\overline{X}$	Mdn	Range
1	25	23	18–29
2	28	29	17–39
3	31	31	15–45

Two-Way Analysis of Variance

CHAPTER **13**

All the types of experiments we have considered so far have involved two or more groups differing with respect to some single independent variable. In a good many experiments, however, the investigator studies simultaneously the effects of more than one independent variable. For example, in a study of the effects of hormones on animal development, we might have four groups of male animals who are 1, 7, 14, or 21 days old and divide each of these groups into three subgroups, injected with hormone dosages of 0, 2, or 6 units. This general arrangement, in which we study the effects of two or more variables, is known as a *factorial design*. In this chapter we will be considering the analysis of variance appropriate for data from experiments employing a double classification or *two-way* factorial design—that is, experiments in which two variables are simultaneously manipulated.

We have seen that experiments investigating a single variable need not compare only two groups but may involve any number. Similarly, there is no limit to the number of groups representing each variable in a two-way factorial study. In our animal experiment, for example, we suggested studying four age groups and three dosage levels, a design calling for 4 × 3, or 12, independent groups. We could extend our study to include more conditions within each variable, such as 1, 2, 7, 14, 21, and 28 days of age and dosages of 0, 2, 6, or 10 units, thus calling for 6 × 4, or 24, groups. As we increase the number of specific conditions within each type of variable, the total number of groups that needs to be tested to form a complete factorial design increases quite drastically, requiring considerable time and effort to conduct the experiment. However, the logic of the two-way design and the statistical methods we use to analyze the experimental results remain the same, whatever the number of experimental conditions within each of the variables.

A word now about terminology. We will refer to the two kinds of conditions being investigated as *variables*, using the letters A and B as identifying labels. The specific experimental conditions employed in investigating each of the variables we will identify by the numerals 1, 2, 3, and so on to the mth condition for the A variable or to the nth condition for the B variable.

Many of the variables investigators wish to study involve some kind of quantifiable dimension such as time, distance, or amount, so that the specific conditions being investigated can be rank-ordered in magnitude from least to most, and the numerals 1, 2, 3, and so forth used to reflect this order. Often, however, the variables we study involve two or more discrete categories or types of conditions. We may, for example, be interested in comparing several different kinds of teaching methods, persons of different nationalities, men versus women, and so on. In these instances, the groups or conditions obviously cannot be rank-ordered along some dimension, so our use of identifying numerical labels is purely nominative, the assignment of number to group being completely arbitrary.

MAIN EFFECTS AND INTERACTION IN ANALYSIS OF VARIANCE

Application of the analysis of variance technique to the data from a two-way factorial experiment allows us to make three kinds of statements about our results: (1) the effects on our dependent measure of the different conditions of variable A, independent of variations in B conditions (*main effects of A*), (2) the effects of the different conditions of variable B, independent of variations in A conditions (*main effects of B*), and (3) the joint effects or *interaction* of variables A and B. Rather than proceeding directly to the statistical steps that will allow us to make these comparisons, we first will explain what each of them involves.

Table 13.1 presents the results of three hypothetical experiments employing 2 X 2, 2 X 3, and 3 X 3 factorial designs, respectively, with equal N's per group in each experiment. The numbers entered in the cells are the means obtained for each of the specific

Table 13.1   Two-Way Tables Showing Means for Hypothetical 2 X 2, 2 X 3, and 3 X 3 Factorial Experiments, with Equal N's per Group

I.   2 X 2, N per grp = 20

Variable A		Variable B 1	Variable B 2	$\overline{X}_{row}$	N_{row}
	1	20	27	23.5	40
	2	26	18	22.0	40
$\overline{X}_{col}$		23.0	22.5		
N_{col}		40	40		

II.   2 X 3, N per grp = 10

Variable A		Variable B 1	Variable B 2	Variable B 3	$\overline{X}_{row}$	N_{row}
	1	3	7	8	6	30
	2	5	9	10	8	30
$\overline{X}_{col}$		4	8	9		
N_{col}		20	20	20		

III.   3 X 3, N per grp = 5

Variable A		Variable B 1	Variable B 2	Variable B 3	$\overline{X}_{row}$	N_{row}
	1	29	30	31	30.0	15
	2	30	34	36	33.3	15
	3	31	38	42	37.0	15
$\overline{X}_{col}$		30.0	34.0	36.3		
N_{col}		15	15	15		

groups. For each table we have also reported the mean of all the cases represented in each column and in each row. For example, in the 3×3 experiment whose results are reported in Table 13.1, we added together the scores of all individuals tested under the first A condition (A1), whatever their B condition, and divided the sum by 15, the total number of A1 subjects. This yielded a mean of 30.0, as shown at the right of the A1 row. (Since the N's in each cell of the table are equal, we could also get the row mean by averaging the three A1 cell means.) The means of all the A2 groups and of all the A3 groups, shown at the right of the second and third rows, were obtained in a similar fashion, as were the column means, shown at the bottom of the table. These column values, of course, are the means of all 15 of the cases tested under the B1, B2, and B3 conditions, respectively, whatever their A condition.

Main Effects

Determination of the main effects of variable A and variable B by means of a two-way analysis of variance (ANOVA) involves testing a particular null hypothesis about the row $\overline{X}$'s and about the column $\overline{X}$'s. Each of these hypotheses states that the means of the populations from which the samples were drawn are all equal. Thus, in the case of the A variable, H_0 states that each A group, averaged over all B groups, is drawn from a population with the same mean ($\mu_{A1} = \mu_{A2} = \cdots = \mu_{Am}$) and that therefore any differences among the $\overline{X}$'s of the A groups (row $\overline{X}$'s) are due to chance factors. Similarly, H_0 for variable B states that each B group, averaged over all A groups, is drawn from a population with the same mean ($\mu_{B1} = \mu_{B2} = \cdots = \mu_{Bn}$) and that any differences among the $\overline{X}$'s of the B groups (column $\overline{X}$'s) are therefore due to chance.

A method of testing these null hypotheses may occur to you immediately. In comparing the means of the A groups (row $\overline{X}$'s), why not ignore the B variable and consider all those tested under the same A condition (all those in the same row) as forming a single group? Then we could perform a one-way analysis of variance of the A groups, following the procedures outlined in Chapter 12. Another one-way ANOVA could then be performed in the same way on the column data, allowing us to test the null hypothesis about the means of the B groups. In conducting this type of analysis, you will recall, we find an F ratio that compares the relative size of two measures of variability: the variability between the means of the groups and the variability within each of the groups, the latter being due to uncontrolled, unidentified factors such as original differences in ability among our subjects. If the between-groups measure is sufficiently larger than the within-groups measure, we reject the null hypothesis.

However, if we were to perform these one-way analyses of variance on the data from a two-way experiment, we would run into a major problem. Our measure of within-groups variability for the A (row) variable would reflect differences among the subjects in each row that are due not only to uncontrolled, unidentified factors, but also to any systematic effect of variations in B (column) conditions. In like manner, our within-groups measure for the column data would reflect any differences among subjects due to the effects of variations in A conditions (rows). These systematic effects would inflate each of the measures of within-group variability and hence decrease the size of the F ratio. To avoid this complication, we modify our approach by computing a *single* within-groups measure based on each of our separate experimental groups—that is, within each *cell* of our two-way table. We then obtain our F ratios for the main effects of variables A and B by dividing each of our measures of between-groups variability (one between column $\overline{X}$'s, the other between row $\overline{X}$'s) by this single measure of within-groups (within-cells) variability.

Interaction between Variables

The reason we conduct a factorial study is not simply to investigate the main effects of more than one variable. One of the major advantages of the factorial experiment is that it allows us to study the *joint effects* of the two variables, how the variables combine or *interact* to influence our response measure.

The ways in which variables can combine are numerous. One simple type of combination is an *additive* one—a combination of the form A + B. The data from the hypothetical 2 X 3 experiment reported in Table 13.1 illustrate this type of combination. We can most easily explain an additive relationship by referring to the graphic presentation of these data in Figure 13.1. We have shown the three B conditions along the baseline and have plotted the means of the three A1 groups and the three A2 groups in separate curves. As you can see by inspection, the curves for the A1 and A2 groups are *parallel*. This fact indicates that the magnitude of the effect of the A variable is the *same* for each condition of B. An alternate way of describing this type of relationship is to state that the difference in $\overline{X}$'s between A1 and A2 groups tested under the same B condition is the same for each of the B conditions. In this example, as you can easily see from the values reported in Table 13.1, the difference in $\overline{X}$'s (A2 - A1) at each level of B is 2 (5 - 3 = 2, 9 - 7 = 2, 10 - 8 = 2). We could have graphed the data the other way around, representing the two A conditions along the baseline and plotting the means of the B1, B2, and B3 groups in separate curves. As you may determine for yourself by graphing the data in this manner, the three curves would again turn out to be parallel. You can also see that this would be so by examining the means of groups shown in Table 13.1. The difference between the $\overline{X}$'s of any given pair of B groups (1 versus 2, 2 versus 3, or 1 versus 3) is the same whatever the A condition under which the pair was tested. For B1 versus B2, for example, the difference at both levels of A is 4.

Figure 13.1 Graph of $\overline{X}$'s from 2 X 3 experiment of Table 13.1.

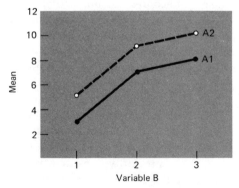

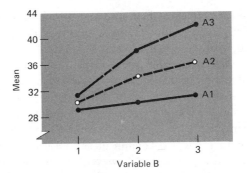

Figure 13.2 Graph of $\overline{X}$'s from 3 × 3 experiment of Table 13.1.

The data from the 2 × 2 and 3 × 3 experiments reported in Table 13.1 are examples of two kinds of *nonadditive* combinations, as can again be conveniently demonstrated by presenting their results graphically. Looking first at the data of the 3 × 3 experiment plotted in Figure 13.2, you will observe that the curves for the three A groups are not parallel; rather, they diverge. That is, the magnitude of the differences between the A groups becomes larger as one goes from B1 to B2 and from B2 to B3. Had the curves for the three B groups been plotted instead, these curves would also have been shown to diverge with changes in the A condition.

An example of a very complex type of nonadditive interaction is found in the data of the 2 × 2 experiment reported in Table 13.1. Scores are affected by both A and B conditions but, as you can see from the plot of the data in Figure 13.3, the *direction* of the effect of variations in the A condition is different for each B condition (and vice versa). Strange as it may seem at first glance, this type of interaction occurs fairly frequently in psychological data. For example, individuals who are given a long list of very simple arithmetic problems to solve with instructions that emphasize the importance of speed may produce not only more answers in a given period of time, but more *correct* answers than individuals given the same problems with instructions that emphasize the importance of accuracy more than speed. The same sets of instructions, however, may produce the *opposite* results with respect to correct responses when a list of complicated problems is used; individuals given speed instructions might make all sorts of errors in their attempt to hurry and therefore arrive at a smaller number of correct solutions than the group trying to be accurate. Thus, speed instructions may lead to either better or

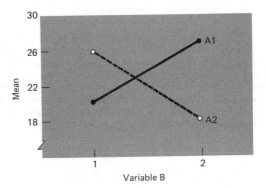

Figure 13.3 Graph of $\overline{X}$'s from 2 × 2 experiment of Table 13.1.

worse performance than accuracy instructions, depending on the type of problems employed; the direction of the effects of variations in A is dependent on the particular B condition under which subjects were tested.

One by-product of this type of complex interaction should be mentioned. Since the direction of the influence of one variable is dependent on the other variable, comparisons of the overall means of the various A groups, ignoring B conditions, or of the overall means of the B groups, ignoring A conditions, may reveal that neither variable has any significant main effect. You can see, for example, that in the 2×2 experiment shown in Table 13.1 the two column means are very similar in value, as are the two row means. We would not conclude, however, that neither variable has any kind of effect but rather, assuming that the relevant statistical term is significant, that the *direction* of the effects of one variable is dependent on the specific condition employed with respect to the second variable.

A major purpose in conducting a two-way factorial study, as we have said, is to determine how the two variables combine. The two-way ANOVA technique is of assistance in this process since it allows us to test a hypothesis about the nature of the interaction between the variables. The specific hypothesis being tested is that the variables combine in an *additive* fashion—that the magnitude of the effects of one variable is constant over all conditions of the second variable. Expressing this in another way, we test the hypothesis that the curves resulting from a plot of the $\overline{X}$'s of the groups are *parallel*. If the F ratio found for the interaction between variables A and B is nonsignificant at a given α level, we *accept* the hypothesis that the variables combine additively. If, on the other hand, the F for the interaction between variables is significant, we *reject* this hypothesis and conclude that the variables combine nonadditively. Versatile as the analysis of variance technique is, however, it cannot tell us in this latter instance precisely what kind of nonadditive relationship has occurred; this we must determine by inspection of our data.

In summary, then, a two-way analysis of variance yields F ratios that allow us to test the null hypothesis that μ's are equal concerning differences between the means of the A groups and between the means of the B groups—the *main effects* of each of the two variables—and a hypothesis concerning the *interaction* between the two variables: that they combine in an additive manner. We turn now to the computational steps involved in conducting such an analysis. We will present formulas only for the case in which all groups have the same N.[1]

---

[1] If the N's are unequal, the most convenient procedure may be to use an appropriate statistical package on a computer. Whenever data analyses require very complicated statistical tests, more complex than any procedures in this book, or many applications of a single test, time can be saved by having a computer perform the analyses. These are manuals for three well-known statistical packages: J. T. Helwig, *SAS introductory guide* (Raleigh, N.C.: SAS Institute, 1978). W. R. Klecka, N. H. Nie, and C. H. Hill, *SPSS primer: Statistical package for the social sciences primer* (New York: McGraw-Hill, 1975). University of California, Los Angeles, Health Science Computing Facility, W. P. Dixon, chief eds.) BMDP statistical software 1981. (Berkeley: University of California Press, 1981).

Table 13.2   Data from the Hypothetical Study of Thresholds Where $N_g = 5$

Neutral (A1)						Unpleasant (A2)					
High (B1)		Medium (B2)		Low (B3)		High (B1)		Medium (B2)		Low (B3)	
X	X^2	X	X^2	X	X^2	X	X^2	X	X^2	X	X^2
13	169	21	441	24	576	16	256	22	484	32	1024
19	361	14	196	15	225	14	196	19	361	27	729
21	441	16	256	22	484	21	441	16	256	25	625
16	256	21	441	19	361	19	361	20	400	27	729
11	121	13	169	17	289	23	529	29	841	21	441
80	1348	85	1503	97	1935	93	1783	106	2342	132	3548

$$\sum_{tot} X = 80 + 85 + 97 + 93 + 106 + 132 = 593$$

$$\sum_{tot} X^2 = 1348 + 1503 + 1935 + 1783 + 2342 + 3548 = 12459$$

$$N_g = 5 \qquad N_{tot} = 30$$

Summary of $\sum X$ values

		Word familiarity (variable B)			
		1 (high)	2 (med.)	3 (low)	Sum
Affectivity	1 (neutral)	80	85	97	262
(variable A)	2 (unpleasant)	93	106	132	331
	Sum	173	191	229	593

COMPUTATIONAL STEPS IN TWO-WAY ANALYSIS OF VARIANCE FOR EQUAL N's

In describing the procedures employed in performing a two-way analysis of variance, we will omit the basic formulas defining the various sums of squares and present only the computational formulas, illustrating their application with the data from the hypothetical experiment shown in Table 13.2. In this experiment, the minimum exposure time (duration threshold) required for correct identification of words was being investigated as a function of several word characteristics. Three groups of 10 individuals were given lists of words that had been judged to be high, medium, or low in familiarity; half the subjects in each of these familiarity groups were given a list containing words that were emotionally unpleasant in meaning and the other half a list of emotionally neutral words. The experiment thus formed a 2 × 3 factorial design with five subjects in each of the six groups. The score obtained for each of the subjects was the sum of minimum exposure times for the various words on the list.

In Table 13.2 we have also reported the ΣX and ΣX^2 for each of the six groups in our hypothetical experiment, since these are the basic data entering into our calculations. We will find it convenient to have the ΣX values in each of the groups set up in a two-way table so that we can determine the sums for each column and row, as well as $\sum_{tot} X$. This type of two-way table is also shown in Table 13.2.

Computational Steps in Finding SS's

You will remember that in performing a one-way analysis of variance on the data from a collection of k groups, we first find the total sum of squares, the SS for all the subjects in the experiment. We then proceed to break down SS_{tot} into its two components: SS_{bg} and SS_{wg}. Our first step in analyzing the results of a two-way factorial experiment is to treat the data as a single collection of k groups and find SS_{tot}, SS_{bg}, and SS_{wg} using exactly the same procedures as we employed in the last chapter. Illustrating these procedures[2] with the data from our 2×3 experiment in Table 13.2 (where $k = mn = 6$), we thus have:

$$SS_{tot} = \left(\sum_{tot} X^2\right) - \frac{\left(\sum_{tot} X\right)^2}{N_{tot}} \tag{13.1}$$

$$= 12459 - \frac{(593)^2}{30} = 12459 - 11721.63$$

$$= 737.37$$

$$SS_{bg} = \sum_{g} \left[\frac{(\sum X_g)^2}{N_g}\right] - \frac{\left(\sum_{tot} X\right)^2}{N_{tot}} \tag{13.2}$$

Since the N's in each group are the same, Formula 13.2 can be rewritten for our 2×3 experiment as:

$$SS_{bg} = \left[\frac{(\sum X_{A1B1})^2 + (\sum X_{A1B2})^2 + (\sum X_{A1B3})^2 + \cdots + (\sum X_{A2B3})^2}{N_g}\right]$$

$$- \frac{\left(\sum_{tot} X\right)^2}{N_{tot}}$$

where N_g = the number of subjects in each cell.

---

[2]Note that Formulas 13.1–13.4 are almost the same as Formulas 12.4–12.7 in Chapter 12.

Thus:

$$SS_{bg} = \left[\frac{(80)^2 + (85)^2 + (97)^2 + (93)^2 + (106)^2 + (132)^2}{5} \right] - \frac{(593)^2}{30} \qquad \textbf{(13.3)}$$

$$= 12068.60 - 11721.63 = 346.97$$

We now proceed to find SS_{wg}:

$$SS_{wg} = SS_{tot} - SS_{bg}$$

$$= 737.37 - 346.97 = 390.40$$

We remind you that we could also have computed SS_{wg} directly rather than by subtraction, using the formula:

$$SS_{wg} = \sum_{g} \left[\left(\sum X_g^2 \right) - \frac{\left(\sum X_g \right)^2}{N_g} \right] \qquad \textbf{(13.4)}$$

The SS_{wg} we have just determined is based on the variability within each of our six experimental groups—within each *cell* of our two-way table. As we discussed earlier, we will use this SS_{wg} to obtain a measure of within-groups variability against which to assess the main effects of both variable A and variable B. This same measure is also used in determining the F for the interaction of the variables.

The SS_{bg} that we have computed, however, is not of direct interest to us, since it is based on the variability among the means of all of our independent groups, considered as a single collection of k groups. This variability among the k means comes from three sources: (a) differences in the $\overline{X}$'s of the A groups (row means); (b) differences in the $\overline{X}$'s of the B groups (column means); and (c) the interaction between variables A and B—that is, any departure from an additive combination of the two variables.

Since it is these sources of variability that concern us, we now proceed to break down SS_{bg} and find the SS's for its three components: SS_A, SS_B, and the interaction, $SS_{A \times B}$.

Sum of Squares for Variable A (SS_A). The sum of squares for variable A, assuming an equal N in each A condition, is determined by the formula:

$$SS_A = \frac{\left(\sum X_{A1} \right)^2 + \left(\sum X_{A2} \right)^2 + \cdots + \left(\sum X_{Am} \right)^2}{N_A} - \left[\frac{\left(\sum\limits_{tot} X \right)^2}{N_{tot}} \right] \qquad \textbf{(13.5)}$$

where N_A = number in each A condition.

This formula indicates that we first find the sum of all the individuals tested under the A1 condition (the $\sum X$ of the A1 row) and square this sum. We then find the same quantity for each of the other A groups, on to the last or mth A group. We then add these quantities together and divide the sum by N_A, the number of cases in each of the A conditions. Then we subtract the value $\left(\sum_{tot} X\right)^2 \Big/ N$. Be sure to note that the latter quantity appears in the formulas for SS_{tot} and SS_{bg} so that its value has already been determined.

In our illustrative experiment, there were two A conditions (two degrees of affectivity), each based on a total of 15 cases, 5 each from the B1, B2, and B3 conditions. Thus:

$$SS_A = \frac{\left(\sum X_{A1}\right)^2 + \left(\sum X_{A2}\right)^2}{N_A} - \left[\frac{\left(\sum_{tot} X\right)^2}{N_{tot}}\right]$$

$$= \frac{(262)^2 + (331)^2}{15} - [11721.63]$$

$$= 11880.33 - 11721.63$$

$$= 158.70$$

Sum of Squares for Variable B (SS_B).   The SS_B is determined in a parallel manner:

$$SS_B = \frac{\left(\sum X_{B1}\right)^2 + \left(\sum X_{B2}\right)^2 + \cdots + \left(\sum X_{Bn}\right)^2}{N_B} - \left[\frac{\left(\sum_{tot} X\right)^2}{N_{tot}}\right] \qquad \textbf{(13.6)}$$

where N_B = number in each B condition.

For the three B (word familiarity) groups of our experiment, we thus have:

$$SS_B = \frac{(173)^2 + (191)^2 + (229)^2}{10} - [11721.63]$$

$$= 11885.10 - 11721.63$$

$$= 163.47$$

Sum of Squares for Interaction ($SS_{A \times B}$).   You will recall that SS_{bg} can be divided into three components, representing the SSs for variable A, variable B, and the A × B interaction. Since we have already computed SS_{bg}, SS_A, and SS_B, we can obtain the SS for interaction by subtraction:

$$SS_{A \times B} = SS_{bg} - SS_A - SS_B \qquad \textbf{(13.7)}$$

For our experiment:

$$SS_{A \times B} = 346.97 - 158.70 - 163.47$$

$$= 24.80$$

Although SS for interaction is typically obtained by subtraction, it may be computed directly by the following basic formula:

$$SS_{A \times B} = N_g \left[\sum_g (\overline{X}_{AB} - \overline{X}_A - \overline{X}_B + \overline{X}_{tot})^2 \right] \tag{13.8}$$

where N_g = number of cases per group, $\overline{X}_{AB}$ = the mean of a particular group, $\overline{X}_A$ = the mean of all the cases tested under the same A condition as this group (row $\overline{X}$), $\overline{X}_B$ = the mean of all cases tested under the same B condition as this group (column $\overline{X}$), and $\overline{X}_{tot}$ = the mean of all the cases. We will leave the application of this formula to the data of Table 13.2 as an exercise for those curious enough to find out that it does indeed yield the same value as we obtained by subtraction. Remember that the hypothesis being tested concerning the A $\times$ B interaction is that the two variables combine additively. If the variables combined in perfect additive fashion, we would expect no variability due to interaction and hence that $SS_{A \times B}$ would be equal to zero. This implies in turn that the term $(\overline{X}_{AB} - \overline{X}_A - \overline{X}_B + \overline{X}_{tot})^2$ in our formula for $SS_{A \times B}$ would be zero for each AB group. We may demonstrate this by applying the formula to the data from the 2 $\times$ 3 experiment of Table 13.1, which, as we saw earlier, were an example of a perfect additive relationship between variables. The grand mean for all of them is 7. Applying our formula to these data gives us:

$$SS_{A \times B} = N_g \left[\sum_g (\overline{X}_{AB} - \overline{X}_A - \overline{X}_B + \overline{X}_{tot})^2 \right]$$

$$= 10 \left[(3 - 6 - 4 + 7)^2 + (7 - 6 - 8 + 7)^2 + (8 - 6 - 9 + 7)^2 \right.$$
$$\left. + (5 - 8 - 4 + 7)^2 + (9 - 8 - 8 + 7)^2 + (10 - 8 - 9 + 7)^2 \right]$$

$$= 10(0)$$

$$= 0$$

Computation of MS's and *df*'s

Having determined the SS within groups and the SS's between A groups, between B groups, and the A $\times$ B interaction, the three components making up SS_{bg}, we now can find the variance or mean square (MS) for each of these sources of variability. Just as we stated in the last chapter, these MS's are found by dividing each SS by its *df*. Thus:

$$MS_A = \frac{SS_A}{df_A} \tag{13.9}$$

where $df_A = m - 1$ (the number of A groups minus one).

$$MS_B = \frac{SS_B}{df_B}$$

where $df_B = n - 1$ (the number of B groups minus one).

$$MS_{A \times B} = \frac{SS_{A \times B}}{df_{A \times B}}$$

where $df_{A \times B} = (m - 1)(n - 1)$ or $(df_A)(df_B)$.

$$MS_{wg} = \frac{SS_{wg}}{df_{wg}}$$

where $df_{wg} = N_{tot} - (m)(n)$ [the total number of cases minus the total number of groups—that is, A groups $\times$ B groups].

We should also note that the df for SS_{tot} is equal to $N_{tot} - 1$. Since SS_{tot} is composed of SS_{wg}, SS_A, SS_B, and $SS_{A \times B}$, the dfs for these four components, when summed, equal the total df or $N_{tot} - 1$.

Applying these formulas to the data of our illustrative experiment shown in Table 13.2, we find the following df's:

$$df_A = m - 1 = 2 - 1 = 1$$
$$df_B = n - 1 = 3 - 1 = 2$$
$$df_{A \times B} = (m - 1)(n - 1) = (1)(2) = 2$$
$$df_{wg} = N_{tot} - (m)(n) = 30 - (2)(3) = 24$$

The df_{tot} in our experiment is equal to $30 - 1$ or 29, and you will observe that the four df's obtained above do indeed add up to this number.

The MS's are thus:

$$MS_A = \frac{SS_A}{df_A} = \frac{158.70}{1} = 158.70$$

$$MS_B = \frac{SS_B}{df_B} = \frac{163.47}{2} = 81.74$$

$$MS_{A \times B} = \frac{SS_{A \times B}}{df_{A \times B}} = \frac{24.80}{2} = 12.40$$

$$MS_{wg} = \frac{SS_{wg}}{df_{wg}} = \frac{390.40}{24} = 16.27$$

Computation of F Ratios

Finally we compute F ratios by dividing the MS's for A, B, and the A $\times$ B interaction by MS_{wg}. To illustrate:

$$F_A = \frac{MS_A}{MS_{wg}} = \frac{158.70}{16.27} = 9.75 \tag{13.10}$$

$$F_B = \frac{MS_B}{MS_{wg}} = \frac{81.74}{16.27} = 5.02$$

$$F_{A \times B} = \frac{MS_{A \times B}}{MS_{wg}} = \frac{12.40}{16.27} = .76$$

The results of our calculations of the various values of SS, *df*, MS, and F are summarized in Table 13.3. Now all that remains is to evaluate each F ratio by comparing it with the value of F at the 5% or the 1% level of significance associated with the appropriate *df*'s, as listed in Table F. Looking at this table, we find that with 1 and 24 *df*, the *df*'s associated with the F for variable A, a value of 7.82 is required at $\alpha = .01$. Since our obtained F is 9.75, we conclude that the A (affectivity) groups differ significantly at the 1% level. For 2 and 24 *df*, the *df*'s associated with the F for variable B, a value of 3.40 is required at $\alpha = .05$ and of 5.61 at $\alpha = .01$. Since we obtained an F of 5.02 for the B (word familiarity) groups, we conclude that these groups differ significantly at the 5% level.

However, the F for the A × B interaction is less than one, so obviously it is not significant. Thus, we conclude that while both word familiarity and affectivity are significantly related to duration threshold, the magnitude of the effects of one variable is the same whatever the specific condition of the other variable. The two variables, in other words, combine additively. The *p* value of each F, in terms of whether it reaches or fails to reach the 1% level or the 5% level, is shown in the final column of Table 13.3.

A Step-by-Step Computational Procedure

We will now summarize the steps taken in computing a two-way ANOVA, starting first with a description of the basic data that will enter into our calculations.

Basic data
1. Find the ΣX of each group and enter them in a two-way table such as is illustrated in the lower portion of Table 13.2. By adding the appropriate sums, find the ΣX for each column, each row, and for the total experiment.
2. Find the ΣX^2 for each group and by adding these values, for the total experiment.

Table 13.3 Summary of the Analysis of Variance of Experiment on Duration Thresholds as Related to Word Familiarity and Affectivity

Source of Variation	SS	df	MS	F	p
Affectivity (A)	158.70	1	158.70	9.75	$< .01$
Familiarity (B)	163.47	2	81.74	5.02	$< .05$
Interaction (A × B)	24.80	2	12.40	.76	$> .05$
Within groups	390.40	24	16.27		
Total	737.37	29			

Calculate SS's by the following formulas:

3. $SS_{tot} = \left(\sum_{tot} X^2\right) - \dfrac{\left(\sum_{tot} X\right)^2}{N_{tot}}$

4. $SS_{bg} = \sum_{g} \left[\dfrac{\left(\sum X_g\right)^2}{N_g}\right] - \dfrac{\left(\sum_{tot} X\right)^2}{N_{tot}}$

5. $SS_{wg} = SS_{tot} - SS_{bg}$

   Check: $SS_{wg} = \sum_{g} \left[\left(\sum X_g^2\right) - \dfrac{\left(\sum X_g\right)^2}{N_g}\right]$

6. $SS_A = \dfrac{\left(\sum X_{A1}\right)^2 + \left(\sum X_{A2}\right)^2 + \cdots + \left(\sum X_{Am}\right)^2}{N_A} - \dfrac{\left(\sum_{tot} X\right)^2}{N_{tot}}$

7. $SS_B = \dfrac{\left(\sum X_{B1}\right)^2 + \left(\sum X_{B2}\right)^2 + \cdots + \left(\sum X_{Bn}\right)^2}{N_B} - \dfrac{\left(\sum_{tot} X\right)^2}{N_{tot}}$

8. $SS_{A \times B} = SS_{bg} - SS_A - SS_B$

   Check: $SS_{A \times B} = N_g \left[\sum_{g} (\overline{X}_{AB} - \overline{X}_A - \overline{X}_B + \overline{X}_{tot})^2\right]$

Determine *df*'s by the formulas:

9. $df_A = m - 1$

10. $df_B = n - 1$

11. $df_{A \times B} = (m - 1)(n - 1)$

12. $df_{wg} = N_{tot} - (m)(n)$

Calculate MS's by dividing each SS by its *df*:

13. $MS_A = \dfrac{SS_A}{df_A}$

14. $MS_B = \dfrac{SS_B}{df_B}$

15. $MS_{A \times B} = \dfrac{SS_{A \times B}}{df_{A \times B}}$

16. $MS_{wg} = \dfrac{SS_{wg}}{df_{wg}}$

Calculate F ratios by dividing the MS's for A, B, and A $\times$ B by MS_{wg}:

17. $F_A = \dfrac{MS_A}{MS_{wg}}$

18. $F_B = \dfrac{MS_B}{MS_{wg}}$

19. $F_{A \times B} = \dfrac{MS_{A \times B}}{MS_{wg}}$

Determine the significance of F's:

20. For each of the F ratios found in steps 17–19 find the values required for significance at the 5% and 1% level by entering Table F with the df's associated with the numerator and the denominator of each of these F's.

21. If a given F exceeds the required F at either the 5% or 1% level, reject the null hypothesis at that level; if it falls short, accept the null hypothesis.

Summarize the results:

22. Prepare a summary table, similar to the one in Table 13.3, that contains the values of the SS, df, MS, and F for the main effects of variables A and B and for the A $\times$ B interaction.

If one or more of the F values is significant, we may want to make further comparisons to determine precisely which groups or conditions contributed to the significant differences. This may be done by comparing pairs of $\overline{X}$ values by the Tukey *hsd* test discussed in Chapter 12. The only new idea to note here is that in Table G the number of groups in the experiment, k, is equal to m times n for any *hsd* test we are likely to use with a two-way factorial design.

This concludes our discussion of the analysis of variance technique—one of the most useful tools of statistics. Still further extensions allow us to analyze the data from factorial designs involving three or more separate variables. Modifications also have been developed that are appropriate, for example, for analyzing matched groups, such as might be involved in experiments in which a number of measures are obtained from the same subjects rather than from independent groups. Accordingly, the one-way and two-way analysis of variance techniques that we have discussed are the first steps toward the statistical analysis of many experimental problems more complicated than those discussed in this book.

DEFINITIONS OF TERMS AND SYMBOLS

Two-way factorial design. An experimental design in which two variables, A and B, are manipulated and in which all possible combinations of A conditions and B conditions are represented. For example, with three A conditions and four B conditions, the factorial design calls for $3 \times 4 = 12$ groups.

Two-way ANOVA. Application of the analysis of variance technique to the data from a two-way factorial design.

Main effects. In a two-way design, the effects of variations in A conditions, independent of variations in B conditions, and the effects of variations in B conditions, independent of variations in A conditions.

H_0 **for main effects.** The null hypothesis for variable A is that the population μs from which the A groups are drawn are all equal. Similarly, the null hypothesis for variable B is that the population μ's from which the B groups are drawn are all equal.

A $\times$ B interaction. The manner in which variables A and B combine or interact to determine the dependent variable.

H_0 **for A $\times$ B interaction.** The null hypothesis that the combination of A and B is additive. An additive combination is reflected graphically in a set of parallel curves when the means of the A conditions for each B condition (or the means of the B conditions for each A condition) are plotted.

PROBLEMS

1. A series of experiments employing a two-way factorial design is described below. For each experiment, do the following: I. Plot the $\overline{X}$'s in a single graph, like those in Figures 13.1–13.3 in the text. II. Present the $\overline{X}$'s in a two-way table, such as the one in Table 13.1 in the text, then find and enter in the table the row and column means. (Assume equal N's in each group.) III. After inspecting your graph and table, state which F's would probably be significant if a two-way ANOVA were performed and what you would conclude about the main effects and the interaction between the variables. Be specific in describing your conclusions.

   (a) A company introduced an experimental "flextime" program in one of its plants in which employees were expected to work 40 hours per week but had the option of working more than 8 hours on some days and less on others. Members of two departments, one consisting of assembly line workers and the other of office workers, took part in the program. Members of two other departments of assembly and office workers served as a control group, continuing to work 8 hours per day. After the program had been in operation for six months, a measure of morale and job satisfaction was given to all four groups. The means are as follows, high values indicating high satisfaction.

   Exp.–Assembly = 35   Control–Assembly = 25
   Exp.–Office = 37   Control–Office = 32

   (b) An agriculturist performed an experiment to compare the yield of three different types of hybrid corn in two different soil conditions. In conducting his study, he divided each of two plots, differing in type of soil, into 15 rectangular strips of equal size and planted the three types of seed in alternating strips. The mean yields (in bushels) were as follows:

   Plot 1–H_1 = 61   Plot 1–H_2 = 71   Plot 1–H_3 = 66
   Plot 2–H_1 = 72   Plot 2–H_2 = 81   Plot 2–H_3 = 75.

(c) In a study of the effects of success and failure, a psychologist gave subjects a timed digit-substitution task and afterward told subjects assigned to the success condition that they had performed very well in comparison to others and subjects in the failure condition that they had not performed well. Then all subjects were given a second example of the task and the time to completion was recorded. Subjects had previously been classified as high or low in self-esteem and self-confidence, and half in each self-esteem group had been assigned to the success and failure conditions. The mean time to completion (in seconds) for the groups on the second task were as follows:

Success–High self-esteem = 63   Failure–High self-esteem = 57
Success–Low self-esteem = 58   Failure–Low self-esteem = 64

2. One hundred twenty male students, 40 majoring in business administration, 40 majoring in education, and 40 majoring in psychology, were shown one of two videotapes of a man being interviewed. (Actually, the interview was staged.) Twenty in each of the three groups viewed a tape in which the man portrayed himself as being very competent in a number of areas and as hard-working and ambitious to be highly successful vocationally. The other 20 in each group viewed a tape in which the man portrayed himself as competent in the same ways but was more relaxed about his work and had only moderate vocational ambitions. After watching the tape, the subjects rated the man on a 7-point scale to indicate how personally likable they found him. A rating of 7 indicated "very likable" and a rating of 1 indicated "not at all likable." Shown below are the sums and sums of squares for the several groups (N per group = 20).

		Business	Major Education	Psychology
Ambitious	ΣX	117	67	81
	ΣX^2	703	299	387
Unambitious	ΣX	60	95	78
	ΣX^2	212	499	358

(a) Find the $\overline{X}$'s of the groups and graph them.
(b) Specify the null hypotheses that are to be tested.
(c) Conduct the appropriate ANOVA. To shorten your labors, round off SS's and MS's to one decimal place and F's to two decimal places. Summarize the results in a table patterned after Table 13.3 in the text.
(d) Describe the outcomes of the experiment.
(e) Perform a series of three Tukey *hsd* tests to determine whether the ratings of the two tapes differ significantly within each group of majors.

Chi Square

CHAPTER 14

Most of the statistical procedures we have discussed in previous chapters have involved *measurement* data—that is, characteristics that may vary in amount. We therefore assign each individual in the group we are measuring a number reflecting how much of the characteristic that individual exhibits. Techniques are also available for analyzing *categorical data.* Categorical data, you recall, involve the assignment of each member of a group of individuals to one of two or more discrete categories. The group of individuals is described by reporting the number or proportion of the group falling into each category.

For example, we might ask each of 200 people whether they favor a constitutional amendment permitting classroom prayers in public schools. Assuming that everyone had an opinion, we would in this case have only two categories of responses, Yes or No. We would not measure for each individual the "amount" of his or her response, but rather classify the response as belonging to one of the two categories, and for the group as a whole, record the number of people (the frequency) within each category. Note that all the persons included in each category have the same "score"; there is no variability in the response as classified *within* a category.

Our purpose in questioning this sample of 200 people might be to determine whether in the population as a whole, there is a majority sentiment for or against the constitutional amendment. We might therefore wish to test the hypothesis that in the population the split is 50:50, and to determine whether it is reasonable to reject this hypothesis. In Chapter 9, we discussed a method for testing a hypothesis of this sort. That method, which involves converting the difference between a sample proportion and a hypothetical population proportion into a z score, is applicable to the binomial case in which there are only *two* mutually exclusive categories.

In this chapter we discuss an alternative technique for analyzing categorical data. The statistic we compute is called *chi square*, symbolized as χ^2. One of the desirable features of χ^2 is that it permits hypotheses to be tested about the distribution of individuals among any number of mutually exclusive categories rather than being restricted to the two-category case. Chi square can also be used for another purpose that will be explained later in this chapter.

CHI SQUARE APPLIED TO A SINGLE SAMPLE

When members of a sample have been assigned to categories and the frequency of cases in each category has been determined, the question we typically want to answer is whether the frequencies observed in our sample of N individuals deviate significantly from some *theoretical* or *expected frequencies* based on N and the population frequencies. We have some hypothesis about the proportion in the population that falls into each category and therefore what frequencies should theoretically be expected in our sample of N individuals. In some cases, we may hypothesize equal frequencies in each category, so we want to determine whether the observed frequencies deviate significantly from what would be expected from a hypothesis of equal likelihood. In other cases, we may adopt some other hypothesis.

Having classified a sample of individuals, we have the problem of evaluating the deviations of the *observed* frequencies from theoretically *expected* frequencies if our hypothesis about the population proportions were true. Should our hypothesis about population frequencies be accepted and the deviations attributed to sampling error? Or is it more reasonable to conclude, at a certain probability level, that our hypothesis is false and a nonchance factor was operating? Chi square is used for this purpose. The formula for χ^2 is:

$$\chi^2 = \sum \left[\frac{(O - E)^2}{E} \right]$$ (14.1)

where   O = observed frequency

E = the corresponding expected frequency

According to this formula, we first get the deviation of each observed frequency from its corresponding expected frequency and then square the deviations. Next, each squared deviation is divided by the appropriate expected frequency. The remaining step is to sum these quotients; the sum is the value of χ^2.

Let us see how this works with our example of 200 people questioned about prayers in the public schools. We will test the hypothesis that opinion neither favors nor opposes the constitutional amendment, that in the population from which our sample was drawn opinion is equally divided. The expected frequency is, therefore, 100 in each category. Table 14.1, which shows the computation, indicates that for a 50:50 hypothesis with these data, $\chi^2 = 5.12$. How is this value to be interpreted? The interpretation of a value of χ^2 is basically the same as for *t*. We assume a certain null hypothesis—for the data in Table 14.1 the hypothesis that there is no difference in the population between the two categories of response. Then we ask: If the hypothesis is true, how often would values of χ^2 as large as or larger than our obtained value arise by random sampling error? If we find that values of χ^2 as large as or larger than the value obtained would occur by sampling error only 5 percent or 1 percent of the time, we reject the hypothesis at the 5% or 1% level of significance.

Table 14.1   A 50:50 Hypothesis Tested by χ^2

	Favor	*Do Not Favor*	*Total*
Observed (O)	84	116	200
Expected (E)	100	100	200
O - E	-16	+16	
(O - E)^2	256	256	
$\frac{(O - E)^2}{E}$	2.56	2.56	

$$\chi^2 = \sum \left[\frac{(O - E)^2}{E} \right] = 2.56 + 2.56 = 5.12 \quad df = 1$$

In order to evaluate obtained values of χ^2 we need to know the distribution of χ^2. Table I in Appendix II shows for each df value from 1 to 30 the value of χ^2 that would occur by sampling error at both the .05 and .01 levels of probability. Thus, for 1 df (first row in Table I) a value of χ^2 as large as 6.64 occurs 1 percent of the time by sampling error alone. Similarly, with 6 df, an obtained χ^2 would have to be at least 12.59 to be called significant at the 5% level. Note that, unlike F or t, values of χ^2 necessary for significance at a given level *increase* with increasing df.

Degrees of Freedom. The df for a particular χ^2 does *not* depend upon the number of individuals in the sample. For χ^2, the df is determined by the number of *deviations*, between observed and expected frequencies, that are *independent;* that is, are free to vary. For example, what is the df for Table 14.1? First, we have to note a condition that must always be true for any χ^2: *The sum of the expected frequencies must equal the sum of the observed frequencies.* In Table 14.1 this is true; both observed and expected frequencies sum to 200. Now if, in Table 14.1, we write down 100 as the expected frequency for one category, the expected frequency for the other category (and, therefore, the deviation) is *not* free to vary; it must also be 100, in order that the expected frequencies sum to the same total as the observed frequencies. Therefore, there is 1 df for the data in Table 14.1, because only one deviation is free to vary.

What if the number of categories were four instead of two? The number of expected frequencies that are free to vary and still have the sum of the expected frequencies equal the sum of the observed frequencies is three. Generalizing from these examples, it can be seen that degrees of freedom is equal to the number of categories minus one:

$$df = \text{number of categories} - 1 \qquad\qquad (14.2)$$

Returning to Table 14.1, in which $df = 1$, we can now use Table I to evaluate the calculated χ^2. Table I shows that for 1 df, $\chi^2 = 3.84$ at the 5% level and 6.64 at the 1% level. Our obtained value of 5.12 exceeds the value at the 5% but not at the 1% level, so we may reject our null hypothesis at the 5% level of significance. We conclude that opinion is quite probably not evenly split in the population from which our sample was drawn.

We have noted that chi square is not restricted to testing the hypothesis of equally distributed frequencies or to the two-category situation, but may be used to test any *a priori* hypothesis about any number of categories. By an *a priori* hypothesis we mean one the investigator has formulated before the research is done. For example, the ads for Zippy cola brag that in blind tasting tests of Zippy and two competing brands, 50 percent of adults prefer Zippy, whereas only 30 percent prefer Brand X, and 20 percent prefer Brand Y. A consumer organization decides to investigate this claim and conducts a test with 150 people. The results show that 54 prefer Zippy, 51 prefer Brand X, and 45 prefer Brand Y. If the advertising claim is true, the expected frequencies are 75, 45, 30, respectively; that is, 50 percent of 150 people, 30 percent of 150, and 20 percent of 150. After these three pairs of observed and expected frequencies have been determined, χ^2 can be computed, as in Table 14.1. With three categories, $df = 2$.

Before taking up the use of χ^2 in more complex cases, we shall note some of the conditions that must be met before data may appropriately be analyzed by χ^2. Like any statistical method, the use of χ^2 is subject to certain limitations; if these are kept in mind, many incorrect applications will be avoided.[1]

First, χ^2 can be used only with frequency data. It is not correct, for example, to test by χ^2 the deviation between the obtained mean score on some trait for a group, and some expected or predicted mean. These are measurement data, and one of the reasons why χ^2 cannot be used with such data is that the value of χ^2 varies with the size of the unit of measurement. Thus, weight may be measured either in pounds or in kilograms. But the value of χ^2 between observed and expected weights will be different, even for the same subjects, if the score is recorded in pounds than if it is recorded in kilograms. Who is to say which value of χ^2 is "correct"?

Second, the individual events or observations must be independent of each other. The data for the example on prayers in school may be used as an illustration. There is no reason to think that the answer of any one of the subjects in the sample is dependent upon or correlated with the answer of any other subject; the individual responses are independent. But it would be incorrect, for example, to ask each of 50 people to make guesses as to what card will be drawn from a deck on each of five successive draws and then to claim the total frequency was 250 events. In such a case we would not have 250 independent responses; successive responses by the same individual are very probably related.

Third, no theoretical frequency should be smaller than 5. The distribution of χ^2, part of which is given in Table I, may take different values from those shown in the table if any expected frequency is less than 5. The discrepancy is not large, however, when χ^2 is computed from contingency tables with a fairly large number of cells (more than 4, at a minimum) and only a few theoretical frequencies are less than 5. In this latter instance, it is permissible to relax the stringent rule that *no* theoretical frequency may be less than 5 and to go ahead and compute χ^2.

Fourth, there must be some basis for the way the data are categorized. Preferably, the categories should have some logical basis, or be based on previous acceptable research, and should be set up before the data are collected. Suppose, for example, we asked a large random sample of people what magazine they would take if they could subscribe to only one. We find that 31 magazines were named, ranging from a few mentioned very frequently to several preferred by one person each. We now categorize the data into several "types" of magazines in order to apply χ^2. Such an application would be questionable, because we have not demonstrated that the types (categories) we set up are any more defensible than any others that might be used. And it would be even worse to continue reclassifying the data until we got categories that would give a significant χ^2. Our categories must be set up beforehand, and it must be demonstrated that the categories

---

[1] D. Lewis and C. J. Burke, The use and misuse of the chi-square test. *Psychological Bulletin*, 1949, *46*, 433–487.

are reliable. Reliability would be demonstrated if there were a high degree of agreement among a group of qualified judges as to what magazines should be assigned to what categories.

Finally, recall we said above that *the sum of expected and the sum of observed frequencies must be the same.* This requirement is related to another: *If we are recording whether or not an event occurs, we must include in our data, and use in computing* χ^2, *both the frequency of occurrence and the frequency of nonoccurrence.* For example, suppose we roll a die 120 times and note the frequency with which a one-spot turns up. The expected frequency is 20, since a die has six sides. It would not be correct to test by χ^2 only whether the observed frequency of occurrence of the one-spot deviated significantly from the expected frequency. The χ^2 must be based *both* on the deviation of observed frequency of occurrence from expected frequency of occurrence, and the deviation of observed frequency of nonoccurrence from expected frequency of nonoccurrence. If we ignore the frequency of nonoccurrence, the *sums* of observed and expected frequencies will not be the same, except by chance. The value of χ^2 must be based on the deviation of observed from expected frequency for both the category of occurrence and the category of nonoccurrence.

TESTING SIGNIFICANCE OF DIFFERENCE BY χ^2

Chi square probably has its greatest usefulness in testing for significance of differences between groups. Although there may be any number of groups and any number of categories, most often in research we have two groups and two categories of response; the data are expressed in a 2 X 2 table. Our illustration of χ^2 as it is used to test for significance of difference therefore will be based on a 2 X 2 table, although the method is the same for any number of groups and categories.

Suppose we want to determine whether there is any relation between the amount of education a person has had and method of keeping up with the news. We ask a random sample of college graduates and a random sample of high school graduates whether they keep up with the news mostly by reading a newspaper or mostly by watching television. Table 14.2 shows the data that might have been obtained, arranged in a 2 X 2 table.

Table 14.2 shows that 47 of the 109 college graduates in the sample said they got news mostly by watching television; the other 62 relied mostly on the newspaper. Among the 97 high school graduates, 58 said television, 39 the newspaper. We want to know whether these frequencies indicate a significant difference between the two groups.

When χ^2 is used to test for the significance of a difference between two or more groups, it is said to be a test of *independence.* Perhaps this can best be explained by pointing out that the observed frequencies in Table 14.2 are classified two ways. One way is television versus newspaper; the other way is by educational level. Our problem, stated above in terms of significance of difference, can therefore be restated as the question: Are the two ways of classifying the observed frequencies independent of each other? If they *are* independent, then preference for television or newspaper does *not* depend on educational level and we could combine the educational groups, categorizing

Table 14.2 Testing Significance of Difference by χ^2 Test of Independence

	Television	Newspaper	Total
College graduates	47(55.6)	62(53.4)	109
High school graduates	58(49.4)	39(47.6)	97
Total	105	101	206

O	E	O - E	$(O - E)^2$	$\dfrac{(O - E)^2}{E}$
47	55.6	-8.6	73.96	1.33
62	53.4	+8.6	73.96	1.39
58	49.4	+8.6	73.96	1.50
39	47.6	-8.6	73.96	1.55
				$\chi^2 = 5.77$

merely on the basis of television versus newspaper. This is the same as saying the groups do *not* differ significantly. On the other hand, if the classifications are *not* independent—that is, if they are correlated—then when we categorize the data one way, we must categorize the other way too. In our example this would mean that the educational groups *differ significantly* in their preference for television or newspaper.

Let us now see how we use χ^2 to test for significance of difference, using our example in Table 14.2. We start by assuming the null hypothesis that our two groups were drawn from the same population with respect to method of keeping up with the news. Note that we are not concerned, for either group, with whether television is preferred over the newspaper. If we were, we could use χ^2, as explained earlier, to test an *a priori* hypothesis in each group separately.

Our next step is to determine the expected frequency for each of the four cells of the table. Since we have no *a priori* hypothesis, we reason as follows: if our null hypothesis is correct—if the true (expected) frequencies are the same for both samples—then combining the two samples should give us a better estimate of the true frequencies than we could get from either sample alone. Thus we add, within each column, the observed frequencies for the samples to get the marginal column totals of 105 and 101. We use these values as the best estimates of the division in the single assumed population between television preference and newspaper preference. Since our total sample is 206 cases, the expected percent frequency in the population for television preference is $\frac{105}{206}$, or 50.97 percent. In other words, our best estimate of what we would obtain if we measured the whole population is 50.97 percent getting news mostly by television. In similar fashion, $\frac{101}{206}$, or 49.03 percent of the total frequency, is the expected percent frequency for the newspaper category.

All we have to do now, to get the expected raw frequencies for each sample, is divide the sample N on the basis of these expected percent frequencies. There are 109 college graduates; since we expect 50.97 percent of them to fall in the television category, we have 50.97 percent of 109, or 55.6, as the expected frequency for the upper left-hand cell of the table. This value is shown in parentheses in that cell in Table 14.2. Similarly, 50.97 percent of the 97 high school graduates gives us 49.4 as the expected frequency in the television category for those subjects. The expected frequencies for the newspaper cells are determined by taking 49.03 percent of 109 and of 97.

Now that we have stated the logic for the method of determining the expected frequencies, we can state a simple rule that may be more easily remembered: Simply multiply the marginal total for any column by the marginal total for any row and divide the product by the total N. The result is the expected frequency for the cell common to the row and column whose totals were multiplied together. Thus, in our example, $(105)(109)/206 = 55.6, (105)(97)/206 = 49.4$ and so forth.

Returning to our problem, now that we have the expected frequencies, we compute χ^2 in the usual manner: Each deviation of an observed frequency from its expected frequency is squared, divided by the expected frequency, and the quotients summed. We find that $\chi^2 = 5.77$ for this example. Now, what are the *df*? We will state a general rule for determining the *df* for any table that has at least two rows and two columns, and in which the marginal totals are used in determining the expected frequencies:

$$df = (\text{number of columns} - 1)(\text{number of rows} - 1) \tag{14.3}$$

The rule applies to tables of any number of rows and columns, as long as there are at least two of each. Thus in a 2 × 2 table, the $df = (2 - 1)(2 - 1)$, or 1.

With 1 *df*, our χ^2 value of 5.77 is larger than the value given in Table I for the 5% level. We shall therefore reject the hypothesis at the 5% level of significance that the two groups were drawn from the same population with respect to method of getting the news. We could say the same thing by stating that we reject the hypothesis of independence; the frequencies in the television versus newspaper categories are probably *not* independent of educational level.

Before we leave this example we may note that χ^2 can be computed for a 2 × 2 table without computation of expected frequencies.[2] Let us schematize a 2 × 2 table as follows, where the letters in the cells indicate the *observed* frequencies:

A	B	A + B
C	D	C + D
A + C	B + D	N

With this schema, χ^2 can be computed directly as follows:

$$\chi^2 = \frac{N(AD - BC)^2}{(A + B)(C + D)(A + C)(B + D)} \tag{14.4}$$

---

[2]Reprinted with permission from Q. McNemar, *Psychological Statistics* (New York: Wiley, 1949).

Thus for the data in Table 14.2:

$$\chi^2 = \frac{206\;[(47)(39) - (62)(58)]^2}{(109)(97)(105)(101)} = 5.77$$

A Sidelight on Chi Square. The critical values of χ^2's used to assess hypotheses about expected frequencies have been derived from theoretical distributions that involve continuous rather than discrete values of χ^2. Calculated values of χ^2, on the other hand, vary in discrete fashion. This fact may create a problem in interpreting χ^2, particularly when *df* and expected frequencies are small. Statisticians have therefore historically advised introducing what is called a correction for continuity into the calculational steps for computing χ^2 when *df* and expected frequencies are small. This correction results in reducing the value of χ^2 in comparison to the value obtained with the formulas presented above. Recently there has been controversy among statisticians about the suitability of using this correction with the kinds of data usually obtained in psychology.[3] In light of its currently controversial status, it seems inappropriate to present the correction for continuity here, beyond this brief mention of its existence.

Significance of Differences among Several Groups

The method of applying χ^2 to test for significance of differences among several groups is identical with the method illustrated in the previous section with two groups. As an example, consider the data in Table 14.3. The data represent the number of bachelor's degrees awarded by a university in 1955, 1965, and 1975 in four different fields: social sciences, natural sciences, humanities, and education.

Let us apply χ^2 to the data in Table 14.3. We will be testing for any overall significance of differences among the groups. The procedure is the same as that previously illustrated with a 2 X 2 table. We obtain the theoretical frequency for each cell by multiplying the total for the row containing the cell by the total for the column containing the cell and dividing by the total N, which here is 1400. For example, for the cell in the row for social sciences and the column for the year 1955 we have 308(300)/1400 = 66.00. In Table 14.3 the theoretical frequencies are shown in the E column. Then χ^2 is computed in the usual way: In each cell subtract the theoretical frequency from the observed frequency to get the deviation (O – E), square the deviation, and divide by the theoretical frequency of the cell; the sum of the quotients is the value of χ^2. For the data in Table 14.3, $\chi^2 = 27.56$. Since this is a 3 X 4 table, there are 6 *df*: (3 – 1)(4 – 1). Looking at Table I, we see that for 6 *df* a χ^2 of 16.81 is required for significance at the 1% level. Our obtained χ^2 is far greater than this value. We therefore reject the hypothesis that the pro-

---

[3]See W. J. Conover, Some reasons for not using the Yates continuity correction on 2 X 2 contingency tables (with comments and rejoinder). *Journal of the American Statistical Association*, 1974, *69*, 374–382; and G. Camilli and K. D. Hopkins, Applicability of chi square to 2 X 2 contingency tables with small expected frequencies. *Psychological Bulletin*, 1978, *85*, 163–167. The continuity correction is defended in the comments on Conover's article and in other articles, such as N. Mantel and B. J. Hankey, The odds ratios of a 2 X 2 contingency table. *American Statistician*, 1975, *29*, 143–145.

Table 14.3 Testing Overall Significance of Difference among Several Groups by χ^2

Degree	1955	1965	1975	Total
Social sciences	46	67	195	308
Humanities	51	53	88	192
Natural sciences	85	100	154	339
Education	118	163	280	561
Total	300	383	717	1400

O	E	O - E	$(O - E)^2$	$\dfrac{(O - E)^2}{E}$
46	66.00	-20.00	400.00	6.06
51	41.14	9.86	97.22	2.36
85	72.64	12.36	152.77	2.10
118	120.21	- 2.21	4.88	.04
67	84.26	-17.26	297.91	3.54
53	52.53	.47	.22	.00
100	92.74	7.26	52.71	.57
163	153.47	9.53	90.82	.59
195	157.74	37.26	1388.31	8.80
88	98.33	-10.33	106.71	1.09
154	173.62	-19.62	384.94	2.22
280	287.31	- 7.31	53.44	.19
				$\chi^2 = 27.56$

portion of degrees awarded was constant in the three years for which data were collected. Inspection of Table 14.3 suggests that social sciences and education were gaining in popularity while the natural sciences and, even more, the humanities were losing.

Chi Square with More than 30 df.   Occasionally we may have so many groups and categories that there are more than 30 *df*. This would be the case in a 6 X 8 table (35 *df*), a 9 X 9 table (64 *df*), and so on. Table I does not go beyond 30 *df*. To evaluate a χ^2 based on more than 30 *df*, we convert the χ^2 value to an x/σ or z-score value and look up the probability in the normal curve table (Table B). The following expression is used to convert a χ^2 based on more than 30 *df* to an x/σ value:

$$\frac{x}{\sigma} = \sqrt{2\chi^2} - \sqrt{2n - 1} \tag{14.5}$$

where n = the number of *df* for the χ^2 value.

Let us conclude by reemphasizing the importance of the restrictions on the use of χ^2. These restrictions are not mere hairsplitting, since the use of χ^2 when it is inappropriate may lead us to erroneous conclusions about our data. The same statement can be made about any statistical test; we must be sure before applying a statistical test to any set of data that the assumptions underlying its use are not violated.

In the next chapter we will consider several additional statistical techniques that may be appropriate when the assumptions of some of the tests we have discussed so far cannot be met.

DEFINITIONS OF TERMS AND SYMBOLS

Frequency data. Data showing the number of cases (frequency) in each category in a set of two or more mutually exclusive categories.

Observed frequency (O). Actual number of cases in a sample observed to fall in a certain category.

Expected frequency (E). Theoretically expected number of cases in a certain category, according to some null hypothesis about population proportions that is to be tested.

Chi square (χ^2). A statistic used to test a null hypothesis, usually based on the comparison of observed and expected frequencies. Observed values of chi square are compared to critical values at a given α level that presume the null hypothesis is true.

A priori hypothesis. A hypothesis about expected frequencies, based on a related hypothesis about population proportions, that is set up by the investigator beforehand. This hypothesis is the H_0 to be tested by χ^2.

Test of independence. The use of χ^2 to test the hypothesis that individuals' classification according to one property is unrelated to their classification according to another property. For example, a test of the hypothesis that voters' sex (male or female) is unrelated to whether or not they voted Republican or Democratic in the last primary election.

PROBLEMS

In all problems requiring that you compute χ^2, specify the α level you will use to evaluate the hypothesis you are testing, give the df, and state whether or not χ^2 is significant at the given α level.

1. A psychology professor, interested in various kinds of "response bias," asked students in one of his classes to act as subjects in an investigation of bias. He flipped a fair coin and then asked the students to write down their guess about whether the coin had come up a head or a tail. Of the 200 students, 118 guessed heads and 82 guessed tails.

(a) In order to determine whether or not students exhibit a guessing bias, what statistical hypothesis should be tested using these data?

(b) Test the hypothesis specified in (a) by computing χ^2. What do you conclude?

2. In the investigation of response bias described in problem 1, the students were also asked to guess the suit of a card the professor drew at random from a bridge deck of 52 cards. Of the 200 students, 43 guessed a club, 51 guessed a spade, 54 a heart, and 52 a diamond.

(a) Compute χ^2 to determine whether or not there is a response bias.

(b) What do you conclude?

3. Random samples of 20 physicians and 20 lawyers were asked whether, if they had to do it all over again, they would enter the same or a different occupation. The observed frequencies of their responses are shown below. Do members of the two professions differ?

	Same	Different
Physicians	14	6
Lawyers	12	8

(a) What kind of a χ^2 test is called for?

(b) Compute χ^2 by Formula 14.1.

(c) Recompute χ^2 using Formula 14.4 involving only observed frequencies.

4. Three-year-old and 6-year-old children were repeatedly presented with a row of three circles differing in size. (On each trial, the same three circles were presented, but they appeared in different order.) On each occasion, the child was asked to pick one and was given a colored sticker whenever he or she chose the circle of intermediate size. Presentations continued until the child chose the intermediate circle five times in succession. After reaching this criterion, the child was presented with three triangles, also differing in size, and again asked to choose one. The experimenter was interested in determining whether the child's age was related to the degree to which he or she generalized from training to pick the intermediate-size circle to choice of intermediate-size triangle on the test trial.

Shown below are the numbers of children in each age group who selected each size of triangle. Also shown, in parentheses, are the expected frequencies.

	Small	Medium	Large
Three-year-olds	11(7.97)	13(18.06)	10(7.97)
Six-year-olds	4(7.03)	21(15.94)	5(7.03)

(a) Compute χ^2.

(b) What do you conclude about the effects of age on choice of triangle on the test trial following training on the circles?

5. A random sample of high school seniors planning to go to college has been surveyed in each of five geographical regions. One of the questions asks the students to select the subject matter in which they think they will major. Ten categories are listed (modern languages, natural science, education, social science, and so forth). The numbers of students who selected each category in each region are determined and placed in a 5×10 table. A χ^2 is performed to determine whether or not the proportion of seniors choosing each category of college major is related to the part of the country in which they live. The χ^2 is 55.05. Using $\alpha = .05$, determine whether this χ^2 is significant.

Nonparametric Tests

CHAPTER **15**

When we have measured two or more samples, we may determine the significance of the difference between or among their $\overline{X}$'s by means of a t test or of the F in an analysis of variance. These statistical techniques are examples of what are called *parametric* tests, since they require that we estimate from the sample data the value of at least one population parameter, such as its standard deviation. One of the assumptions we make in applying these parametric techniques to sample data is that the variable we have measured is normally distributed in the populations from which the samples were obtained.

In this chapter we will consider several of the *nonparametric* techniques statisticians have devised, techniques whose theory makes no use of parameter values. A second attribute of these nonparametric methods is that they are based on less restricting assumptions than those underlying parametric ones concerning the shape of the distribution of the characteristic being measured. Thus an alternate label for these methods is *distribution-free*.

NONPARAMETRIC VERSUS PARAMETRIC TESTS

Quite obviously, nonparametric techniques become valuable when the samples we have measured suggest that the assumptions underlying an otherwise appropriate parametric test cannot be met. Nonparametric tests, however, can be applied not only to measurement data, but also to other types such as sets of ranks or classified frequencies. For example, we might ask a factory foreman to rank 20 men working under his supervision, from the best to the poorest worker, and then compare the ranks received by 10 of the men who had been given a training course before being assigned to the job with the ranks of the 10 men who had been given no special training. It is unnecessary to give an illustration of classified frequencies, since we have just dealt with such data in Chapter 14 on χ^2. We should mention, however, that χ^2 is itself an example of a nonparametric technique and that several of the tests we will be discussing in this chapter involve converting our raw data into classified frequencies and then computing a χ^2.

When we have measurement data for which it is permissible to apply a parametric test, we often find it more time-consuming and computationally involved to compute the t, analysis of variance, or whatever, than a parallel nonparametric test. However, whenever possible we continue to use parametric tests in preference to nonparametric ones, for the following reason: If we apply both to the same data, we usually find that a nonparametric test is less powerful in terms of rejecting the null hypothesis. Other things being equal, a larger sample difference is needed between groups to reject the null hypothesis at a given significance level for a nonparametric than for a parametric test. Since, as research workers, we are particularly concerned with minimizing the probability that we will accept our null hypothesis when it is false, we use the more powerful parametric tests whenever we can.

Nonparametric techniques, then, are reserved for situations in which the nature of our data makes the use of a parametric test inappropriate. The latter occurs either because we are not dealing with measurement data or because our measurement data suggest that the assumptions underlying the use of parametric tests are not met and that important errors may result from their use.

SOME NONPARAMETRIC TESTS

Median Test with Two Independent Samples

The first nonparametric procedure we will consider, the *median test*, is appropriate when we wish to compare two independent, random samples. With this technique we test the null hypothesis that the two samples were drawn from populations with the same *median*.

In conducting the test, we first combine the two samples and find the median of the total, combined group. Then we count the number of cases in each sample that fall below and those that fall at or above the overall median.

To illustrate these initial steps, let us assume we have two independent samples with the following scores:

Sample 1 (N = 20)

6	10	3	4	8	2	9	2	7	4
5	3	6	1	9	11	4	8	9	3

Sample 2 (N = 22)

3	15	11	9	14	15	11	4	10	12	15
14	11	12	6	13	6	12	7	14	8	15

When all 42 cases are pooled, the median for the total group is found to be 8.5. When we count the number of cases in each sample that are at or above and below the median of 8.5, we find:

	At or above median	Below median	Total
Sample 1	5	15	20
Sample 2	16	6	22
Total	21	21	42

Now if the samples came from populations with the same median, we would expect that except for sampling error, half of the cases in each sample would fall at or above the median of the total group and half below it. To determine whether there is a significant departure from a 50:50 split within each of our samples, we now perform a χ^2 test with $df = 1$, following the procedures outlined early in the last chapter.

In Table 15.1 we have computed the χ^2 value for the data presented above. Since we were testing a 50:50 hypothesis, we obtained the expected frequencies for the cases above and below the median in each sample simply by dividing the sample N by 2. The χ^2 we found, 9.54, is greater than the value of 6.64 required for significance at the 1% level with 1 df. We therefore reject the null hypothesis and conclude that the samples came from populations with different medians.

Table 15.1   Median Test: χ^2 Testing the 50:50 Hypothesis for Two Groups

	At or Above Median	Below Median	Total
Sample 1	5(10)	15(10)	20
Sample 2	16(11)	6(11)	22
	$\overline{21}$	$\overline{21}$	

O	E	O - E	$(O - E)^2$	$\dfrac{(O - E)^2}{E}$
5	10	-5	25	2.50
16	11	5	25	2.27
15	10	5	25	2.50
6	11	-5	25	2.27
				$\chi^2 = 9.54$

Median Test with More Than Two Independent Samples

The median test we have just described may actually be used to compare any number of independent samples, not merely two. To review our method, the k samples are first combined and the median for the total group of cases determined. We then set up a $2 \times k$ table, entering the number of cases in each of the k samples that fall at or above and below the overall median. Using these data, we then compute a χ^2 testing the hypothesis that there is a 50:50 split in each of the samples. In interpreting this χ^2 we use $df = (2 - 1)(k - 1)$, which, of course, reduces to $k - 1$, or one less than the number of samples. If our computed χ^2 exceeds the tabled χ^2 value for $df = k - 1$ at the 5% or 1% level, we reject the null hypothesis that the samples all came from populations with the same median.

You will note that in conducting a median test, scores taking the same value as the median are classified with the scores falling *above* the median. In situations in which some scores actually do fall exactly at the median, chi square tends to be more significant, even if there is no real effect. In this instance, the median test may be improved upon by obtaining the observed frequencies, as we have just described, but computing χ^2 using the methods outlined later in Chapter 14. More specifically, the method shown in Table 14.2 should be used in the case of two samples, and the method used in Table 14.3 should be used in the case of more than two samples.

A Rank Test for Two Independent Samples (Mann-Whitney U)

The U test, devised by Mann and Whitney and also, independently, by Wilcoxon, is a technique based on score ranks that is appropriate for comparing two independent groups. Our first step is to combine the two samples and rank the total set of scores, assigning the rank of 1 to the largest score, 2 to the second largest, and so on. (The same final result would be obtained if scores were instead ranked from *smallest* to largest.) If tied scores occur, we handle them in the usual way: Each is assigned the average of the ranks the tied scores jointly occupy; for example, if two scores are tied for fourth place, each is assigned a rank of 4.5, the average of 4 and 5.[1]

Having assigned a rank to each score, we find the sum of the ranks for each sample, identifying these sums by R_1 and R_2. Next we calculate a statistic designated as U, one for each sample. (Actually, we need only one of these U values but can use the other as a check on the accuracy of our calculations.) For sample 1, this statistic is determined by the following formula:

$$U_1 = N_1 N_2 + \frac{N_1(N_1 + 1)}{2} - R_1 \tag{15.1}$$

where R_1 is the sum of the ranks assigned to the first sample.

Assuming the null hypothesis, which states, in effect, that the samples come from the same population with respect to the measured characteristic, the sampling distribution of U has a mean (expected U or U_E) equal to the value given by the formula:

$$U_E = \frac{N_1 N_2}{2} \tag{15.2}$$

and a standard deviation whose value is given by the formula:

$$\sigma_U = \sqrt{\frac{N_1 N_2 (N_1 + N_2 + 1)}{12}} \tag{15.3}$$

We now convert our obtained U into a *z* score:

$$z = \frac{x}{\sigma} = \frac{U_1 - U_E}{\sigma_U} \tag{15.4}$$

The U for the second sample could be obtained by a parallel formula, with R_1 being replaced by R_2 and, in the second term of the formula, the N_1 values by N_2 values. We could then find *z* by subtracting U_E from U_2 instead of U_1 in the numerator of the formula, since the two resulting differences are opposite in sign but of the same absolute value; that is, U_1 and U_2 deviate equally from U_E but in the opposite direction. The

---

[1] No problem results from this method of assigning ranks to tied scores when they all occur in a single sample, since the sum of ranks is unaffected. When ties occur across groups, the sum for each group *is* affected, being determined by the particular rank given to each tied score. This, in turn, affects U_1 (or U_2) and *z*. One statistician, J. V. Bradley, *Distribution-free statistical tests* (Englewood Cliffs, N.J.: Prentice-Hall, Inc., 1968), has recommended that this problem be resolved by considering all possible rankings of tied values and computing and reporting the *z*'s based on the U's of all the various combinations. This recommendation seems reasonable in instances in which the conclusion one reaches about the significance of the findings is likely to be affected.

scores of the sample whose U is less than U_E are, of course, lower than those of the other sample and vice versa.

If the N of each of our samples is 8 or more, the sampling distribution of U is normal or approximately so, and we may use the normal curve table to assess the null hypothesis. Thus, if we are testing a bidirectional H_1 and the z we have calculated is greater in absolute value than 1.96 (or 2.58), we reject the null hypothesis at the 5% (or 1%) level and conclude that the samples differed significantly. If z is less than this value, we accept the null hypothesis. When the N of one or both of the samples is less than 8, the sampling distribution of U is not normal and we must use a special table, which is not included in this book, to interpret z.

Table 15.2 Illustration of the Calculation of the Mann-Whitney U Test[*]

Sample 1 (N = 11)		Sample 2 (N = 10)		
X	*Rank*	*X*	*Rank*	

$$U_1 = N_1 N_2 + \frac{N_1(N_1 + 1)}{2} - R_1$$

X	*Rank*	*X*	*Rank*
35	1	34	2
33	3	31	5
32	4	26	9
30	6	25	11
27	7	23	13
26	9	22	14
26	9	21	15
24	12	18	19
20	16.5	17	20
20	16.5	16	21
19	18		$R_2 = 129$
	$R_1 = 102.0$		

$$= (11)(10) + \frac{11(11 + 1)}{2} - 102$$

$$= 176 - 102 = 74$$

OR

$$U_2 = N_1 N_2 + \frac{N_2(N_2 + 1)}{2} - R_2$$

$$= (11)(10) + \frac{10(10 + 1)}{2} - 129$$

$$= 36$$

$$U_E = \frac{N_1 N_2}{2} = \frac{(11)(10)}{2} = 55$$

$$\sigma_U = \sqrt{\frac{N_1 N_2 (N_1 + N_2 + 1)}{12}}$$

$$= \sqrt{\frac{(11)(10)(11 + 10 + 1)}{12}}$$

$$= \sqrt{201.67} = 14.20$$

$$z = \frac{U_1 - U_E}{\sigma_U} \qquad \textbf{OR} \qquad z = \frac{U_2 - U_E}{\sigma_U}$$

$$= \frac{74 - 55}{14.20} = \frac{19}{14.20} \qquad\qquad = \frac{36 - 55}{14.20} = \frac{-19}{14.20}$$

$$= 1.34 \qquad\qquad\qquad\qquad = -1.34$$

[*]Each listed rank is the overall rank of the associated X in the total group of twenty-one cases.

In Table 15.2 we have performed a Mann-Whitney U test on the hypothetical data from two independent samples. The z has been calculated twice, first using U_1 and then U_2 to demonstrate that z will be the same except for sign. Since the sample Ns are greater than 8, we use the normal curve to assess our calculated z of 1.34 and see that it falls far short of 1.96, the value needed for significance at the 5% level. We therefore accept the null hypothesis and conclude that the two samples could have come from the same population.

In both our discussion and illustration of the Mann-Whitney U test, we have assumed that the raw scores were measurement data of some kind and that we first had to convert them into ranks in order to perform the test. Sometimes, however, the raw scores *are* the ranks we need, as in the example given earlier in which the work efficiency of 20 men had been ranked, and we wished to compare the efficiency of those in the total group who had been given a special training course with the efficiency of those who had not. Except for omitting the initial step of converting raw scores to ranks, we would compute the Mann-Whitney U in the usual manner with data of this type. The same remarks hold true for the other rank tests we will discuss.

A Rank Test for Two or More Independent Groups (Kruskal-Wallis One-Way Analysis of Variance by Ranks)

The Kruskal-Wallis one-way analysis of variance by ranks is an extension of the Mann-Whitney U test that allows us to compare any number of independent groups rather than only two. We first pool all the k groups, arrange the total group of scores in order, and assign the *smallest* score a rank of 1, the next smallest a rank of 2, and so on, handling ties in the usual manner. (It is essential in this test that the *smallest* score be given a rank of 1, and so on down to the *largest* score.) Having ranked the total group, we then find the sum of the ranks (R_i) for each of the k samples.

If each of the samples came from the same population, then each of these sums of ranks (each R_i) should be equal except for sampling error. We test this null hypothesis by computing a statistic H by the following formula:

$$H = \frac{12}{N(N + 1)} \sum \left(\frac{R_i^2}{n_i} \right) - 3(N + 1) \qquad (15.5)$$

where: N = total number of cases in the k groups

n_i = number of cases in a given sample

R_i = sum of the ranks in a given sample

The expression $\sum (R_i^2/n_i)$, then, indicates that, for each group, we are to square the sum of ranks (R), divide this R^2 by the number of cases in the group, and finally to sum these quantities for the k groups. An illustration of these procedures involving four groups is shown in Table 15.3.

Table 15.3 Illustration of the Calculation of the Kruskal-Wallis One-Way Analysis of Variance for Ranks*

Group I (n = 6)		Group II (n = 8)		Group III (n = 7)		Group IV (n = 7)	
X	Rank	X	Rank	X	Rank	X	Rank
99	28	85	18.5	75	9	91	23
98	27	82	16	70	6	89	22
96	26	81	14	68	5	86	20
95	25	81	14	67	4	85	18.5
93	24	78	11	65	3	84	17
88	21	76	10	62	2	81	14
	$R_1 = 151$	73	8	59	1	79	12
		72	7		$R_3 = 30$		$R_4 = 126.5$
			$R_2 = 98.5$				

$$H = \frac{12}{N(N+1)} \sum \left(\frac{R_i^2}{n_i} \right) - 3(N+1)$$

$$= \left[\frac{12}{28(28+1)} \right] \left[\frac{151^2}{6} + \frac{98.5^2}{8} + \frac{30^2}{7} + \frac{126.5^2}{7} \right] - 3(28+1)$$

$$= (.0148)(7427.56) - 87 = 22.93$$

*Each listed rank is the overall rank of the associated X in the total group of 28 cases.

When each of the samples has five or more cases, the statistic H takes the chi-square distribution with *df* equal to *k* - 1, the number of groups minus one. In our example where *k* = 4, therefore, we enter Table I with 4 - 1 or 3 *df* and find that a χ^2 of 11.34 is required for significance at the 1% level. The H of 22.93 we obtained in Table 15.3 is greater than this value, so we are able to reject the hypothesis that all of our samples came from the same population.

A Rank Test for Two Matched Samples (Wilcoxon Signed-Ranks)

The nonparametric techniques we have discussed so far are used to compare two or more *independently* selected groups. The test we take up now was designed to allow us to compare distributions consisting of matched pairs: the same individual tested under two conditions, or pairs of different individuals matched on some basis of similarity before being tested. The procedure, devised by Wilcoxon, is identified as a matched-pairs signed-ranks test or, more simply, as the Wilcoxon signed-ranks test.

To illustrate, assume we have tested each member of a group under two different experimental conditions. Our first step is to find the difference (D) between the pair of scores (I and II) for each subject. We now discard from further consideration any pair of scores for which the D is 0. We then rank the remaining D's in order of *absolute* magnitude, assigning the rank of 1 to the *smallest* absolute D, 2 to the next *smallest*, and so on. As always, ties are handled by assigning to each tied D the average of the ranks the ties jointly occupy. Next we give to each of these absolute ranks the sign of the associated D score, giving us a set of *signed ranks*.

These procedures are illustrated in Table 15.4, in which, for convenience, we ordered the 10 subjects in terms of the absolute magnitude of their D scores, shown in the fourth column. Since the first subject received the same score in both conditions, so that $D = 0$, no rank was assigned to the D for this individual. For the nine remaining D's the *smallest* absolute D was given a rank of 1, and so on, down to rank 9 for the largest absolute D. Finally, in the last column we have given each of the absolute ranks the same sign, + or −, as its D.

Our next step is to sum the positive signed ranks and then to sum the negative signed ranks. If the two distributions of scores come from the same population—that is, if the null hypothesis that our two experimental conditions have no differential effects is correct—we would expect that the positive and negative sums would have the same absolute value except for sampling error.

For small samples a special table has been set up that allows us to test the null hypothesis directly from the sums of our signed ranks. This table is reproduced in Table J of Appendix II for N's up to 25, where N is the *total number of signed ranks* (N_{S-R}). Since some of our D's may be 0 and thus discarded, N_{S-R} may be less than the total number of original pairs of scores. The table is used as follows. The *smaller* (in absolute value) of the two sums of signed ranks is designated as T. We then look up in Table J the *maximum* value T may take and still be significant at the 5% or 1% level (two-tailed test) for our N_{S-R}. (Unlike t or F, the smaller the T the more significant it is.) In our example in Table 15.4 the number of signed ranks is 9 and T is 3.5. Looking at Table J, we find that for $N_{S-R} = 9$, a T of 6 *or less* is required for significance at the 5% level and a T

Table 15.4   Illustration of the Calculation of the Wilcoxon Signed-Ranks Test[*]

Subject	Score I	Score II	D	Rank of $|D|$	Signed Rank
1	39	39	0	—	—
2	47	48	−1	1	−1
3	29	27	2	2.5	2.5
4	33	35	−2	2.5	−2.5
5	36	33	3	4	4
6	48	43	5	5	5
7	38	31	7	6.5	6.5
8	43	36	7	6.5	6.5
9	32	23	9	8	8
10	52	40	12	9	9
					$\sum + = \overline{41.5}$
$N_{S-R} = 9$	T = 3.5				$\sum - = -3.5$

[*]The raw data are from 10 subjects tested under two experimental conditions.

of 2 *or less* at the 1% level. Since our obtained T of 3.5 is less than 6 but not less than 2, we reject the null hypothesis at the 5% level.

For larger samples the sampling distribution of T is approximately normal with a mean (expected T or T_E) given by the formula:

$$T_E = \frac{N(N + 1)}{4} \tag{15.6}$$

and a standard deviation given by the formula:

$$\sigma_T = \sqrt{\frac{N(N + 1)(2N + 1)}{24}} \tag{15.7}$$

where N = the number of *signed ranks*. We can then determine the z for our calculated T:

$$z = \frac{T - T_E}{\sigma_T} \tag{15.8}$$

We assess the null hypothesis by comparing our calculated z with the values of 1.96 and 2.58 required for significance in a normal distribution at the 5% and 1% levels, respectively.

A Rank Test for Two or More Matched Groups (Friedman Two-Way Analysis of Variance by Ranks)

On occasion we might want to test the same subjects not merely twice but three or more times. We might, for example, give each of a group of students three passages to memorize that contain political opinions with which they agree strongly, disagree strongly, or have no reactions, in order to determine whether this content factor affects their memorization. Or we might train a group of animals to run to a 10-unit food reward, then give 20 trials with a 5-unit reward to see if there is any systematic decline in running speeds to the reduced reward over the 20 trials. If the underlying assumptions are met, the data obtained by repeated measurement of the same subject may be analyzed by extensions of the analysis of variance techniques discussed in earlier chapters—extensions that will not be presented in this text.

If the assumptions for the analysis of variance are not fulfilled by our data, a nonparametric test that may be applied to data involving two or more measures from each subject is the Friedman two-way analysis of variance for ranks. In this analysis one variable is considered to be subjects and the other variable is measures.

In applying the Friedman rank test we first set up a table such as the one at the upper left of Table 15.5, in which each of the rows contains the k scores earned by each of the N subjects. In our example each subject was measured three times. As shown at the upper right of Table 15.5 we now rank, from 1 to k, the scores in each *row*—the

Table 15.5   Illustration of the Calculation of $\chi_R{}^2$ in the Friedman Two-Way Analysis of Variance by Ranks*

Raw Scores				Scores Ranked by Rows			
	Measure				*Measure*		
Subject	*I*	*II*	*III*	*Subject*	*I*	*II*	*III*
1	12	14	11	1	2	1	3
2	13	19	16	2	3	1	2
3	15	18	19	3	3	2	1
4	17	22	20	4	3	1	2
5	17	15	22	5	2	3	1
6	19	21	24	6	3	2	1
7	21	19	18	7	1	2	3
					17	12	13

$$\chi_R{}^2 = \frac{12}{Nk(k+1)} \left(\sum R_i{}^2 \right) - 3N(k+1)$$

$$= \left[\frac{12}{(7)(3)(3+1)} \right] [17^2 + 12^2 + 13^2] - 3(7)(3+1)$$

$$= 86 - 84 = 2 \qquad df = k - 1 = 2$$

*The raw data are from seven subjects tested under three experimental conditions.

scores of each *subject*. We have chosen in our example to go from largest to smallest in ranking the scores of each subject but could have done it the opposite way if we wished.

The next step is to find the sums of the ranks for each of the k measures—that is, for each of the *columns*. If our experimental conditions had no differential effects, any difference in the rank of the scores for a given subject is due to chance factors, and across subjects, any given rank is therefore as likely to be found with one measure as another. This implies, of course, that if the null hypothesis is true, each of the column sums of ranks would be equal except for sampling error.

We test the null hypothesis by computing a χ^2 based on the column sums. This χ^2 is given by the formula:

$$\chi_R{}^2 = \frac{12}{Nk(k+1)} \left(\sum R_i{}^2 \right) - 3N(k+1) \qquad \qquad (15.9)$$

where k = number of measures

N = number of subjects

R_i = a column sum of ranks

The resulting $\chi_R{}^2$ approximates the chi-square distribution with $df = k - 1$. In Table 15.5 we calculated $\chi_R{}^2$ to be 2.00 and df to be 2. This value is far less than the χ^2 of

5.99 found in Table I to be associated with the 5% level for 2 *df*. We therefore accept the null hypothesis and conclude that there was no significant difference among the three measures.

RELATIVE EFFICIENCY OF PARAMETRIC AND NONPARAMETRIC TESTS

We stress once more that a nonparametric test may not be as powerful as the corresponding parametric test such as *t* or F. One measure of the relative usefulness of these types of tests is the *relative efficiency* of two tests. Relative efficiency is defined as the *ratio of the sample sizes required by each test to obtain the same power value.* For example, suppose that for a given $\mu_1 - \mu_2$ value a *t* test had a power of .70 when N = 10 and a certain nonparametric test had the same power with an N of 12. According to our definition, the relative efficiency of the nonparametric test (compared to the *t* test) is $\frac{10}{12}$ or .833. This example illustrates the greater efficiency of the parametric test, the usual finding when sample N's are moderate or large in size. However, if very small samples are used (total N's of about 6), many nonparametric tests have relative efficiencies nearly equal to 1, compared to the corresponding parametric tests.

The relative efficiencies just mentioned were computed under the assumption that the data came from populations and samples satisfying the assumptions of the parametric test involved. Such data are precisely those for which parametric tests such as the *t* and the F test were developed. So nonparametric tests often work quite well even in cases where parametric tests are preferable. But once assumptions such as normality, which are required by the *t* and F tests, are known to be false, we know that *t* or F may lead to inappropriate conclusions about significance levels. In that case a nonparametric test will probably be preferable, even if it has low relative efficiency, because we can then be sure about our significance level.

Unfortunately, relative efficiency is not a constant that can be determined for any pair of tests. Rather, its value changes as the size of the sample or samples for one test changes. Relative efficiency also changes as other factors change, such as the α level desired and the value of the population difference (such as the value of $\mu_1 - \mu_2$) to be detected. Although we may want to know the relative efficiency for small-sample situations, it is more convenient to find a related measure, called the *asymptotic relative efficiency* (or ARE), which is constant in value despite variations in such factors as we have mentioned.

What is asymptotic relative efficiency? *Asymptotic* means the final level reached or almost reached at the end or *asymptote* of a process, such as the level of proficiency reached on a motor task so well practiced that no further improvement is noticeable. Similarly, *asymptotic relative efficiency* is attained after N becomes large enough that relative efficiency does not change appreciably with further increases in the N for each group. The asymptotic relative efficiencies for the five nonparametric tests described in this chapter are given in Table 15.6.

The tabled values help us decide what nonparametric test to use in a given situation. If we have two independent groups of scores with no tied ranks between groups, for ex-

Table 15.6   Asymptotic Relative Efficiencies (ARE's) of Five Nonparametric
Tests Compared to the Corresponding Parametric Tests

Test	ARE
Median test (two or more independent samples)	.637
Mann-Whitney U test (two independent samples)	.955
Kruskal-Wallis H test (two or more independent samples)	.955
Wilcoxon signed-rank test	.955
Friedman rank test*	.637 for $k = 2$
	.716 for $k = 3$
	. . .
	.910 for $k = 20$
	. . .
	.955 for $k = \infty$

*Where k = number of treatments given to each of N subjects.

ample, we would prefer the Mann–Whitney U test to the median test because of the latter's inefficiency, particularly with larger N's. Similarly, the Kruskal-Wallis test, which can be used with more than two groups, is more efficient than the median test and should be used in preference to the latter. However, if there are several cases of tied ranks across groups, complications from these ties may make the median test preferable to the Kruskal-Wallis or Mann-Whitney, even though its ARE is lower.

Nonparametric, distribution-free tests are a relatively recent development in statistics, and new techniques of this type continue to be devised to handle different kinds of statistical problems presented by empirical data. Although parametric tests continue to be preferable if the assumptions underlying their use can be met, nonparametric tests, a selected group of which we have presented in this chapter, are often of invaluable assistance to the research worker.

DEFINITIONS OF TERMS AND SYMBOLS

Nonparametric or distribution-free test.   A statistical test that does not require estimation of population parameters and makes few assumptions about the shape of the distribution from which the data to be analyzed come.

Median test.   A technique for testing the null hypothesis that two or more independent random samples come from populations with the same median. The test involves classifying members of each sample as falling above or below the overall median for the total group and then performing a chi-square test to determine whether all groups have equivalent proportions of scores at or above the median.

Mann-Whitney U test. A substitute for the *t* test with two independent groups, useful for situations in which normal distributions cannot be assumed. All scores in the two groups are ranked together, and separate sums of ranks are obtained for each group. Then either sum may be used with other information to calculate first a U and then a *z* to be compared to entries in Table B to determine the significance of the difference between the two groups. For small samples, a special table must be used in place of Table B.

Kruskal-Wallis one-way analysis of variance by ranks. The generalization of the Mann-Whitney U test for the case of two or more independent groups.

Wilcoxon signed-ranks test. A distribution-free analog of the *t* test for matched pairs, for use when the parent populations are not normal. A statistic called T, based on a ranking of the differences between pairs of scores, is calculated and its value is compared to entries in Table J in order to determine the significance of the difference between the two conditions being compared.

Friedman two-way analysis of variance by ranks. A chi-square test using ranks to compare the effects of treatments in two or more matched groups.

Relative efficiency. A method of comparing the power of two tests to detect a false null hypothesis. Relative efficiency is equal to the ratio of the sample sizes required by each test to obtain the same power value with the N for a nonparametric test being evaluated going in the denominator.

Asymptotic relative efficiency (ARE). Relative efficiency when N is very large for each group.

PROBLEMS

1. Randomly selected groups of male professional football players, college professors, and business executives were given a test of their enjoyment of interpersonal competition. The results are shown below, with high scores indicating high competitiveness.

Football players	College professors	Businessmen
35	31	35
35	28	34
35	26	31
34	22	27
34	21	26
34	20	24
33	19	23
32	19	21
30	18	21
28	18	19

   (a) If the groups were compared by means of a one-way analysis of variance, what assumptions underlying this parametric test appear to be violated?

   (b) Compare the groups by a median test, using $\alpha = .05$. What do you conclude about the competitiveness of the three groups?

2. Twenty students participated in an experiment in which they studied a prose passage for 3 minutes and then were asked to recall its content. The passage was an essay on the merits of the capitalistic economic system. The students had previously responded in their class to a political attitudes questionnaire and on the basis of the class median, were divided into those who were below the median (liberal) or above the median (conservative) in their attitudes. The recall data for the two groups are reported below. Perform a Mann-Whitney U test on these data, using $\alpha = .05$. What do you conclude about the effects on recall of material that agrees or disagrees with one's attitudes?

Conservative: 25, 23, 21, 21, 20, 19, 14, 11
Liberal: 24, 22, 18, 17, 16, 15, 15, 14, 13, 12, 11, 10

3. Laboratory rats prefer a sucrose solution to plain water and will run down a straight alley faster when they anticipate a sucrose reward. In a study of taste threshold, one group of thirsty rats was given 50 trials in a straight alley to a water reward, a second group was tested with a very weak sucrose solution, and a third with a slightly stronger solution. The running times in the last block of 10 trials are shown below. Analyze the data by the Kruskal-Wallis test. What do you conclude about the capacity of the animals to detect the sucrose?

Water: 30, 29, 26, 21, 18
Lower sucrose: 31, 28, 27, 22, 20
Higher sucrose: 33, 33, 30, 25, 19

4. In a word recognition study, subjects were shown a list of words for $\frac{1}{100}$ second each and asked to guess what the word was. Before each of these test words, they were shown another word for 1 second. For half of the list, the word shown first was associated with the test word (such as "table" followed by "chair") and for the other half, the first word had no discernible association with the test word ("green" followed by "horse"). The number of correct guesses made by each subject for test words preceded by associated and nonassociated words is shown below. Perform a Wilcoxon signed-ranks test of these data, using $\alpha = .05$. What do you conclude about the effects of the experimental conditions?

Subject	Associated	Nonassociated
1	19	17
2	18	19
3	18	13
4	17	11
5	16	12
6	15	15
7	15	12
8	12	10
9	10	8
10	9	10

5. A child psychologist has devised a behavior modification program to be applied by parents whose young children frequently throw temper tantrums. He asks parents to record the number of tantrums exhibited by their children for a period of one week during which they are being trained how to treat their children, for another week one month later, and a final week six months later. The following results are obtained from five children. Compare the distribution by means of the Friedman two-way analysis of variance by rank. What do you conclude about the effect of the training program?

Child	Week I	Week II	Week III
1	18	8	2
2	17	6	0
3	13	14	3
4	11	3	4
5	10	11	5

List of Formulas

APPENDIX I

Formula		Page

(4.1) $\overline{X} = \dfrac{\sum X}{N}$

(4.2) $\overline{X} = \dfrac{\sum (fX)}{N}$

(4.3) $\mu = \dfrac{\sum X}{N}$

where N is the total number of cases in the population

(4.4) Score (percentile) = LRL + i $\dfrac{[n\text{th case}] - [cum.\ f \text{ int. below}]}{\text{no. of cases in int.}}$

where LRL = lower real limit of the interval and nth case = number of cases corresponding to given percent

(4.5) Percentile rank = $100 \left(\dfrac{\text{no. of cases}}{N} \right)$

where no. of cases
= $cum.\ f$ in int. below + (f in int.) $\left(\dfrac{\text{score} - \text{LRL}}{i} \right)$

(5.1) Semi-interquartile range = $\dfrac{Q_3 - Q_1}{2}$

(5.2) $S^2 = \dfrac{\sum x^2}{N} = \dfrac{\sum (X - \overline{X})^2}{N}$

(5.3) $S = \sqrt{\dfrac{\sum x^2}{N}} = \sqrt{\dfrac{\sum (X - \overline{X})^2}{N}}$

(5.4) $s^2 = \dfrac{\sum x^2}{N - 1}$

(5.5) $s = \sqrt{\dfrac{\sum x^2}{N - 1}} = \sqrt{\dfrac{\sum (X - \overline{X})^2}{N - 1}}$

(5.6) $\sigma = \sqrt{\dfrac{\sum x^2}{N}} = \sqrt{\dfrac{\sum (X - \mu)^2}{N}}$

where N is the total number of cases in the population

(5.7) $S = \sqrt{\dfrac{\sum X^2}{N} - \overline{X}^2}$

Formula		Page

(5.8) $s = \sqrt{\dfrac{\sum X^2 - \left(\sum X\right)^2 / N}{N-1}}$ ⟶ 66

(5.9) $S = \sqrt{\dfrac{\sum f X^2}{N} - \left(\dfrac{\sum f X}{N}\right)^2}$ ⟶ 66

(5.10) $s = \sqrt{\dfrac{\sum f X^2 - \left(\sum f X\right)^2 / N}{N-1}}$ ⟶ 66

(5.11) $z = \dfrac{X - \overline{X}}{S}$ (based on sample statistics) ⟶ 70

$z = \dfrac{X - \mu}{\sigma}$ (based on population values) ⟶ 75

(5.12) $X = \overline{X} + z(S)$ ⟶ 71

(5.13) $\bar{z} = \dfrac{\sum z}{N} = 0$ ⟶ 72

(5.14) $S_z{}^2 = S_z = 1$ ⟶ 72

(5.15) Standard score = (specified σ) (z) + (specified μ) ⟶ 73

(6.1) $r = \dfrac{\sum z_X z_Y}{N}$ where N = no. of pairs, $df = N - 2$ ⟶ 82

$z_X = \dfrac{X - \overline{X}}{S_X}$ and $z_Y = \dfrac{Y - \overline{Y}}{S_Y}$

(6.2) $r = \dfrac{\sum XY - N\overline{X}\,\overline{Y}}{\sqrt{\sum X^2 - N\overline{X}^2}\,\sqrt{\sum Y^2 - N\overline{Y}^2}}$ ⟶ 85

(6.3) $a = r\left(\dfrac{s_Y}{s_X}\right)$ ⟶ 93

(6.4) $b = \overline{Y} - r\left(\dfrac{s_Y}{s_X}\right)\overline{X}$ ⟶ 93

(6.5) $Y_{pred} = r\left(\dfrac{s_Y}{s_X}\right)X - r\left(\dfrac{s_Y}{s_X}\right)\overline{X} + \overline{Y}$ ⟶ 94

(6.6) $X_{pred} = r\left(\dfrac{s_X}{s_Y}\right)Y - r\left(\dfrac{s_X}{s_Y}\right)\overline{Y} + \overline{X}$ ⟶ 94

273

Tables

APPENDIX II

CONTENTS

ACKNOWLEDGMENTS

The authors thank the authors, editors, and publishers listed below for permission to reprint materials from their publications in the tables of this appendix.

Table A. The Rand Corporation, *A Million Random Digits with 100,000 Normal Deviates*, p. 243 (Glencoe, Ill.: The Free Press, 1955). By permission of the Rand Corporation.

Table B. The original data for Table B came from *Tables for Statisticians and Biometricians*, edited by Karl Pearson, published by Cambridge University Press, and are used here by permission of the publisher. The adaptation of these data is taken from E. F. Lindquist, *A First Course in Statistics* (revised edition), with permission of the publisher, Houghton Mifflin Company.

Table C. Donald B. Owen, *Handbook of Statistical Tables*, © 1962. U.S. Department of Energy. Table 2.1. Published by Addison-Wesley Publishing Company, Inc., Reading, Massachusetts. Reprinted with permission of the publisher.

Table D. Table D is taken from Table VI of Fisher & Yates: *Statistical Tables for Biological, Agricultural and Medical Research*, published by Longman Group Ltd. London (previously published by Oliver & Boyd Ltd. Edinburgh) and by permission of the authors and publishers. Part of this table is reprinted by permission from *Statistical Methods* by George W. Snedecor and William G. Cochran, sixth edition © 1967 by Iowa State University Press, Ames, Iowa.

Table E. Computed from E. G. Olds, Distribution of the sum of squares of rank differences for small numbers of individuals, *Annals of Mathematical Statistics*, 1938, *9*, 133–148, and, The 5% significance levels for sums of squares of rank differences and a correction, *Annals of Mathematical Statistics*, 1949, *20*, 117–118, by permission of the author and the Institute of Mathematical Statistics.

Table F. M. Merrington and C. M. Thompson, Tables of percentage points of the inverted beta (F) distribution. *Biometrika*, 1943, *33*, 73–78. By permission of the Biometrika Trustees.

Table G. Adapted from Table 29 of E. S. Pearson and H. O. Hartley, (eds.), *Biometrika Tables for Statisticians.* Vol. I, 3rd ed. (Cambridge: Published for the Biometrika Trustees at the University Press, 1966). By permission of the Biometrika Trustees.

Table H. J. Sandler, A test of the significance of the difference between the means of correlated measures, based on a simplification of Student's *t*. *British Journal of Psychology*, 1955, *46*, 225–226. By permission of the author and the British Psychological Society.

Table I. ⸝Table I is taken from Table IV of Fisher & Yates: *Statistical Tables for Biological, Agricultural, and Medical Research*, published by Longman Group Ltd. London (previously published by Oliver & Boyd Ltd. Edinburgh) and by permission of the authors and publishers.

Table J. Adapted from Table 2 of F. Wilcoxon and R. A. Wilcox, *Some Rapid Approximate Statistical Procedures* (Pearl River, N.Y.: Lederle Laboratories, 1964.) Reproduced with the permission of the American Cyanamid Company.

Table K. E. F. Lindquist, *A First Course in Statistics* (rev. ed.), (Boston: Houghton Mifflin). By permission of the publisher.

Table A   Random Numbers

51645	98661	88422	26428	45198	97112	45964	90089	33098	97772
40550	53760	85879	20708	45861	56334	25416	50149	98693	55835
74535	68977	91426	42636	37570	46742	24752	95509	91849	57785
95421	88518	29311	11910	52808	78534	33741	42239	57559	61282
82073	36081	48877	06766	41307	17197	61531	53971	59280	42311
63183	66223	19937	66014	60160	47714	99520	59522	94782	40116
26780	26262	06010	86755	02820	27335	73665	49199	06690	99820
56432	62150	56003	61085	78492	46768	90508	71745	89590	42715
24380	65621	11298	07821	32725	54464	15733	72871	40300	23936
84752	94790	55405	21262	26036	09293	27952	69253	02076	08067
25691	31722	21715	01059	30389	64211	85558	24496	71369	97464
75959	66324	72381	43577	92298	10094	67799	49994	31886	66773
22564	64528	96588	75722	41460	88123	53507	93663	56136	99007
92388	60971	68037	35312	33843	23649	83106	87798	70699	01294
83497	98701	96442	60874	82103	41350	52770	50953	57656	94747
74469	91798	65091	22432	88248	71914	40804	43110	46960	94124
77841	37239	28922	77915	50378	54787	17897	04064	97463	68605
39642	31647	82528	75480	01718	39615	83303	27843	90364	38249
08456	44896	96370	01327	66730	74615	51954	21716	45985	15824
72338	98537	51112	71728	83864	72256	90992	93280	21458	62702
47770	38347	32186	08650	45220	36517	95131	83329	59349	76898
80836	30521	56718	81620	41500	52293	41362	77342	03121	11008
18352	95404	86417	06379	62319	21663	13592	56974	87497	53620
26778	64000	90764	38105	76210	30287	87823	65748	21821	61610
13716	62729	33795	87160	98516	76882	49223	25004	88223	64613
32985	26814	51833	57363	04067	84648	85505	41465	71769	99550
86675	38742	39639	67704	03454	22996	92696	35795	77185	45750
60438	60026	72498	71331	01121	72275	71667	38987	72864	02379
75188	55511	94496	01498	15935	45228	62286	91563	35899	92588
74107	65389	41614	16567	03801	58808	34609	37869	62248	12103
55904	45492	93459	97948	26999	63484	22844	26405	51929	89432
51722	49793	75248	49617	29035	04950	98602	47715	85247	73738
22701	43144	30281	94746	55393	51025	39757	51729	68780	71962
10119	94807	23987	99253	02173	63107	75148	10911	40047	94712
12717	94106	62367	71530	94688	68341	00996	64154	21125	04256
28516	45108	31054	51464	12364	19560	27992	23960	59285	01302
49961	51838	48316	20781	28576	18868	61128	74465	50206	60516
74018	61495	56744	30314	01001	41509	23222	29999	35083	13110
48999	29283	83865	22337	07139	95196	82131	13811	40242	20798
84394	31145	33299	87324	85165	94951	10944	93668	11797	18025

table of Z scores

Table B Percentage of Total Area under the Normal Curve between Mean Ordinate and Ordinate at Any Given Sigma Distance from the Mean

$\frac{x}{\sigma}$	.00	.01	.02	.03	.04	.05	.06	.07	.08	.09
0.0	00.00	00.40	00.80	01.20	01.60	01.99	02.39	02.79	03.19	03.59
0.1	03.98	04.38	04.78	05.17	05.57	05.96	06.36	06.75	07.14	07.53
0.2	07.93	08.32	08.71	09.10	09.48	09.87	10.26	10.64	11.03	11.41
0.3	11.79	12.17	12.55	12.93	13.31	13.68	14.06	14.43	14.80	15.17
0.4	15.54	15.91	16.28	16.64	17.00	17.36	17.72	18.08	18.44	18.79
0.5	19.15	19.50	19.85	20.19	20.54	20.88	21.23	21.57	21.90	22.24
0.6	22.57	22.91	23.24	23.57	23.89	24.22	24.54	24.86	25.17	25.49
0.7	25.80	26.11	26.42	26.73	27.04	27.34	27.64	27.94	28.23	28.52
0.8	28.81	29.10	29.39	29.67	29.95	30.23	30.51	30.78	31.06	31.33
0.9	31.59	31.86	32.12	32.38	32.64	32.90	33.15	33.40	33.65	33.89
1.0	34.13	34.38	34.61	34.85	35.08	35.31	35.54	35.77	35.99	36.21
1.1	36.43	36.65	36.86	37.08	37.29	37.49	37.70	37.90	38.10	38.30
1.2	38.49	38.69	38.88	39.07	39.25	39.44	39.62	39.80	39.97	40.15
1.3	40.32	40.49	40.66	40.82	40.99	41.15	41.31	41.47	41.62	41.77
1.4	41.92	42.07	42.22	42.36	42.51	42.65	42.79	42.92	43.06	43.19
1.5	43.32	43.45	43.57	43.70	43.83	43.94	44.06	44.18	44.29	44.41
1.6	44.52	44.63	44.74	44.84	44.95	45.05	45.15	45.25	45.35	45.45
1.7	45.54	45.64	45.73	45.82	45.91	45.99	46.08	46.16	46.25	46.33
1.8	46.41	46.49	46.56	46.64	46.71	46.78	46.86	46.93	46.99	47.06
1.9	47.13	47.19	47.26	47.32	47.38	47.44	47.50	47.56	47.61	47.67
2.0	47.72	47.78	47.83	47.88	47.93	47.98	48.03	48.08	48.12	48.17
2.1	48.21	48.26	48.30	48.34	48.38	48.42	48.46	48.50	48.54	48.57
2.2	48.61	48.64	48.68	48.71	48.75	48.78	48.81	48.84	48.87	48.90
2.3	48.93	48.96	48.98	49.01	49.04	49.06	49.09	49.11	49.13	49.16
2.4	49.18	49.20	49.22	49.25	49.27	49.29	49.31	49.32	49.34	49.36
2.5	49.38	49.40	49.41	49.43	49.45	49.46	49.48	49.49	49.51	49.52
2.6	49.53	49.55	49.56	49.57	49.59	49.60	49.61	49.62	49.63	49.64
2.7	49.65	49.66	49.67	49.68	49.69	49.70	49.71	49.72	49.73	49.74
2.8	49.74	49.75	49.76	49.77	49.77	49.78	49.79	49.79	49.80	49.81
2.9	49.81	49.82	49.82	49.83	49.84	49.84	49.85	49.85	49.86	49.86
3.0	49.87									
3.5	49.98									
4.0	49.997									
5.0	49.99997									

(handwritten margin notes: "x", "x/s", "z"; "how many z scores fall between s"; "95% of the cases")

t test for one mean table

Table C Critical Values of *t*

use 1 tail only when you're sure variance can only go in one Direction — ex. Wt training when you know they'll get stronger

df	Level of Significance for One-Tailed Test			
	5%	2.5%	1%	.5%
	Level of Significance for Two-Tailed Test			
	10%	5%	2%	1%
1	6.3138	12.7062	31.8207	63.6574
2	2.9200	4.3027	6.9646	9.9248
3	2.3534	3.1824	4.5407	5.8409
4	2.1318	2.7764	3.7469	4.6041
5	2.0150	2.5706	3.3649	4.0322
6	1.9432	2.4469	3.1427	3.7074
7	1.8946	2.3646	2.9980	3.4995
8	1.8595	2.3060	2.8965	3.3554
9	1.8331	2.2622	2.8214	3.2498
10	1.8125	2.2281	2.7638	3.1693
11	1.7959	2.2010	2.7181	3.1058
12	1.7823	2.1788	2.6810	3.0545
13	1.7709	2.1604	2.6503	3.0123
14	1.7613	2.1448	2.6245	2.9768
15	1.7531	2.1315	2.6025	2.9467
16	1.7459	2.1199	2.5835	2.9208
17	1.7396	2.1098	2.5669	2.8982
18	1.7341	2.1009	2.5524	2.8784
19	1.7291	2.0930	2.5395	2.8609
20	1.7247	2.0860	2.5280	2.8453
21	1.7207	2.0796	2.5177	2.8314
22	1.7171	2.0739	2.5083	2.8188
23	1.7139	2.0687	2.4999	2.8073
24	1.7109	2.0639	2.4922	2.7969
25	1.7081	2.0595	2.4851	2.7874
26	1.7056	2.0555	2.4786	2.7787
27	1.7033	2.0518	2.4727	2.7707
28	1.7011	2.0484	2.4671	2.7633
29	1.6991	2.0452	2.4620	2.7564
30	1.6973	2.0423	2.4573	2.7500
35	1.6869	2.0301	2.4377	2.7238
40	1.6839	2.0211	2.4233	2.7045
45	1.6794	2.0141	2.4121	2.6896
50	1.6759	2.0086	2.4033	2.6778

Critical values of *t, continued*

df	Level of Significance for One-Tailed Test			
	5%	2.5%	1%	.5%
	Level of Significance for Two-Tailed Test			
df	10%	5%	2%	1%
60	1.6706	2.0003	2.3901	2.6603
70	1.6669	1.9944	2.3808	2.6479
80	1.6641	1.9901	2.3739	2.6387
90	1.6620	1.9867	2.3685	2.6316
100	1.6602	1.9840	2.3642	2.6259
110	1.6588	1.9818	2.3607	2.6213
120	1.6577	1.9799	2.3598	2.6174
∞	1.6449	1.9600	2.3263	2.5758

Table D Values of *r* at the 5% and 1% Levels of Significance (Two-Tailed Test)

✳ Degrees of = n - 2

Degrees of Freedom (df)	5%	1%	Degrees of Freedom (df) = h - 2	5%	1%
1	.997	1.000	24	.388	.496
2	.950	.990	25	.381	.487
3	.878	.959	26	.374	.478
4	.811	.917	27	.367	.470
5	.754	.874	28	.361	.463
6	.707	.834	29	.355	.456
7	.666	.798	30	.349	.449
8	.632	.765	35	.325	.418
9	.602	.735	40	.304	.393
10	.576	.708	45	.288	.372
11	.553	.684	50	.273	.354
12	.532	.661	60	.250	.325
13	.514	.641	70	.232	.302
14	.497	.623	80	.217	.283
15	.482	.606	90	.205	.267
16	.468	.590	100	.195	.254
17	.456	.575	125	.174	.228
18	.444	.561	150	.159	.208
19	.433	.549	200	.138	.181
20	.423	.537	300	.113	.148
21	.413	.526	400	.098	.128
22	.404	.515	500	.088	.115
23	.396	.505	1000	.062	.081

as df are fewer, required r goes up

So ... go to next hardest r to get

ei: - n of 33, go to 30

- 124 go to 100

✸ Product moment Correlation table

interval/ratio Data

Table E   Values of r_S (Rank-Order Correlation Coefficient) at the 5% and 1% Levels of Significance (Two-Tailed Test)

N	5%	1%
5	1.000	—
6	.886	1.000
7	.786	.929
8	.738	.881
9	.683	.833
10	.648	.794
12	.591	.777
14	.544	.714
16	.506	.665
18	.475	.625
20	.450	.591
22	.428	.562
24	.409	.537
26	.392	.515
28	.377	.496
30	.364	.478

- greater the N the greater chance you have of getting a significant result.

- ordinal Data (ranked) where you don't know distance between ranks   table

Analysis of Variance

Table F   Values of F at the 5% and 1% Significance Levels

(df Associated with the Denominator)		(df Associated with the Numerator)								
		1	2	3	4	5	6	7	8	9
1	5%	161	200	216	225	230	234	237	239	241
	1%	4052	5000	5403	5625	5764	5859	5928	5982	6022
2	5%	18.5	19.0	19.2	19.2	19.3	19.3	19.4	19.4	19.4
	1%	98.5	99.0	99.2	99.2	99.3	99.3	99.4	99.4	99.4
3	5%	10.1	9.55	9.28	9.12	9.01	8.94	8.89	8.85	8.81
	1%	34.1	30.8	29.5	28.7	28.2	27.9	27.7	27.5	27.3
4	5%	7.71	6.94	6.59	6.39	6.26	6.16	6.09	6.04	6.00
	1%	21.2	18.0	16.7	16.0	15.5	15.2	15.0	14.8	14.7
5	5%	6.61	5.79	5.41	5.19	5.05	4.95	4.88	4.82	4.77
	1%	16.3	13.3	12.1	11.4	11.0	10.7	10.5	10.3	10.2
6	5%	5.99	5.14	4.76	4.53	4.39	4.28	4.21	4.15	4.10
	1%	13.7	10.9	9.78	9.15	8.75	8.47	8.26	8.10	7.98
7	5%	5.59	4.74	4.35	4.12	3.97	3.87	3.79	3.73	3.68
	1%	12.2	9.55	8.45	7.85	7.46	7.19	6.99	6.84	6.72
8	5%	5.32	4.46	4.07	3.84	3.69	3.58	3.50	3.44	3.39
	1%	11.3	8.65	7.59	7.01	6.63	6.37	6.18	6.03	5.91
9	5%	5.12	4.26	3.86	3.63	3.48	3.37	3.29	3.23	3.18
	1%	10.6	8.02	6.99	6.42	6.06	5.80	5.61	5.47	5.35
10	5%	4.96	4.10	3.71	3.48	3.33	3.22	3.14	3.07	3.02
	1%	10.0	7.56	6.55	5.99	5.64	5.39	5.20	5.06	4.94
11	5%	4.84	3.98	3.59	3.36	3.20	3.09	3.01	2.95	2.90
	1%	9.65	7.21	6.22	5.67	5.32	5.07	4.89	4.74	4.63
12	5%	4.75	3.89	3.49	3.26	3.11	3.00	2.91	2.85	2.80
	1%	9.33	6.93	5.95	5.41	5.06	4.82	4.64	4.50	4.39
13	5%	4.67	3.81	3.41	3.18	3.03	2.92	2.83	2.77	2.71
	1%	9.07	6.70	5.74	5.21	4.86	4.62	4.44	4.30	4.19
14	5%	4.60	3.74	3.34	3.11	2.96	2.85	2.76	2.70	2.65
	1%	8.86	6.51	5.56	5.04	4.70	4.46	4.28	4.14	4.03
15	5%	4.54	3.68	3.29	3.06	2.90	2.79	2.71	2.64	2.59
	1%	8.68	6.36	5.42	4.89	4.56	4.32	4.14	4.00	3.89
16	5%	4.49	3.63	3.24	3.01	2.85	2.74	2.66	2.59	2.54
	1%	8.53	6.23	5.29	4.77	4.44	4.20	4.03	3.89	3.78
17	5%	4.45	3.59	3.20	2.96	2.81	2.70	2.61	2.55	2.49
	1%	8.40	6.11	5.18	4.67	4.34	4.10	3.93	3.79	3.68
18	5%	4.41	3.55	3.16	2.93	2.77	2.66	2.58	2.51	2.46
	1%	8.29	6.01	5.09	4.58	4.25	4.01	3.84	3.71	3.60

Values of F at the 5% and 1% significance levels, *continued*

(df Associated with the Denominator)		1	2	3	4	5	6	7	8	9
		\multicolumn{9}{c}{(df Associated with the Numerator)}								
19	5%	4.38	3.52	3.13	2.90	2.74	2.63	2.54	2.48	2.42
	1%	8.18	5.93	5.01	4.50	4.17	3.94	3.77	3.63	3.52
20	5%	4.35	3.49	3.10	2.87	2.71	2.60	2.51	2.45	2.39
	1%	8.10	5.85	4.94	4.43	4.10	3.87	3.70	3.56	3.46
21	5%	4.32	3.47	3.07	2.84	2.68	2.57	2.49	2.42	2.37
	1%	8.02	5.78	4.87	4.37	4.04	3.81	3.64	3.51	3.40
22	5%	4.30	3.44	3.05	2.82	2.66	2.55	2.46	2.40	2.34
	1%	7.95	5.72	4.82	4.31	3.99	3.76	3.59	3.45	3.35
23	5%	4.28	3.42	3.03	2.80	2.64	2.53	2.44	2.37	2.32
	1%	7.88	5.66	4.76	4.26	3.94	3.71	3.54	3.41	3.30
24	5%	4.26	3.40	3.01	2.78	2.62	2.51	2.42	2.36	2.30
	1%	7.82	5.61	4.72	4.22	3.90	3.67	3.50	3.36	3.26
25	5%	4.24	3.39	2.99	2.76	2.60	2.49	2.40	2.34	2.28
	1%	7.77	5.57	4.68	4.18	3.86	3.63	3.46	3.32	3.22
26	5%	4.23	3.37	2.98	2.74	2.59	2.47	2.39	2.32	2.27
	1%	7.72	5.53	4.64	4.14	3.82	3.59	3.42	3.29	3.18
27	5%	4.21	3.35	2.96	2.73	2.57	2.46	2.37	2.31	2.25
	1%	7.68	5.49	4.60	4.11	3.78	3.56	3.39	3.26	3.15
28	5%	4.20	3.34	2.95	2.71	2.56	2.45	2.36	2.29	2.24
	1%	7.64	5.45	4.57	4.07	3.75	3.53	3.36	3.23	3.12
29	5%	4.18	3.33	2.93	2.70	2.55	2.43	2.35	2.28	2.22
	1%	7.60	5.42	4.54	4.04	3.73	3.50	3.33	3.20	3.09
30	5%	4.17	3.32	2.92	2.69	2.53	2.42	2.33	2.27	2.21
	1%	7.56	5.39	4.51	4.02	3.70	3.47	3.30	3.17	3.07
40	5%	4.08	3.23	2.84	2.61	2.45	2.34	2.25	2.18	2.12
	1%	7.31	5.18	4.31	3.83	3.51	3.29	3.12	2.99	2.89
60	5%	4.00	3.15	2.76	2.53	2.37	2.25	2.17	2.10	2.04
	1%	7.08	4.98	4.13	3.65	3.34	3.12	2.95	2.82	2.72
120	5%	3.92	3.07	2.68	2.45	2.29	2.18	2.09	2.02	1.96
	1%	6.85	4.79	3.95	3.48	3.17	2.96	2.79	2.66	2.56

Table G   Values of q_α (The Studentized Range Statistic) at the 5% and 1% Levels of Significance

df_{wg}	α	2	3	4	5	6	7	8	9	10
					k = Number of Means					
1	.05	17.97	26.98	32.82	37.08	40.41	43.12	45.40	47.36	49.07
	.01	90.03	135.00	164.30	185.60	202.20	215.80	227.20	237.00	245.60
2	.05	6.08	8.33	9.80	10.88	11.74	12.44	13.03	13.54	13.99
	.01	14.04	19.02	22.29	24.72	26.63	28.20	29.53	30.68	31.69
3	.05	4.50	5.91	6.82	7.50	8.04	8.48	8.85	9.18	9.46
	.01	8.26	10.62	12.17	13.33	14.24	15.00	15.64	16.20	16.69
4	.05	3.93	5.04	5.76	6.29	6.71	7.05	7.35	7.60	7.83
	.01	6.51	8.12	9.17	9.96	10.58	11.10	11.55	11.93	12.27
5	.05	3.64	4.60	5.22	5.67	6.03	6.33	6.58	6.80	6.99
	.01	5.70	6.98	7.80	8.42	8.91	9.32	9.67	9.97	10.24
6	.05	3.46	4.34	4.90	5.30	5.63	5.90	6.12	6.32	6.49
	.01	5.24	6.33	7.03	7.56	7.97	8.32	8.61	8.87	9.10
7	.05	3.34	4.16	4.68	5.06	5.36	5.61	5.82	6.00	6.16
	.01	4.95	5.92	6.54	7.01	7.37	7.68	7.94	8.17	8.37
8	.05	3.26	4.04	4.53	4.89	5.17	5.40	5.60	5.77	5.92
	.01	4.75	5.64	6.20	6.62	6.96	7.24	7.47	7.68	7.86
9	.05	3.20	3.95	4.41	4.76	5.02	5.24	5.43	5.59	5.74
	.01	4.60	5.43	5.96	6.35	6.66	6.91	7.13	7.33	7.49
10	.05	3.15	3.88	4.33	4.65	4.91	5.12	5.30	5.46	5.60
	.01	4.48	5.27	5.77	6.14	6.43	6.67	6.87	7.05	7.21
11	.05	3.11	3.82	4.26	4.57	4.82	5.03	5.20	5.35	5.49
	.01	4.39	5.15	5.62	5.97	6.25	6.48	6.67	6.84	6.99
12	.05	3.08	3.77	4.20	4.51	4.75	4.95	5.12	5.27	5.39
	.01	4.32	5.05	5.50	5.84	6.10	6.32	6.51	6.67	6.81
13	.05	3.06	3.73	4.15	4.45	4.69	4.88	5.05	5.19	5.32
	.01	4.26	4.96	5.40	5.73	5.98	6.19	6.37	6.53	6.67
14	.05	3.03	3.70	4.11	4.41	4.64	4.83	4.99	5.13	5.25
	.01	4.21	4.89	5.32	5.63	5.88	6.08	6.26	6.41	6.54
15	.05	3.01	3.67	4.08	4.37	4.59	4.78	4.94	5.08	5.20
	.01	4.17	4.84	5.25	5.56	5.80	5.99	6.16	6.31	6.44
16	.05	3.00	3.65	4.05	4.33	4.56	4.74	4.90	5.03	5.15
	.01	4.13	4.79	5.19	5.49	5.72	5.92	6.08	6.22	6.35
17	.05	2.98	3.63	4.02	4.30	4.52	4.70	4.86	4.99	5.11
	.01	4.10	4.74	5.14	5.43	5.66	5.85	6.01	6.15	6.27

Values of q_α (the Studentized range statistic) at the 5% and 1% levels of significance, *continued.*

df_{wg}	α	2	3	4	5	6	7	8	9	10
					k = Number of Means					
18	.05	2.97	3.61	4.00	4.28	4.49	4.67	4.82	4.96	5.07
	.01	4.07	4.70	5.09	5.38	5.60	5.79	5.94	6.08	6.20
19	.05	2.96	3.59	3.98	4.25	4.47	4.65	4.79	4.92	5.04
	.01	4.05	4.67	5.05	5.33	5.55	5.73	5.89	6.02	6.14
20	.05	2.95	3.58	3.96	4.23	4.45	4.62	4.77	4.90	5.01
	.01	4.02	4.64	5.02	5.29	5.51	5.69	5.84	5.97	6.09
24	.05	2.92	3.53	3.90	4.17	4.37	4.54	4.68	4.81	4.92
	.01	3.96	4.55	4.91	5.17	5.37	5.54	5.69	5.81	5.92
30	.05	2.89	3.49	3.85	4.10	4.30	4.46	4.60	4.72	4.82
	.01	3.89	4.45	4.80	5.05	5.24	5.40	5.54	5.65	5.76
40	.05	2.86	3.44	3.79	4.04	4.23	4.39	4.52	4.63	4.73
	.01	3.82	4.37	4.70	4.93	5.11	5.26	5.39	5.50	5.60
60	.05	2.83	3.40	3.74	3.98	4.16	4.31	4.44	4.55	4.65
	.01	3.76	4.28	4.59	4.82	4.99	5.13	5.25	5.36	5.45
120	.05	2.80	3.36	3.68	3.92	4.10	4.24	4.36	4.47	4.56
	.01	3.70	4.20	4.50	4.71	4.87	5.01	5.12	5.21	5.30
∞	.05	2.77	3.31	3.63	3.86	4.03	4.17	4.29	4.39	4.47
	.01	3.64	4.12	4.40	4.60	4.76	4.88	4.99	5.08	5.16

Table H   Critical Values of Sandler's *A* Statistic

A is significant at a given level if it is equal to or *less than* the value listed in the table.

	Level of Significance for One-Tailed Test				
	5%	2.5%	1%	.5%	.05%
	Level of Significance for Two-Tailed Test				
df = *N* - 1*	10%	5%	2%	1%	.1%
1	.5125	.5031	.50049	.50012	.5000012
2	.412	.369	.347	.340	.334
3	.385	.324	.286	.272	.254
4	.376	.304	.257	.238	.211
5	.372	.293	.240	.218	.184
6	.370	.286	.230	.205	.167
7	.369	.281	.222	.196	.155
8	.368	.278	.217	.190	.146
9	.368	.276	.213	.185	.139
10	.368	.274	.210	.181	.134
11	.368	.273	.207	.178	.130
12	.368	.271	.205	.176	.126
13	.368	.270	.204	.174	.124
14	.368	.270	.202	.172	.121
15	.368	.269	.201	.170	.119
16	.368	.268	.200	.169	.117
17	.368	.268	.199	.168	.116
18	.368	.267	.198	.167	.114
19	.368	.267	.197	.166	.113
20	.368	.266	.197	.165	.112
21	.368	.266	.196	.165	.111
22	.368	.266	.196	.164	.110
23	.368	.266	.195	.163	.109
24	.368	.265	.195	.163	.108
25	.368	.265	.194	.162	.108
26	.368	.265	.194	.162	.107
27	.368	.265	.193	.161	.107
28	.368	.265	.193	.161	.106
29	.368	.264	.193	.161	.106
30	.368	.264	.193	.160	.105
40	.368	.263	.191	.158	.102
60	.369	.262	.189	.155	.099
120	.369	.261	.187	.153	.095
∞	.370	.260	.185	.151	.092

*N = number of pairs.

Table I Values of Chi Square (χ^2) at the 5% and 1% Levels of Significance

Degrees of Freedom (df)	5%	1%
1	3.84	6.64
2	5.99	9.21
3	7.82	11.34
4	9.49	13.28
5	11.07	15.09
6	12.59	16.81
7	14.07	18.48
8	15.51	20.09
9	16.92	21.67
10	18.31	23.21
11	19.68	24.72
12	21.03	26.22
13	22.36	27.69
14	23.68	29.14
15	25.00	30.58
16	26.30	32.00
17	27.59	33.41
18	28.87	34.80
19	30.14	36.19
20	31.41	37.57
21	32.67	38.93
22	33.92	40.29
23	35.17	41.64
24	36.42	42.98
25	37.65	44.31
26	38.88	45.64
27	40.11	46.96
28	41.34	48.28
29	42.56	49.59
30	43.77	50.89

Table J Values of T at the 5% and 1% Levels of Significance in the Wilcoxon Signed-Ranks Test

N_{S-R}	5%	1%
6	1	—
7	2	—
8	4	0
9	6	2
10	8	3
11	11	5
12	14	7
13	17	10
14	21	13
15	25	16
16	30	19
17	35	23
18	40	28
19	46	32
20	52	37
21	59	43
22	66	49
23	73	55
24	81	61
25	90	68

Table K   Table of Squares and Square Roots of the Numbers from 1 to 1000

Number	Square	Square Root	Number	Square	Square Root
1	1	1.000	41	16 81	6.403
2	4	1.414	42	17 64	6.481
3	9	1.732	43	18 49	6.557
4	16	2.000	44	19 36	6.633
5	25	2.236	45	20 25	6.708
6	36	2.449	46	21 16	6.782
7	49	2.646	47	22 09	6.856
8	64	2.828	48	23 04	6.928
9	81	3.000	49	24 01	7.000
10	1 00	3.162	50	25 00	7.071
11	1 21	3.317	51	26 01	7.141
12	1 44	3.464	52	27 04	7.211
13	1 69	3.606	53	28 09	7.280
14	1 96	3.742	54	29 16	7.348
15	2 25	3.873	55	30 25	7.416
16	2 56	4.000	56	31 36	7.483
17	2 89	4.123	57	32 49	7.550
18	3 24	4.243	58	33 64	7.616
19	3 61	4.359	59	34 81	7.681
20	4 00	4.472	60	36 00	7.746
21	4 41	4.583	61	37 21	7.810
22	4 84	4.690	62	38 44	7.874
23	5 29	4.796	63	39 69	7.937
24	5 76	4.899	64	40 96	8.000
25	6 25	5.000	65	42 25	8.062
26	6 76	5.099	66	43 56	8.124
27	7 29	5.196	67	44 89	8.185
28	7 84	5.292	68	46 24	8.246
29	8 41	5.385	69	47 61	8.307
30	9 00	5.477	70	49 00	8.367
31	9 61	5.568	71	50 41	8.426
32	10 24	5.657	72	51 84	8.485
33	10 89	5.745	73	53 29	8.544
34	11 56	5.831	74	54 76	8.602
35	12 25	5.916	75	56 25	8.660
36	12 96	6.000	76	57 76	8.718
37	13 69	6.083	77	59 29	8.775
38	14 44	6.164	78	60 84	8.832
39	15 21	6.245	79	62 41	8.888
40	16 00	6.325	80	64 00	8.944

Table of squares and square roots, *continued*

Number	Square	Square Root	Number	Square	Square Root
81	65 61	9.000	121	1 46 41	11.000
82	67 24	9.055	122	1 48 84	11.045
83	68 89	9.110	123	1 51 29	11.091
84	70 56	9.165	124	1 53 76	11.136
85	72 25	9.220	125	1 56 25	11.180
86	73 96	9.274	126	1 58 76	11.225
87	75 69	9.327	127	1 61 29	11.269
88	77 44	9.381	128	1 63 84	11.314
89	79 21	9.434	129	1 66 41	11.358
90	81 00	9.487	130	1 69 00	11.402
91	82 81	9.539	131	1 71 61	11.446
92	84 64	9.592	132	1 74 24	11.489
93	86 49	9.644	133	1 76 89	11.533
94	88 36	9.695	134	1 79 56	11.576
95	90 25	9.747	135	1 82 25	11.619
96	92 16	9.798	136	1 84 96	11.662
97	94 09	9.849	137	1 87 69	11.705
98	96 04	9.899	138	1 90 44	11.747
99	98 01	9.950	139	1 93 21	11.790
100	1 00 00	10.000	140	1 96 00	11.832
101	1 02 01	10.050	141	1 98 81	11.874
102	1 04 04	10.100	142	2 01 64	11.916
103	1 06 09	10.149	143	2 04 49	11.958
104	1 08 16	10.198	144	2 07 36	12.000
105	1 10 25	10.247	145	2 10 25	12.042
106	1 12 36	10.296	146	2 13 16	12.083
107	1 14 49	10.344	147	2 16 09	12.124
108	1 16 64	10.392	148	2 19 04	12.166
109	1 18 81	10.440	149	2 22 01	12.207
110	1 21 00	10.488	150	2 25 00	12.247
111	1 23 21	10.536	151	2 28 01	12.288
112	1 25 44	10.583	152	2 31 04	12.329
113	1 27 69	10.630	153	2 34 09	12.369
114	1 29 96	10.677	154	2 37 16	12.410
115	1 32 25	10.724	155	2 40 25	12.450
116	1 34 56	10.770	156	2 43 36	12.490
117	1 36 89	10.817	157	2 46 49	12.530
118	1 39 24	10.863	158	2 49 64	12.570
119	1 41 61	10.909	159	2 52 81	12.610
120	1 44 00	10.954	160	2 56 00	12.649

Table of squares and square roots, *continued*

Number	Square	Square Root	Number	Square	Square Root
161	2 59 21	12.689	201	4 04 01	14.177
162	2 62 44	12.728	202	4 08 04	14.213
163	2 65 69	12.767	203	4 12 09	14.248
164	2 68 96	12.806	204	4 16 16	14.283
165	2 72 25	12.845	205	4 20 25	14.318
166	2 75 56	12.884	206	4 24 36	14.353
167	2 78 89	12.923	207	4 28 49	14.387
168	2 82 24	12.961	208	4 32 64	14.422
169	2 85 61	13.000	209	4 36 81	14.457
170	2 89 00	13.038	210	4 41 00	14.491
171	2 92 41	13.077	211	4 45 21	14.526
172	2 95 84	13.115	212	4 49 44	14.560
173	2 99 29	13.153	213	4 53 69	14.595
174	3 02 76	13.191	214	4 57 96	14.629
175	3 06 25	13.229	215	4 62 25	14.663
176	3 09 76	13.266	216	4 66 56	14.697
177	3 13 29	13.304	217	4 70 89	14.731
178	3 16 84	13.342	218	4 75 24	14.765
179	3 20 41	13.379	219	4 79 61	14.799
180	3 24 00	13.416	220	4 84 00	14.832
181	3 27 61	13.454	221	4 88 41	14.866
182	3 31 24	13.491	222	4 92 84	14.900
183	3 34 89	13.528	223	4 97 29	14.933
184	3 38 56	13.565	224	5 01 76	14.967
185	3 42 25	13.601	225	5 06 25	15.000
186	3 45 96	13.638	226	5 10 76	15.033
187	3 49 69	13.675	227	5 15 29	15.067
188	3 53 44	13.711	228	5 19 84	15.100
189	3 57 21	13.748	229	5 24 41	15.133
190	3 61 00	13.784	230	5 29 00	15.166
191	3 64 81	13.820	231	5 33 61	15.199
192	3 68 64	13.856	232	5 38 24	15.232
193	3 72 49	13.892	233	5 42 89	15.264
194	3 76 36	13.928	234	5 47 56	15.297
195	3 80 25	13.964	235	5 52 25	15.330
196	3 84 16	14.000	236	5 56 96	15.362
197	3 88 09	14.036	237	5 61 69	15.395
198	3 92 04	14.071	238	5 66 44	15.427
199	3 96 01	14.107	239	5 71 21	15.460
200	4 00 00	14.142	240	5 76 00	15.492

Table of squares and square roots, *continued*

Number	Square	Square Root	Number	Square	Square Root
241	5 80 81	15.524	281	7 89 61	16.763
242	5 85 64	15.556	282	7 95 24	16.793
243	5 90 49	15.588	283	8 00 89	16.823
244	5 95 36	15.620	284	8 06 56	16.852
245	6 00 25	15.652	285	8 12 25	16.882
246	6 05 16	15.684	286	8 17 96	16.912
247	6 10 09	15.716	287	8 23 69	16.941
248	6 15 04	15.748	288	8 29 44	16.971
249	6 20 01	15.780	289	8 35 21	17.000
250	6 25 00	15.811	290	8 41 00	17.029
251	6 30 01	15.843	291	8 46 81	17.059
252	6 35 04	15.875	292	8 52 64	17.088
253	6 40 09	15.906	293	8 58 49	17.117
254	6 45 16	15.937	294	8 64 36	17.146
255	6 50 25	15.969	295	8 70 25	17.176
256	6 55 36	16.000	296	8 76 16	17.205
257	6 60 49	16.031	297	8 82 09	17.234
258	6 65 64	16.062	298	8 88 04	17.263
259	6 70 81	16.093	299	8 94 01	17.292
260	6 76 00	16.125	300	9 00 00	17.321
261	6 81 21	16.155	301	9 06 01	17.349
262	6 86 44	16.186	302	9 12 04	17.378
263	6 91 69	16.217	303	9 18 09	17.407
264	6 96 96	16.248	304	9 24 16	17.436
265	7 02 25	16.279	305	9 30 25	17.464
266	7 07 56	16.310	306	9 36 36	17.493
267	7 12 89	16.340	307	9 42 49	17.521
268	7 18 24	16.371	308	9 48 64	17.550
269	7 23 61	16.401	309	9 54 81	17.578
270	7 29 00	16.432	310	9 61 00	17.607
271	7 34 41	16.462	311	9 67 21	17.635
272	7 39 84	16.492	312	9 73 44	17.664
273	7 45 29	16.523	313	9 79 69	17.692
274	7 50 76	16.553	314	9 85 96	17.720
275	7 56 25	16.583	315	9 92 25	17.748
276	7 61 76	16.613	316	9 98 56	17.776
277	7 67 29	16.643	317	10 04 89	17.804
278	7 72 84	16.673	318	10 11 24	17.833
279	7 78 41	16.703	319	10 17 61	17.861
280	7 84 00	16.733	320	10 24 00	17.889

Table of squares and square roots, *continued*

Number	Square	Square Root	Number	Square	Square Root
321	10 30 41	17.916	361	13 03 21	19.000
322	10 36 84	17.944	362	13 10 44	19.026
323	10 43 29	17.972	363	13 17 69	19.053
324	10 49 76	18.000	364	13 24 96	19.079
325	10 56 25	18.028	365	13 32 25	19.105
326	10 62 76	18.055	366	13 39 56	19.131
327	10 69 29	18.083	367	13 46 89	19.157
328	10 75 84	18.111	368	13 54 24	19.183
329	10 82 41	18.138	369	13 61 61	19.209
330	10 89 00	18.166	370	13 69 00	19.235
331	10 95 61	18.193	371	13 76 41	19.261
332	11 02 24	18.221	372	13 83 84	19.287
333	11 08 89	18.248	373	13 91 29	19.313
334	11 15 56	18.276	374	13 98 76	19.339
335	11 22 25	18.303	375	14 06 25	19.363
336	11 28 96	18.330	376	14 13 76	19.391
337	11 35 69	18.358	377	14 21 29	19.416
338	11 42 44	18.385	378	14 28 84	19.442
339	11 49 21	18.412	379	14 36 41	19.468
340	11 56 00	18.439	380	14 44 00	19.494
341	11 62 81	18.466	381	14 51 61	19.519
342	11 69 64	18.493	382	14 59 24	19.545
343	11 76 49	18.520	383	14 66 89	19.570
344	11 83 36	18.547	384	14 74 56	19.596
345	11 90 25	18.574	385	14 82 25	19.621
346	11 97 16	18.601	386	14 89 96	19.647
347	12 04 09	18.628	387	14 97 69	19.672
348	12 11 04	18.655	388	15 05 44	19.698
349	12 18 01	18.682	389	15 13 21	19.723
350	12 25 00	18.708	390	15 21 00	19.748
351	12 32 01	18.735	391	15 28 81	19.774
352	12 39 04	18.762	392	15 36 64	19.799
353	12 46 09	18.788	393	15 44 49	19.824
354	12 53 16	18.815	394	15 52 36	19.849
355	12 60 25	18.841	395	15 60 25	19.875
356	12 67 36	18.868	396	15 68 16	19.900
357	12 74 49	18.894	397	15 76 09	19.925
358	12 81 64	18.921	398	15 84 04	19.950
359	12 88 81	18.947	399	15 92 01	19.975
360	12 96 00	18.974	400	16 00 00	20.000

Table of squares and square roots, *continued*

Number	Square	Square Root	Number	Square	Square Root
401	16 08 01	20.025	441	19 44 81	21.000
402	16 16 04	20.050	442	19 53 64	21.024
403	16 24 09	20.075	443	19 62 49	21.048
404	16 32 16	20.100	444	19 71 36	21.071
405	16 40 25	20.125	445	19 80 25	21.095
406	16 48 36	20.149	446	19 89 16	21.119
407	16 56 49	20.174	447	19 98 09	21.142
408	16 64 64	20.199	448	20 07 04	21.166
409	16 72 81	20.224	449	20 16 01	21.190
410	16 81 00	20.248	450	20 25 00	21.213
411	16 89 21	20.273	451	20 34 01	21.237
412	16 97 44	20.298	452	20 43 04	21.260
413	17 05 69	20.322	453	20 52 09	21.284
414	17 13 96	20.347	454	20 61 16	21.307
415	17 22 25	20.372	455	20 70 25	21.331
416	17 30 56	20.396	456	20 79 36	21.354
417	17 38 89	20.421	457	20 88 49	21.378
418	17 47 24	20.445	458	20 97 64	21.401
419	17 55 61	20.469	459	21 06 81	21.424
420	17 64 00	20.494	460	21 16 00	21.448
421	17 72 41	20.518	461	21 25 21	21.471
422	17 80 84	20.543	462	21 34 44	21.494
423	17 89 29	20.567	463	21 43 69	21.517
424	17 97 76	20.591	464	21 52 96	21.541
425	18 06 25	20.616	465	21 62 25	21.564
426	18 14 76	20.640	466	21 71 56	21.587
427	18 23 29	20.664	467	21 80 89	21.610
428	18 31 84	20.688	468	21 90 24	21.633
429	18 40 41	20.712	469	21 99 61	21.656
430	18 49 00	20.736	470	22 09 00	21.679
431	18 57 61	20.761	471	22 18 41	21.703
432	18 66 24	20.785	472	22 27 84	21.726
433	18 74 89	20.809	473	22 37 29	21.749
434	18 83 56	20.833	474	22 46 76	21.772
435	18 92 25	20.857	475	22 56 25	21.794
436	19 00 96	20.881	476	22 65 76	21.817
437	19 09 69	20.905	477	22 75 29	21.840
438	19 18 44	20.928	478	22 84 84	21.863
439	19 27 21	20.952	479	22 94 41	21.886
440	19 36 00	20.976	480	23 04 00	21.909

Table of squares and square roots, *continued*

Number	Square	Square Root	Number	Square	Square Root
481	23 13 61	21.932	521	27 14 41	22.825
482	23 23 24	21.954	522	27 24 84	22.847
483	23 32 89	21.977	523	27 35 29	22.869
484	23 42 56	22.000	524	27 45 76	22.891
485	23 52 25	22.023	525	27 56 25	22.913
486	23 61 96	22.045	526	27 66 76	22.935
487	23 71 69	22.068	527	27 77 29	22.956
488	23 81 44	22.091	528	27 87 84	22.978
489	23 91 21	22.113	529	27 98 41	23.000
490	24 01 00	22.136	530	28 09 00	23.022
491	24 10 81	22.159	531	28 19 61	23.043
492	24 20 64	22.181	532	28 30 24	23.065
493	24 30 49	22.204	533	28 40 89	23.087
494	24 40 36	22.226	534	28 51 56	23.108
495	24 50 25	22.249	535	28 62 25	23.130
496	24 60 16	22.271	536	28 72 96	23.152
497	24 70 09	22.293	537	28 83 69	23.173
498	24 80 04	22.316	538	28 94 44	23.195
499	24 90 01	22.338	539	29 05 21	23.216
500	25 00 00	22.361	540	29 16 00	23.238
501	25 10 01	22.383	541	29 26 81	23.259
502	25 20 04	22.405	542	29 37 64	23.281
503	25 30 09	22.428	543	29 48 49	23.302
504	25 40 16	22.450	544	29 59 36	23.324
505	25 50 25	22.472	545	29 70 25	23.345
506	25 60 36	22.494	546	29 81 16	23.367
507	25 70 49	22.517	547	29 92 09	23.388
508	25 80 64	22.539	548	30 03 04	23.409
509	25 90 81	22.561	549	30 14 01	23.431
510	26 01 00	22.583	550	30 25 00	23.452
511	26 11 21	22.605	551	30 36 01	23.473
512	26 21 44	22.627	552	30 47 04	23.495
513	26 31 69	22.650	553	30 58 09	23.516
514	26 41 96	22.672	554	30 69 16	23.537
515	26 52 25	22.694	555	30 80 25	23.558
516	26 62 56	22.716	556	30 91 36	23.580
517	26 72 89	22.738	557	31 02 49	23.601
518	26 83 24	22.760	558	31 13 64	23.622
519	26 93 61	22.782	559	31 24 81	23.643
520	27 04 00	22.804	560	31 36 00	23.664

Table of squares and square roots, *continued*

Number	Square	Square Root	Number	Square	Square Root
561	31 47 21	23.685	601	36 12 01	24.515
562	31 58 44	23.707	602	36 24 04	24.536
563	31 69 69	23.728	603	36 36 09	24.556
564	31 80 96	23.749	604	36 48 16	24.576
565	31 92 25	23.770	605	36 60 25	24.597
566	32 03 56	23.791	606	36 72 36	24.617
567	32 14 89	23.812	607	36 84 49	24.637
568	32 26 24	23.833	608	36 96 64	24.658
569	32 37 61	23.854	609	37 08 81	24.678
570	32 49 00	23.875	610	37 21 00	24.698
571	32 60 41	23.896	611	37 33 21	24.718
572	32 71 84	23.917	612	37 45 44	24.739
573	32 83 29	23.937	613	37 57 69	24.759
574	32 94 76	23.958	614	37 69 96	24.779
575	33 06 25	23.979	615	37 82 25	24.799
576	33 17 76	24.000	616	37 94 56	24.819
577	33 29 29	24.021	617	38 06 89	24.839
578	33 40 84	24.042	618	38 19 24	24.860
579	33 52 41	24.062	619	38 31 61	24.880
580	33 64 00	24.083	620	38 44 00	24.900
581	33 75 61	24.104	621	38 56 41	24.920
582	33 87 24	24.125	622	38 68 84	24.940
583	33 98 89	24.145	623	38 81 29	24.960
584	34 10 56	24.166	624	38 93 76	24.980
585	34 22 25	24.187	625	39 06 25	25.000
586	34 33 96	24.207	626	39 18 76	25.020
587	34 45 69	24.228	627	39 31 29	25.040
588	34 57 44	24.249	628	39 43 84	25.060
589	34 69 21	24.269	629	39 56 41	25.080
590	34 81 00	24.290	630	39 69 00	25.100
591	34 92 81	24.310	631	39 81 61	25.120
592	35 04 64	24.331	632	39 94 24	25.140
593	35 16 49	24.352	633	40 06 89	25.159
594	35 28 36	24.372	634	40 19 56	25.179
595	35 40 25	24.393	635	40 32 25	25.199
596	35 52 16	24.413	636	40 44 96	25.219
597	35 64 09	24.434	637	40 57 69	25.239
598	35 76 04	24.454	638	40 70 44	25.259
599	35 88 01	24.474	639	40 83 21	25.278
600	36 00 00	24.495	640	40 96 00	25.298

Table of squares and square roots, *continued*

Number	Square	Square Root	Number	Square	Square Root
641	41 08 81	25.318	681	46 37 61	26.096
642	41 21 64	25.338	682	46 51 24	26.115
643	41 34 49	25.357	683	46 64 89	26.134
644	41 47 36	25.377	684	46 78 56	26.153
645	41 60 25	25.397	685	46 92 25	26.173
646	41 73 16	25.417	686	47 05 96	26.192
647	41 86 09	25.436	687	47 19 69	26.211
648	41 99 04	25.456	688	47 33 44	26.230
649	42 12 01	25.475	689	47 47 21	26.249
650	42 25 00	25.495	690	47 61 00	26.268
651	42 38 01	25.515	691	47 74 81	26.287
652	42 51 04	25.534	692	47 88 64	26.306
653	42 64 09	25.554	693	48 02 49	26.325
654	42 77 16	25.573	694	48 16 36	26.344
655	42 90 25	25.593	695	48 30 25	26.363
656	43 03 36	25.612	696	48 44 16	26.382
657	43 16 49	25.632	697	48 58 09	26.401
658	43 29 64	25.652	698	48 72 04	26.420
659	43 42 81	25.671	699	48 86 01	26.439
660	43 56 00	25.690	700	49 00 00	26.458
661	43 69 21	25.710	701	49 14 01	26.476
662	43 82 44	25.729	702	49 28 04	26.495
663	43 95 69	25.749	703	49 42 09	26.514
664	44 08 96	25.768	704	49 56 16	26.533
665	44 22 25	25.788	705	49 70 25	26.552
666	44 35 56	25.807	706	49 84 36	26.571
667	44 48 89	25.826	707	49 98 49	26.589
668	44 62 24	25.846	708	50 12 64	26.608
669	44 75 61	25.865	709	50 26 81	26.627
670	44 89 00	25.884	710	50 41 00	26.646
671	45 02 41	25.904	711	50 55 21	26.665
672	45 15 84	25.923	712	50 69 44	26.683
673	45 29 29	25.942	713	50 83 69	26.702
674	45 42 76	25.962	714	50 97 96	26.721
675	45 56 25	25.981	715	51 12 25	26.739
676	45 69 76	26.000	716	51 26 56	26.758
677	45 83 29	26.019	717	51 40 89	26.777
678	45 96 84	26.038	718	51 55 24	26.796
679	46 10 41	26.058	719	51 69 61	26.814
680	46 24 00	26.077	720	51 84 00	26.833

Table of squares and square roots, *continued*

Number	Square	Square Root	Number	Square	Square Root
721	51 98 41	26.851	761	57 91 21	27.586
722	52 12 84	26.870	762	58 06 44	27.604
723	52 27 29	26.889	763	58 21 69	27.622
724	52 41 76	26.907	764	58 36 96	27.641
725	52 56 25	26.926	765	58 52 25	27.659
726	52 70 76	26.944	766	58 67 56	27.677
727	52 85 29	26.963	767	58 82 89	27.695
728	52 99 84	26.981	768	58 98 24	27.713
729	53 14 41	27.000	769	59 13 61	27.731
730	53 29 00	27.019	770	59 29 00	27.749
731	53 43 61	27.037	771	59 44 41	27.767
732	53 58 24	27.055	772	59 59 84	27.785
733	53 72 89	27.074	773	59 75 29	27.803
734	53 87 56	27.092	774	59 90 76	27.821
735	54 02 25	27.111	775	60 06 25	27.839
736	54 16 96	27.129	776	60 21 76	27.857
737	54 31 69	27.148	777	60 37 29	27.875
738	54 46 44	27.166	778	60 52 84	27.893
739	54 61 21	27.185	779	60 68 41	27.911
740	54 76 00	27.203	780	60 84 00	27.928
741	54 90 81	27.221	781	60 99 61	27.946
742	55 05 64	27.240	782	61 15 24	27.964
743	55 20 49	27.258	783	61 30 89	27.982
744	55 35 36	27.276	784	61 46 56	28.000
745	55 50 25	27.295	785	61 62 25	28.018
746	55 65 16	27.313	786	61 77 96	28.036
747	55 80 09	27.331	787	61 93 69	28.054
748	55 95 04	27.350	788	62 09 44	28.071
749	56 10 01	27.368	789	62 25 21	28.089
750	56 25 00	27.386	790	62 41 00	28.107
751	56 40 01	27.404	791	62 56 81	28.125
752	56 55 04	27.423	792	62 72 64	28.142
753	56 70 09	27.441	793	62 88 49	28.160
754	56 85 16	27.459	794	63 04 36	28.178
755	57 00 25	27.477	795	63 20 25	28.196
756	57 15 36	27.495	796	63 36 16	28.213
757	57 30 49	27.514	797	63 52 09	28.231
758	57 45 64	27.532	798	63 68 04	28.249
759	57 60 81	27.550	799	63 84 01	28.267
760	57 76 00	27.568	800	64 00 00	28.284

Table of squares and square roots, *continued*

Number	Square	Square Root	Number	Square	Square Root
801	64 16 01	28.302	841	70 72 81	29.000
802	64 32 04	28.320	842	70 89 64	29.017
803	64 48 09	28.337	843	71 06 49	29.034
804	64 64 16	28.355	844	71 23 36	29.052
805	64 80 25	28.373	845	71 40 25	29.069
806	64 96 36	28.390	846	71 57 16	29.086
807	65 12 49	28.408	847	71 74 09	29.103
808	65 28 64	28.425	848	71 91 04	29.120
809	65 44 81	28.443	849	72 08 01	29.138
810	65 61 00	28.460	850	72 25 00	29.155
811	65 77 21	28.478	851	72 42 01	29.172
812	65 93 44	28.496	852	72 59 04	29.189
813	66 09 69	28.513	853	72 76 09	29.206
814	66 25 96	28.531	854	72 93 16	29.223
815	66 42 25	28.548	855	73 10 25	29.240
816	66 58 56	28.566	856	73 27 36	29.257
817	66 74 89	28.583	857	73 44 49	29.275
818	66 91 24	28.601	858	73 61 64	29.292
819	67 07 61	28.618	859	73 78 81	29.309
820	67 24 00	28.636	860	73 96 00	29.326
821	67 40 41	28.653	861	74 13 21	29.343
822	67 56 84	28.671	862	74 30 44	29.360
823	67 73 29	28.688	863	74 47 69	29.377
824	67 89 76	28.705	864	74 64 96	29.394
825	68 06 25	28.723	865	74 82 25	29.411
826	68 22 76	28.740	866	74 99 56	29.428
827	68 39 29	28.758	867	75 16 89	29.445
828	68 55 84	28.775	868	75 34 24	29.462
829	68 72 41	28.792	869	75 51 61	29.479
830	68 89 00	28.810	870	75 69 00	29.496
831	69 05 61	28.827	871	75 86 41	29.513
832	69 22 24	28.844	872	76 03 84	29.530
833	69 38 89	28.862	873	76 21 29	29.547
834	69 55 56	28.879	874	76 38 76	29.563
835	69 72 25	28.896	875	76 56 25	29.580
836	69 88 96	28.914	876	76 73 76	29.597
837	70 05 69	28.931	877	76 91 29	29.614
838	70 22 44	28.948	878	77 08 84	29.631
839	70 39 21	28.965	879	77 26 41	29.648
840	70 56 00	28.983	880	77 44 00	29.665

Table of squares and square roots, *continued*

Number	Square	Square Root	Number	Square	Square Root
881	77 61 61	29.682	921	84 82 41	30.348
882	77 79 24	29.698	922	85 00 84	30.364
883	77 96 89	29.715	923	85 19 29	30.381
884	78 14 56	29.732	924	85 37 76	30.397
885	78 32 25	29.749	925	85 56 25	30.414
886	78 49 96	29.766	926	85 74 76	30.430
887	78 67 69	29.783	927	85 93 29	30.447
888	78 85 44	29.799	928	86 11 84	30.463
889	79 03 21	29.816	929	86 30 41	30.480
890	79 21 00	29.833	930	86 49 00	30.496
891	79 38 81	29.850	931	86 67 61	30.512
892	79 56 64	29.866	932	86 86 24	30.529
893	79 74 49	29.883	933	87 04 89	30.545
894	79 92 36	29.900	934	87 23 56	30.561
895	80 10 25	29.916	935	87 42 25	30.578
896	80 28 16	29.933	936	87 60 96	30.594
897	80 46 09	29.950	937	87 79 69	30.610
898	80 64 04	29.967	938	87 98 44	30.627
899	80 82 01	29.983	939	88 17 21	30.643
900	81 00 00	30.000	940	88 36 00	30.659
901	81 18 01	30.017	941	88 54 81	30.676
902	81 36 04	30.033	942	88 73 64	30.692
903	81 54 09	30.050	943	88 92 49	30.708
904	81 72 16	30.067	944	89 11 36	30.725
905	81 90 25	30.083	945	89 30 25	30.741
906	82 08 36	30.100	946	89 49 16	30.757
907	82 26 49	30.116	947	89 68 09	30.773
908	82 44 64	30.133	948	89 87 04	30.790
909	82 62 81	30.150	949	90 06 01	30.806
910	82 81 00	30.166	950	90 25 00	30.822
911	82 99 21	30.183	951	90 44 01	30.838
912	83 17 44	30.199	952	90 63 04	30.854
913	83 35 69	30.216	953	90 82 09	30.871
914	83 53 96	30.232	954	91 01 16	30.887
915	83 72 25	30.249	955	91 20 25	30.903
916	83 90 56	30.265	956	91 39 36	30.919
917	84 08 89	30.282	957	91 58 49	30.935
918	84 27 24	30.299	958	91 77 64	30.952
919	84 45 61	30.315	959	91 96 81	30.968
920	84 64 00	30.332	960	92 16 00	30.984

Table of squares and square roots, *continued*

Number	Square	Square Root	Number	Square	Square Root
961	92 35 21	31.000	981	96 23 61	31.321
962	92 54 44	31.016	982	96 43 24	31.337
963	92 73 69	31.032	983	96 62 89	31.353
964	92 92 96	31.048	984	96 82 56	31.369
965	93 12 25	31.064	985	97 02 25	31.385
966	93 31 56	31.081	986	97 21 96	31.401
967	93 50 89	31.097	987	97 41 69	31.417
968	93 70 24	31.113	988	97 61 44	31.432
969	93 89 61	31.129	989	97 81 21	31.448
970	94 09 00	31.145	990	98 01 00	31.464
971	94 28 41	31.161	991	98 20 81	31.480
972	94 47 84	31.177	992	98 40 64	31.496
973	94 67 29	31.193	993	98 60 49	31.512
974	94 86 76	31.209	994	98 80 36	31.528
975	95 06 25	31.225	995	99 00 25	31.544
976	95 25 76	31.241	996	99 20 16	31.559
977	95 45 29	31.257	997	99 40 09	31.575
978	95 64 84	31.273	998	99 60 04	31.591
979	95 84 41	31.289	999	99 80 01	31.607
980	96 04 00	31.305	1000	100 00 00	31.623

Answers to
Chapter Problems

APPENDIX III

Answers are given to all problems except those that call for the student to supply examples.

CHAPTER 1

1. (a) Population of all those who voted. Alternatively, a sample of all individuals who were eligible to vote in the election. (b) Sample of those with listings in the residential directory. (c) Sample from students enrolled in the course (assuming some students were absent). (d) Population.

2. (a) Random. (b) Matched. (c) Randomized.

3. Starting at the sixth entry in column 3 and continuing down the column and down column 4, the 10 students chosen are those assigned the numbers 19, 6, 11, 20, 7, 1, 8, 2, 4, and 3.

4. (a) Rank order. (b) Categorical. (c) Measurement. (d) Measurement.

5. (a) Discrete. (b) Continuous. (c) Discrete. (d) Continuous. (e) Continuous.

CHAPTER 2

1. The grouped frequency distribution shown below has $i = 3$. With $i = 2$, too many class intervals would result to reveal the pattern of the scores adequately. Although $i = 4$ would be acceptable, it would result in midpoints with fractional values rather than whole numbers. For this reason, $i = 3$ is preferable to $i = 4$.

Class Interval	f	Class Interval	f	Class Interval	f	Class Interval	f
63–65	1	51–53	5	39–41	7	27–29	0
60–62	0	48–50	4	36–38	3	24–26	1
57–59	2	45–47	6	33–35	5	21–23	2
54–56	3	42–44	9	30–32	2		N = 50

2. The distribution with $i = 3$ is preferable to the one with $i = 5$. The latter distribution is so compact that the group pattern is beginning to be obscured. Furthermore, comparison with the distribution in problem 1 will be easier if both use the same i.

$i = 3$ Class Interval	f	$i = 3$ Class Interval	f	$i = 5$ Class Interval	f
57–59	1	36–38	6	55–59	2
54–56	1	33–35	10	50–54	1
51–53	0	30–32	7	45–49	5
48–50	2	27–29	4	40–44	6
45–47	4	24–26	4	35–39	11
42–44	2	21–23	3	30–34	13
39–41	5	18–20	1	25–29	7
			N = 50	20–24	5
					N = 50

3. **(a)** R = 65 - 21 = 44. **(b)** Real limits of highest and lowest intervals: 62.5–65.5 and 20.5–23.5. **(c)** Midpoints: 64 and 22.

4. The i's shown below yield a desirable number of intervals for the given N. However, in items (2), (3), and (5), slightly different i's would also be acceptable. In grouped distributions, the number of intervals was obtained by dividing the range (R) by i and rounding upward to a whole number.

	i	Number of Intervals	Midpoint	Highest Class Real Limits	Highest Class Apparent Limits
(1) Not useful to group because of small N					
(2)	5	13	73	70.5–75.5	71–75
(3)	10	19	278.5	273.5–283.5	274–283
(4)	1	10	—	24.5–25.5	—
(5)	.10	12	1.475	1.425–1.525	1.43–1.52
(6)	2	16	33.5	32.5–34.5	33–34

CHAPTER 3

1. **(a)** A figure with the two polygons is shown below.

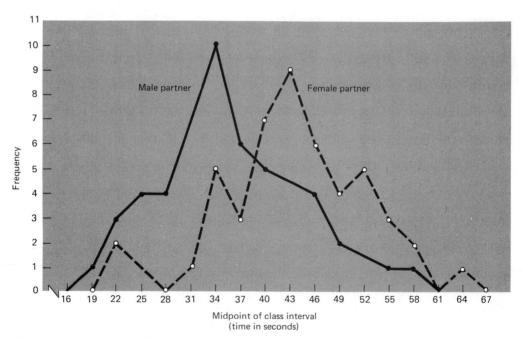

Frequency polygons for male students who believe they are competing with another male or a female on anagram task.

(b) The distribution for the males competing with another male is slightly more skewed than the distribution of males competing with a female. For male partner, the skew is in a positive direction; for female partner, it is in a negative direction. **(c)** Male students competing with another male tend to complete the anagrams in a shorter time than males competing with a female.

2.   **(a)** and **(b)** The cumulative frequencies, cumulative percentages, and the cumulative polygon based on these data are shown below.

Class Interval	f	cum f	cum %
36–40	6	50	100
31–35	9	44	88
26–30	12	35	70
21–25	13	23	46
16–20	6	10	20
11–15	0	4	8
6–10	3	4	8
1–5	1	1	2
	N = 50		

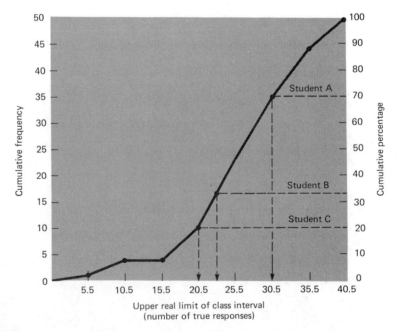

Cumulative polygon for responses to true-false misconceptions test given to introductory psychology students.

(c) Student A: 30.5; student B: 23; student C: 20.5. The procedures used to make these estimates are shown in the cumulative polygon above.

3.   (a) 53. (b) 65. (c) 68. (d) 43. (e) 75, 10.

4.   (a)

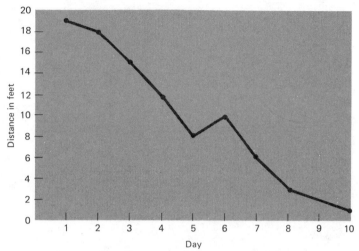

Distance from caged snake at which man stopped on each treatment day.

(b) Over the 10-day treatment period, the man quite regularly came closer and closer to the caged snake, stopping one foot away on the tenth day.

5.

Percentage of students at the university who reside in state, out of state, or outside the United States.

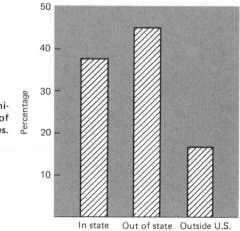

6.

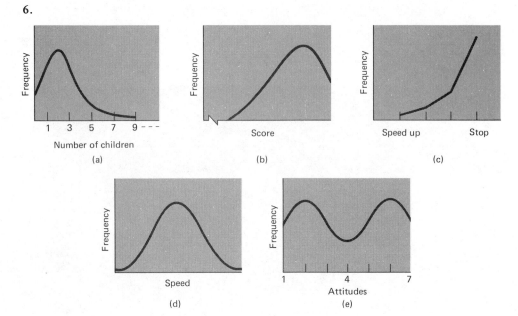

Labels: **(a)** Positively skewed. **(b)** Negatively skewed. **(c)** J curve. **(d)** Symmetrical. **(e)** It is possible that most people who came to the debate already had strong opinions, being either very much for or very much against legalized abortion. On this assumption (and the assumption that proponents and opponents were equally represented in the audience), the curve shown here is bimodal and symmetrical.

CHAPTER 4

1. **(a)** Sum of the X scores. **(b)** Mean of the X scores; $\overline{X} = \Sigma\,X/N$. **(c)** Sum of the Y scores. **(d)** Mean of the Y scores. **(e)** The square of the quantity $\Sigma\,X$. **(f)** The sum of the squared X scores. **(g)** The sum of the deviation of each X from $\overline{X}$; $\Sigma\,x = \Sigma\,(X - \overline{X})$. Except for rounding error in $\overline{X}$, $\Sigma\,x$ always equals 0. **(h)** Multiply each X by the corresponding Y and sum the cross products.

Student	X	X^2	$x = X - \overline{X}$	Y	XY
A	4	16	$-$ .5	5	20
B	1	1	-3.5	2	2
C	7	49	2.5	6	42
D	4	16	$-$ .5	3	12
E	9	81	4.5	8	72
F	2	4	-2.5	2	4
G	6	36	1.5	5	30
H	3	9	-1.5	1	3
	$\sum X = 36$	$\sum X^2 = 212$	$\sum x = 0.0$	$\sum Y = 32$	$\sum XY = 185$

$$\overline{X} = \frac{36}{8} = 4.5 \qquad\qquad \overline{Y} = \frac{32}{8} = 4.0$$

$$\left(\sum X\right)^2 = 36^2 = 1296$$

2. **(a)** Mode = 13, the midpoint of the class interval with the greatest frequency. **(b)** Using Formula 4.2, $\overline{X} = \Sigma(fX)/N = 366/30 = 12.2$. $\Sigma(fX)$ is obtained by multiplying the midpoint (X) of each class interval by its corresponding frequency (f) and summing the resulting fX values. **(c)** The median or 50th percentile is the 15th case from the bottom of the distribution (50% of the N of 30 = 15). Applying Formula 4.4, Median = $11.5 + 3\{(15 - 13)/7\} = 11.5 + .86 = 12.36$. **(d)** Applying Formula 4.5, (1) $25 + 2\{(19 - 17.5)/3\} = 25 + 1.00 = 26.00$; (2) $100(26/30) = 86.67$. The percentile rank of the score 19 is 86.67. **(e)** 10% of 30 = 3. Three cases fall in the bottom interval, 0–2, and there are 0 cases in the interval above. The 10th percentile could therefore take any value between 2.5 (upper real limit of class 0–2) and 5.5 (lower real limit of class 6–8) and meet the definition of that percentile. The midpoint of this range of values (which corresponds to the midpoint of the interval with 0 frequency) is selected as the value of the 10th percentile if a single value is to be specified. This score value is 4.

3. Sample means are unbiased estimates of μ—that is, they systematically tend neither to underestimate or overestimate μ. When only one sample is available, the best single estimate of μ is therefore $\overline{X}$, which in this problem is 63.4.

4. **(a)** For $N = 7$, $(N + 1)/2 = 4$. When the seven numbers are rearranged in order, the value of the 4th (middle) case is 14. **(b)** For $N = 8$, $N/2 = 4$ and $(N + 2)/2 = 5$. When the numbers are ordered, the midpoint between the 4th and 5th case is 18. **(c)** For $N = 6$, $N/2 = 3$ and $(N + 2)/2 = 4$. The midpoint between the 3rd and 4th case is 165.5.

5. **(a)** Positively skewed

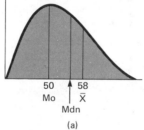

(a)

(b) Approximately symmetrical

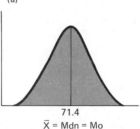

(c) Negatively skewed

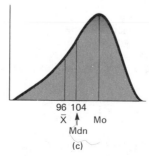

(c)

CHAPTER 5

1. (a) Dist. 2. (b) Dist. 1. (c) Dist. 2. (d) Dist. 1.

2. For distributions 1 and 2, respectively, the values of R are as follows: (a) 4, 7. (b) 28, 14. (c) 11, 28. (d) 9, 9.

3. In (c), the two distributions are quite similar except for the lowest score (5) in dist. 1, which is very different from the rest. If this score were omitted, R for dist. 1 would drop from 28 to 7, a value that is closer to the R of 11 in dist. 2. Thus R is highly influenced by an extreme score, which may be quite deviant. In (d), the two R's are the same, even though the scores in dist. 2 are less variable, more of them clustering around the center of the distribution than in dist. 1. Being based on only the highest score and the lowest score in a distribution, R does not necessarily reflect the variability among the remainder of the scores.

4. Q_1 is the score value of the 10th case (25% of 40) and Q_3 is the score value of the 30th case (75% of 40). After the cumulative frequencies are found, it can be seen by inspection that $Q_1 = 55.5$ and $Q_3 = 58.5$. By Formula 5.1, the semi-interquartile range = $(58.5 - 55.5)/2 = \frac{3}{2} = 1.5$.

5. σ^2, S^2, and s^2 are symbols, respectively, for the variance of a population, the variance of a sample, and an unbiased estimate of the population σ^2 based on the data from a sample. σ, S, and s are symbols, respectively, for the standard deviation of a population, standard deviation of a sample, and an estimate of the population σ. A slight bias remains in s as an estimate of σ.

6.

X	x	x^2	X	x	x^2
11	5	25	5	-1	1
10	4	16	5	-1	1
8	2	4	4	-2	4
7	1	1	2	-4	16
7	1	1	1	-5	25
			$\sum X = 60$	$\sum x = 0$	$\sum x^2 = 94$

$\overline{X} = \frac{60}{10} = 6$ $S^2 = \frac{94}{10} = 9.4$ $S = \sqrt{9.4} = 3.07$

7. $S = \sqrt{454/10 - (6)^2} = \sqrt{45.4 - 36} = \sqrt{9.4} = 3.07$. As it should, the S is the same as was found in problem 6.

8. $s^2 = 94/(10 - 1) = 10.44; s = \sqrt{10.44} = 3.23$.

9. $s = \sqrt{\dfrac{454 - (60)^2/10}{10 - 1}} = \sqrt{(454 - 360)/9} = \sqrt{10.44} = 3.23$.

10. For symmetrical bell-shaped curves, three standard deviation units on either side of $\overline{X}$, for a total of 6, will typically encompass all or almost all the cases. An approximation of S can therefore be obtained by finding $\frac{1}{6}$ of the value of R. (a) R/6 = $\frac{48}{6} = 8$; calculated S of 8.5 is probably accurate. (b) $\frac{21}{6} = 3.5$; calculated $S = 7.2$ is too large. (c) $\frac{59}{6} = 9.8$; $s = 9.3$ is probably accurate. (d) $\frac{46}{6} = 7.7$; $S = 3.8$ is too small. (e) S was calculated to be -3.6, obviously wrong since S always takes a positive value.

11. (a)

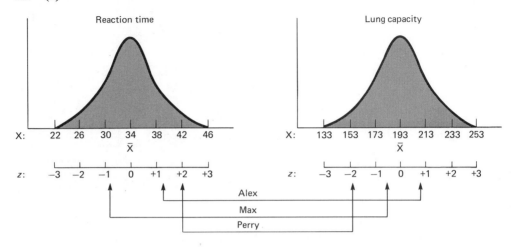

(b)

	z_{RT}	z_{LC}
Alex	$(39–34)/4 = 1.25$	$(210–193)/20 = .85$
Max	$(31–34)/4 = -.75$	$(183–193)/20 = -.50$
Perry	$(42–34)/4 = 2.00$	$(155–193)/20 = -1.90$

(c) Alex and Max each had similar standings on the two tests, Alex being above average on both and Max being below average on both. Perry, however, was much above average on RT and much below average on LC. **(d)** Sam: $X_{RT} = z(S) + \overline{X} = 2.50(4) + 34 = 10 + 34 = 44$; Jim: $X_{RT} = -1.75(4) + 34 = (-7) + 34 = 27$; Boris: $X_{LC} = .25(20) + 193 = 5 + 193 = 198$.

12. (a) Convert $X = 80$ into a z score: $z = (80 - 65)/10 = 1.50$. T scores have a mean of 50 and a standard deviation of 10. Therefore, by formula 5.16, $T = 10(1.5) + 50 = 65$. **(b)** $z = (41 - 65)/10 = -2.40$; standard score $= (100) (-2.4) + 500 = 260$. **(c)** Determine z score equivalent of standard score of 280: $z = (280 - 200)/50 = 1.60$. The raw-score equivalent of $z = 1.6$ is $X = z(S) + \overline{X} = 1.6(10) + 65 = 81$.

CHAPTER 6

1. (a) Linear; negative r. Those who watch TV more tend to have poorer grades. **(b)** Curvilinear. The relationship is likely to take the form of an inverted U, women who are emaciated or who are obese tending to be in poorer health than those of intermediate weights. **(c)** Linear; positive r. Higher level of education is associated with higher incomes. **(d)** Little or no relationship is expected between the two variables; the correlation would be close to zero.

2. (a) In almost all pairs, the z's have the same sign so that their cross-product is positive, and within pairs, the z values tend to be quite similar. A high positive r is expected. **(b)** Inspection of the scatter plot, below, suggests that the relationship is linear.

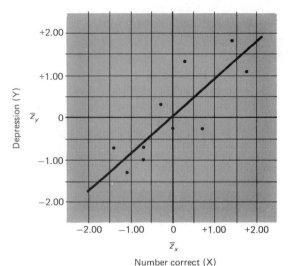

Scatter plot of *z* scores for number correct and depression scores and a straight line that comes close to the points.

(c) $r = \Sigma z_X z_Y / N = 8.25/10 = .825$. **(d)** Obtain ΣXY by multiplying each X by its corresponding Y and summing the resulting cross-products. Then obtain $\overline{X}$ and $\overline{Y}$ by dividing ΣX and ΣY by N. All other values called for in Formula 6.2 are given in the problem. Entering these values into the formula,

$$r = [841 - 10(5)(15)]/\sqrt{330 - 10(5^2)} \ \sqrt{2402 - 10(15^2)}$$

$$= 91/(\sqrt{80} \ \sqrt{152}) = 91/110.26 = .825.$$

3. The X scores and Y scores are first converted to ranks, and the difference (*d*) between pairs of ranks and d^2 obtained, as follows:

Subject	Rank X	Rank Y	*d*	d^2	Subject	Rank X	Rank Y	*d*	d^2
1	5	5.5	.5	.25	6	9	10	1	1
2	7.5	9	-1.5	2.25	7	4	2	2	4
3	2	1	1	1	8	1	3	-2	4
4	10	7.5	2.5	6.25	9	7.5	7.5	0	0
5	6	4	-2	4	10	3	5.5	-2.5	6.25

$r_S = 1 - \{6 \ \Sigma \ d^2/N(N^2 - 1)\} = 1 - \{6(29)/10(10^2 - 1)\} = 1 - \{174/990\} = 1 - .176 = .824$. This value is very close to the Pearson *r* obtained in problem **2(c)**. When measurement data are converted to ranks and r_S is computed, r_S is expected to be similar to *r* in value but not necessarily identical.

4. **(a)** The teacher assumes that lack of playmates causes lack of self-confidence, so that giving children the opportunity to make friends will raise their confidence. However, causality cannot be inferred from the existence of a correlation. X may bring about Y, Y may bring about X, or both X and Y may be "caused" by a simi-

lar set of factors. **(b)** $Y_{pred} = \{.50(1.4)/2.0\} (X) - \{.50(1.4)/2.0\} (16.8) + (5.2) = .35X - .68$.
(c) and (d)

X	Y_{pred}	Actual Y
14	4.22	2
17	5.27	4
19	5.97	7

The predictions correspond only modestly well to actual scores. Since r is not high, these errors in prediction are not surprising. **(e)** With $r = .50$, $r^2 = .25$. Only 25 percent of the variability in one set of scores is explained by variations in the other set of scores.

5. **(a)** $X_{pred} = [(.94)(5.90)/5.64] (Y) - (.983)(58.75) + 47.90 = .983Y - 9.87$. **(b)** For $Y = 50$, $X_{pred} = .983(50) - 9.87 = 39.30$. For $Y = 70$, $X_{pred} = .983(70) - 9.87 = 58.96$.

(c)

Scatter plot of work satisfaction and life satisfaction scores and regression line of X on Y.

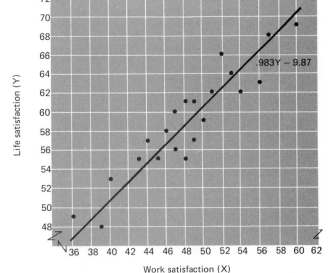

(d) $s_{est\,X} = s_X \sqrt{1 - r^2} = 5.9 \sqrt{1 - (.94)^2} = 5.9 \sqrt{.116} = 5.9(.341) = 2.01$. Thus the standard deviation of actual X points around the regression line at any given Y value should be about 2.01 score units. Inspection of the deviation of the points from the regression line at the various Y values (for example, $Y = 57$) reveals deviations consistent with this statement.

1. **(a)** NME. A given crime may involve more than one type of offense and thus be assigned to more than one category. For example, a murder may be committed in the course of an armed robbery. **(b)** ME, if it is assumed that a student is permitted to be enrolled in only one college. If the university allows dual enrollment, the answer is NME. **(c)** ME. Each student receives only one course grade and thus can be assigned to only one of the five letter-grade categories.

2. **(a)** NI. Once one member of a pair has been assigned to the first condition, the other is automatically assigned to the second condition. Further, the pairs are likely to have similar growth patterns so that their scores on the experimental measures are likely to be correlated. **(b)** NI. The factors that lead an individual to earn good or poor grades in one semester tend to be present on successive semesters as well. **(c)** I. Assuming that the cards have been properly shuffled and dealt, the particular cards that the person is dealt in one hand do not influence the cards in successive hands.

3. **(a)** $p(A) = n_A/N = \frac{60}{200} = .30$. **(b)** $p(B) = .25$; $p(C) = .45$; $p(B \text{ or } C) = p(A) + p(B) = .25 + .45 = .70$. **(c)** $p(\overline{A})$ or complement of A. **(d)** $p(C, B) = p(C) \times p(B) = .45 \times .25 = .1125$. **(e)** $p(A, A, A) = .3 \times .3 \times .3 = .027$.

4. **(a)** $p(A1, \text{Ha}) = p(A1) + p(\text{Ha}) = \frac{1}{50} + \frac{1}{50} = \frac{2}{50} = .04$. **(b)** $\frac{13}{50} = .26$. **(c)** Since the first prize winner has already been selected, 49 contestants remain for the second drawing. Therefore $p(\text{Ca}) = \frac{1}{49}$. **(d)** $p(\text{Ia}, \text{Me}, \text{Wy}) = p(\text{Ia}) \times p(\text{Me}) \times p(\text{Wy}) = \frac{1}{50} \times \frac{1}{49} \times \frac{1}{48} = .0000085$.

5. **(a)**

	Less	Same	More	Total
3-year-olds	.033	.067	.200	.300
5-year-olds	.000	.167	.167	.333
8-year-olds	.000	.300	.067	.367
Total	.033	.534	.434	1.000

Note: Except for rounding errors, the cell entries in each row and column should equal the row and column totals. **(b)** $p \text{ (same)} = \frac{16}{30} = .533$. ($p$ can also be read directly from the table in **(a)**.) Table has slight rounding error.

(c)

	Less	Same	More	Total
3-year-olds	.111	.222	.667	1.00
5-year-olds	.000	.500	.500	1.00
8-year-olds	.000	.818	.182	1.00

(d) $p \text{ (same)} = \frac{5}{10} = .500$. ($p$ can also be read directly from the table in **(c)**.) **(e)** $p \text{ (more)} = \frac{6}{9} = .667$. ($p$ can also be read directly from the table.)

6. $P_N = N! = 4! = 4(3)(2)(1) = 24$

7. **(a)** $P_r^N = \frac{N!}{(N-r)!} = \frac{10!}{(10-2)!} = 90$. **(b)** $C_r^N = \frac{N!}{r!(N-r)!} = \frac{10!}{2!(8!)} = 45$.

1. (a) On each card, there are 4 equally possible arrangements of 1 winner and 3 losers. (That is, the winner is equally likely to appear on line 1, 2, 3, or 4). The number of possible patterns over the three cards is therefore $4 \times 4 \times 4 = 64$ and the probability of each pattern is $(\frac{1}{4})^3$ or $\frac{1}{64}$. (b) $(p + q)^N = (\frac{1}{4} + \frac{3}{4})^3 = (\frac{1}{4})^3 + 3(\frac{1}{4})^2 (\frac{3}{4}) + 3(\frac{1}{4})(\frac{3}{4})^2 + (\frac{3}{4})^3 = \frac{1}{64} + \frac{9}{64} + \frac{27}{64} + \frac{27}{64} = .0156 + .1406 + .4219 + .4219 = 1$. These values correspond, respectively, to the probability of 3, 2, 1 and 0 "winners." (c) $C_1^3 \, pq^2 = \dfrac{3!}{1!(3-1)!} (\frac{1}{4})(\frac{3}{4})^2 = 3(.25)(.5625) = .4219$. This is the same p value obtained in (b). (d) p (2 or 3 winners) $= p$ (2 winners) $+ p$ (3 winners) $= .1406 + .0156 = .1562$. (e) $\mu = Nq = 3(\frac{3}{4}) = \frac{9}{4} = 2.25$. $\sigma = \sqrt{Npq} = \sqrt{3(1/4)(3/4)} = \sqrt{9/16} = \frac{3}{4} = .75$.

2. (a) $z = (63 - 75)/12 = {}^{-}\frac{12}{12} = -1.00$; from Table B, 34.13% of the cases in a normal distribution are between the mean and $z = -1.00$. Therefore $50 - 34.13 = 15.87\%$ score 63 or below. (b) The z for X = 63 was calculated in (a). For X = 87, $z = (87 - 75)/12 = +1.00$. The percentage lying between 63 and 87 is $34.13 + 34.13 = 68.26\%$. (c) 68.26% of 5000 = 3413 students. (d) For X = 81, $z = (81 - 75)/12 = +.5$; Table B shows that 19.15% of the cases lie between the mean and this point. For X = 93, $z = (93 - 75)/12 = +1.5$; 43.32% lie between the mean and this point. Therefore $43.32 - 19.15 = 24.17\%$ lie between these two z scores or between the raw scores of 81 and 93. (e) $z = (66 - 75)/12 = -.75$; 27.34% of the area lies between the mean and this point and $50 - 27.34 = 22.66\%$ below it. The percentile rank of X = 66 is therefore 22.66. (f) The z score at or below which 67% of the cases fall is first determined. Since 50% fall at or below the mean, look up $67 - 50 = 17\%$ in the body of Table B and determine the associated z score. This z is +.44. The raw-score equivalent of the 67th percentile is X $= .44(12) + 75 = 5.28 + 75 = 80.28$.

3. (a) $z = (62 - 58)/10 = +.4$; 15.54% lie between the mean and this point and 50% below the mean. The probability of X = 62 or below is therefore .6554. (b) $z = \pm 5/10 = \pm.50$; 19.15% of the cases lie between the mean and each of these z values and $50 - 19.15 = 30.85\%$ beyond each of them. The probability of a deviation this large or larger in either direction is therefore $.3085 + .3085 = .6170$. (c) From Table B, 47.5% lie between the mean and $z = \pm 1.96$ and 2.5% lie beyond each of these z values. X $= +1.96(10) + 58 = 77.6$ and X $= -1.96(10) + 58 = 38.4$. The extreme 5% had scores of 77.6 or more or 38.4 or less.

CHAPTER 9

1. (a) Sampling distribution of $\overline{X}$'s. (b) 78, the value of μ for the population. (c) Standard error of the mean. By Formula 9.1, $\sigma_{\overline{X}} = 10/\sqrt{25} = 2.0$. (d) The mean would still be equal to μ (i.e., 78). $\sigma_{\overline{X}} = 10/\sqrt{49} = 1.43$. (e) 1.00 (f) Without complex calculations, beyond the scope of this text, your representations cannot be exact. Your figures should have the major features of the following three curves.

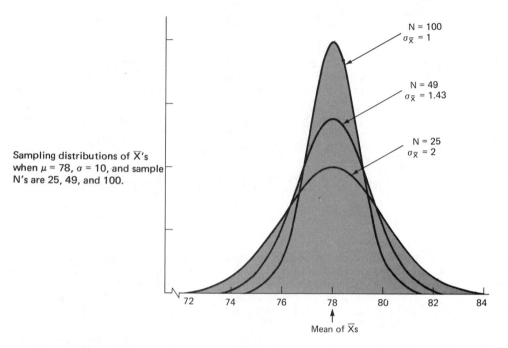

Sampling distributions of $\overline{X}$'s when $\mu = 78$, $\sigma = 10$, and sample N's are 25, 49, and 100.

N = 100
$\sigma_{\overline{X}} = 1$

N = 49
$\sigma_{\overline{X}} = 1.43$

N = 25
$\sigma_{\overline{X}} = 2$

Mean of $\overline{X}$s

(g) As N's increase, variability among $\overline{X}$'s decrease, as reflected in the decreasing value of $\sigma_{\overline{X}}$. The value of $\sigma_{\overline{X}}$ approaches 0 as the sample N's approach the population N.

2. (a) The sampling distribution of $\overline{X}$'s with N = 64 can be expected to have a mean of 150 and $\sigma_{\overline{X}} = 16/\sqrt{64} = 2.00$. The given $\overline{X}$ of 153 must first be converted to a z score: $z = (153 - 150)/2 = 1.5$. From Table B, 43.32% lie between the mean and this z and 6.68% beyond it. The probability of $\overline{X} = 153$ or greater is therefore .0668. (b) $z = \pm 4/2 = \pm 2$. In a normal distribution, 2.28% of the cases fall at or above $z = +2$ and an equal percentage at or below $z = -2$. The probability of deviating ± 4 score units or more is therefore 2(.0228) = .0456. (c) From the body of Table B, it can be seen that 44.95 of the cases fall between the mean and a z of 1.64 and (approximately) 5% beyond this z. Converting this z into its corresponding $\overline{X}$, $\overline{X} = 1.64(2) + 150 = 153.28$. (d) $s_{\overline{X}} = 17/\sqrt{64} = 2.125$. Through errors of random samplings, the estimate of σ was somewhat higher than the actual σ, and the estimate of the standard error of the mean was therefore also slightly higher than the true value.

3. (a) $s_{\overline{X}} = 1.2/\sqrt{144} = .10$. (b) If there is no stereotype, the mean rating in the population would be 4 (any other value of μ would indicate some degree of stereotype). Therefore: H_0: $\mu = 4.00$; H_1: $\mu \neq 4.00$. (c) $t = (4.15 - 4.00)/.10 = 1.50$. With a sample of this size, $t = 1.96$ at $\alpha = .05$ (two-tailed test).* We therefore retain the hypothesis that $\mu = 4.00$. We conclude that 16-year-old girls perceive that being a salesclerk is equally appropriate for men and women. (d) We could have made a Type II

---

*In this and all subsequent problems in which H_1 is bidirectional, the critical value of t at $\alpha = .05$ or .01 is specified without sign. H_0 is rejected if the absolute value of the calculated t equals or exceeds the critical value at the specified α level. When H_1 is unidirectional, the sign (plus or minus) of the critical value of t must be specified and for H_0 to be rejected, the calculated t must have the same sign.

error, falsely accepting the null hypothesis that $\mu = 4.00$ when it is actually some other value. **(e)** $\alpha = 05$; the higher the α level (.04, .03, .02 ....), the greater the probability of a Type II error.

4. **(a)** H_0: $\mu = 4.00$. H_1: $\mu < 4.00$ because law has traditionally been a male-dominated profession and is therefore likely to be rated as more appropriate for men than women (rather than the reverse). **(b)** $s_{\bar{x}} = 1.50/\sqrt{25} = .30$; $t = (3.12 - 4.00)/.30 = -2.93$; $df = 25 - 1 = 24$. Since H_1 is unidirectional, either a two-tailed or a one-tailed test could be made. For 24 df and $\alpha = .01$, the value of t for a two-tailed test is -2.80 and for a one-tailed test, t is -2.49 (Table C). The calculated value of t is higher than both these critical values so that whichever test is chosen, H_0 is rejected at the .01 significance level. We conclude that there is a sex stereotype about lawyers in 16-year-old boys, this profession being regarded as more appropriate for men.

5. For small df's, the t distribution is symmetrical but flatter than the normal distribution in the middle (around $t = 0$) and higher than the normal distribution on the two tails. As df's become larger and larger, the t distribution increasing resembles the normal curve.

6. **(a)** H_0: $\mu = 500$; H_1: $\mu \neq 500$. $s_{\bar{x}} = 50/\sqrt{9} = 16.67$. $t = (480 - 500)/16.67 = -1.20$. With $df = 8$ and $\alpha = .05$, $t = 2.31$. Since $p > .05$, accept H_0 and the validity of the manufacturer's claim. **(b)** With N = 16, $s_{\bar{x}} = 50/\sqrt{16}$ and $t = -20/12.5 = -1.60$; $df = 15$ and $t_{.05} = 2.13$; H_0 is not rejected. **(c)** With N = 49, $s_{\bar{x}} = 50/\sqrt{49}$ and $t = -20/7.14 = -2.80$; $df = 48$. Reject H_0. **(d)** With $\bar{X}$ and s held constant, $s_{\bar{x}}$ decreased and t increased as N increased. With small N's, H_0 was accepted at $\alpha = .05$ but was rejected with a larger N.

7. **(a)** $s_{\bar{x}} = 7.5/\sqrt{144} = .62$. 95% CI: $78.4 \pm (1.96)(.62) = 78.4 \pm 1.22 = 77.18$ to 79.62. **(b)** $78.4 \pm (2.58)(.62) = 76.8$ to 80.0.

8. The proportion who chose "grass" was .55 (p) and the proportion who chose "table" was .45 (q). (To avoid decimals, p and q will be treated as percentages in the solution to the problem.) H_0: $p_t = q_t = 50$; H_1: $p_t \neq 50$. $\sigma_p = \sqrt{\{50(50)\}/40} = 7.91$. $z = (55 - 50)/7.9 = .63$. For $\alpha = .05$, $z = 1.96$. The obtained z is not significant at $\alpha = .05$. It is concluded that the two words are preferred equally in the population.

9. Calling color p and form q, H_0: $p_t = q_t = 50$; H_1: $p_t \neq 50$. $\sigma_p = \sqrt{\{50(50)\}/20} = 11.18$. $z = (75 - 50)/11.18 = 2.24$. For $\alpha = .05$, $z = 1.96$ so H_0 is rejected and it is concluded at the .05 significance level that 5-year-old children are more likely to sort by color than by form. Had α been set at .01, however, H_0 would be retained, since the calculated value of z is less than $z = 2.58$ at $\alpha = .01$.

CHAPTER 10

1. **(a)** Independent variable: type of therapy. Dependent variable: score on the test evaluating emotional status. H_0: The two therapy groups come from populations with the same mean on the test—that is, the population mean is the same after insight therapy as it is after cognitive therapy. **(b)** Independent variable: Type of worker. Dependent variable: job satisfaction score. H_0: The population mean of job satisfaction scores for "blue collar" workers is the same as the population mean for "white collar" workers. **(c)** Independent variable: presence or absence of social ex-

perience given to infant monkeys. Dependent variable: amount of fear in response to a strange toy. H_0: The population mean of fear responses is the same in monkeys previously given social experience as in monkeys previously given no social experience.

2. (a) By Formula 9.1, $\sigma_{\bar{X}WWI} = 3/\sqrt{100} = .3$ and $\sigma_{\bar{X}WWII} = 3/\sqrt{100} = .3$. (b) The mean of the distribution of differences between pairs of $\bar{X}$'s ($\bar{X}_{WWII} - \bar{X}_{WWI}$) takes the same value as the difference between the μ's of the two populations. Thus $68 - 65 = 3$. (c) Standard error of the difference between means ($\sigma_{\bar{X}_1 - \bar{X}_2}$). By Formula 10.1, $\sigma_{\bar{X}WWII - \bar{X}WWI} = \sqrt{(.3)^2 + (.3)^2} = .42$. (d) The mean of the distribution of differences between $\bar{X}$'s is 3 and its standard deviation is .42. The z score of $\bar{X}_{WWII} - \bar{X}_{WWI} = 4$ is therefore $(4 - 3)/.42 = 2.38$. From Table B, it can be determined that a z score of 2.38 or greater has a probability of only .0087.

3. (a) H_0: $\mu_{WWII} - \mu_{WWI} = 0$; H_1: $\mu_{WWII} - \mu_{WWI} \neq 0$. A specific value must be specified in H_0. Since the question is whether the two population μ's are the same or different, H_0 must specify that they are the same since this hypothesis specifies a value, zero, for the difference between μ's. (b) H_1: $\mu_{WWII} > \mu_{WWI}$. Better nutrition and other similar factors have resulted in systematic increases in people's height over the past century or more. We therefore expect that the average recruit in WWII would also be taller. (c) $s_{\bar{X}WWII - \bar{X}WWI} = \sqrt{(3.1)^2/70 + (2.9)^2/70} = \sqrt{.137 + .120} = .51$; $t = (67.2 - 65.8)/.51 = 2.75$. For $\alpha = .01$, $t = 2.58$ for a two-tailed test. Since the calculated t is greater than this value, H_0 can be rejected at the .01 significance level. (d) It is concluded that inductees were taller in the Second than in the First World War.

4. Since the experimenter is uncertain about the direction of any difference that may occur, the H_1 must be bidirectional. Thus: H_0: $\mu_L - \mu_{NL} = 0$; H_1: $\mu_L - \mu_{NL} \neq 0$. $s_{\bar{X}_L - \bar{X}_{NL}} = \sqrt{(4.1)^2/20 + (3.6)^2/20} = 1.22$; $t = (43.2 - 47.7)/1.22 = -3.69$; $df = 20 + 20 - 2 = 38$. Calculated t is greater than 2.72, the t at $\alpha = .01$ for a two-tailed test. ($df = 35$ is used since $df = 38$ does not appear in Table C.) H_0 is rejected at $\alpha = .01$. It is concluded at the .01 significance level that on the average, lonely women students are lower in intimacy of self-disclosure than nonlonely women.

5. H_0: $\mu_E - \mu_S = 0$; H_1: $\mu_E - \mu_S \neq 0$.

$$s_{\bar{X}_E - \bar{X}_S} = \sqrt{\frac{4(5.6)^2 + 9(5.2)^2}{13} \left(\frac{1}{5} + \frac{1}{10}\right)} = \sqrt{8.51} = 2.92.$$

$t = (58.4 - 52.6)/2.92 = 1.99$. For $df = 15 - 2 = 13$; $t = 2.16$ at $\alpha = .05$. The calculated t is below this value, so the null hypothesis cannot be rejected. With $\alpha = .05$, it cannot be concluded that the drugs have any differential effects.

6. (a) $t = (58.9 - 53.1)/1.53 = 3.79$; $df = 50 - 2 = 48$. From Table C, $t = 2.014$ for $\alpha = .05$ and $df = 45$. (48 df is not listed; 45 is the nearest df below this figure.) The calculated t is larger than this value and H_0 is rejected. H_0 could also be rejected at $\alpha = .01$ (for 45 df, $t_{.01} = 2.69$). (b) Three factors influence the power of the t test to detect a false H_0: the size of the difference between μ's (which is reflected in the sizes of the difference between pairs of $\bar{X}$'s), the magnitude of the population σ's (estimated by s), and the sample N's. In the two experiments, the difference between $\bar{X}$'s

turned out to be the same and the s's were comparable. However, the sample N's were larger in the second experiment, which resulted in decreasing $s_{\bar{X}_E - \bar{X}_S}$ and thus in increasing t. We cannot conclude with absolute certainty that a Type II error was made in the first experiment or that the second experiment "proved" the new drug was more beneficial. However, in view of the near significance of t in problem 5, it seems reasonable to conclude that the drugs are differentially effective and that a significant difference was found in the second experiment but not the first because of the larger samples it employed.

CHAPTER 11

1. (a) The moderate correlation suggests that the set of variables on which the pairs of patients were matched had some effect on performance. The matching was thus done on relevant variables. (b) H_0: $\mu_S - \mu_M = 0$; H_1: $\mu_S - \mu_M \neq 0$. $s_{\bar{X}_S - \bar{X}_M} = \sqrt{(2.3)^2 + (2.6)^2} = 3.47$; $t = (84.2 - 91.1)/3.47 = -1.99$. $df = 15 + 15 - 2 = 28$. For 28 df, $t_{.05} = 2.05$. H_0 cannot be rejected.

    (c) $s_{\bar{X}_S - \bar{X}_M} = \sqrt{(2.3)^2 + (2.6)^2 - 2(.45)(2.3)(2.6)} = 2.58$.

    $t = (84.2 - 91.1)/2.58 = -2.67$. $df = N - 1 = 15 - 1 = 14$. For 14 df, $t_{.05} = 2.14$. H_0 is rejected at $\alpha = .05$. It is concluded that schizophrenics do not perform as well on the task as medical patients. H_0 could be rejected because, with matched pairs whose performance is correlated, Formula 11.1 reduces $s_{\bar{X}_1 - \bar{X}_2}$ and permits a more powerful test of H_0. The conclusion that the groups differ significantly is the correct one to draw because it follows from the proper formula for a matched pairs design.

2. (a) H_0: $\mu_{Ch} - \mu_S = 0$; H_1: $\mu_{Ch} - \mu_S \neq 0$. When the differences (D) between pairs of scores and the square of each D are obtained, $\Sigma D = -26$, $\bar{D} = -26/10 = -2.6$, and $\Sigma D^2 = 186$. $s_D^2 = \{(186) - (-26)^2/10\}/10 - 1 = 13.16$. $s_{\bar{D}} = \sqrt{13.16/10} = 1.147$; $t = -2.6/1.147 = -2.27$. $df = 10 - 1 = 9$. For 9 df, $t_{.05} = 2.26$. H_0 is significant at the .05 level. It is concluded that after being made ill, the animals subsequently found the chocolate-flavored mash less palatable. (b) $A = 186/(26)^2 = .275$. This value is slightly smaller than the *maximum* A of .276 allowed at $\alpha = .05$ with $n - 1 = 9$. Therefore, H_0 is significant at the .05 level, the same conclusion reached in (b) with t.

3. (a) The population correlation is equal to 0. (b) For $df = 20 - 2 = 18$, $r_{.05} = .444$. Obtained r is not significantly different from 0. (c) For $df = 28$, $r_{.01} = .463$. Obtained r is significant. (d) For $df = 203$, $r_{.05} = .138$. Obtained r is significant. (e) For $N = 16$, $r_S = .665$ at $\alpha = .01$. Obtained r_S is not significant. (f) For $N = 25$, $r_S = .537$ at $\alpha = .01$. Obtained r_S is not significant.

CHAPTER 12

1. (a) $\bar{X}_1 = 16$, $\bar{X}_2 = 12.8$, $\bar{X}_3 = 10.2$. (b) H_0: $\mu_1 = \mu_2 = \mu_3$. The population mean reaction times are assumed to be equal for tilts of $45°$, $90°$, and $180°$.

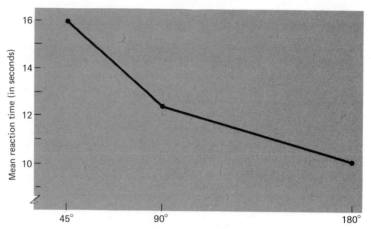

Mean reaction times for the three groups in object orientation study.

(c) $SS_{bg} = \dfrac{(80)^2}{5} + \dfrac{(64)^2}{5} + \dfrac{(51)^2}{5} - \dfrac{(195)^2}{15} = 2619 - 2535 = 84.$

$SS_{tot} = 1300 + 852 + 535 - \dfrac{(195)^2}{15} = 2687 - 2535 = 152.$

$SS_{wg} = 152 - 84 = 68.$

$df_{bg} = 3 - 1 = 2; \quad df_{wg} = 15 - 3 = 12; \quad df_{tot} = 15 - 1 = 14.$

$MS_{bg} = \dfrac{84}{2} = 42.0; \quad MS_{wg} = \dfrac{68}{12} = 5.7; \quad F = \dfrac{42.0}{5.7} = 7.37.$

For $df = 2$ and 12, $F_{.01} = 6.93$; calculated F is significant at $\alpha = .01$.

Source of Variation	SS	df	MS	F	p
Between groups	84	2	42.0	7.37	$<.01$
Within groups	68	12	5.7		
Total	152	14			

(d) There is a significant tendency for mean reaction times to differ as tilt changes. The observed trend is for means to be greatest with the least tilt of one member of a pair and smallest with the most tilt. (A multiple-comparison test such as the Tukey *hsd* test could indicate whether the specific trend observed is present in the populations as well.)

2. Being conservative since $q_{4,116}$ is not tabled for the .05 level, we find $q_{4,60} = 3.74$ for the .05 level. Then $hsd = 3.74 \sqrt{\dfrac{27.30}{30}} = 3.74\,(.9539) = 3.57.$

Absolute Difference between
Pairs of Means

	$\overline{X}_1$	$\overline{X}_2$	$\overline{X}_3$
$\overline{X}_2$	1.2		
$\overline{X}_3$	6.8	8.0	
$\overline{X}_4$	4.4	5.6	2.4

The underlined differences exceed the *hsd* value of 3.57 at $\alpha = .05$. The significant F from the ANOVA leads to the conclusion that there are differences among the means. The follow-up *hsd* test indicates that the specific nature of these differences are such that discussion groups 3 and 4 earned significantly better grades than did groups 1 and 2. Neither groups 1 and 2 nor groups 3 and 4 differ significantly.

3. **(a)** The sample means are all less than the medians, suggesting that non-normality may be present. **(b)** No apparent problem with non-normal data or unequal population variances. **(c)** Since the differences between upper and lower points in the ranges are 11, 22, and 30, respectively, we suspect that the population variances are unequal, contradicting the assumption of equal variances in the three populations.

CHAPTER 13

1. In each of the figures below, one variable was chosen to be represented on the baseline and the other to be represented in separate curves. Choice of which variable is to be shown on the baseline is more or less arbitrary, and you may have made the opposite choice.

(a) I.

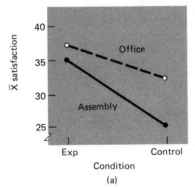

II. Table of $\overline{X}$'s

	Group		
	Exp.	Control	Row $\overline{X}$
Assembly	35	25	30
Office	37	32	34.5
Col. $\overline{X}$	36	28.5	

III. Both variables appear to have main effects. Workers in the experimental flextime program are more satisfied than workers in the control group working set hours, and office workers are more satisfied than assembly line workers. There also appears to be an interaction between the variables, the difference in satisfaction between the two types of working conditions being greater for the assembly line workers.

(b) I.

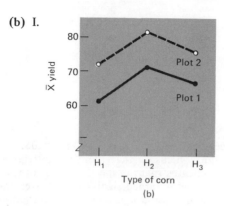

(b)

II. Table of $\overline{X}$'s

	Corn Type			
	H_1	H_2	H_3	Row $\overline{X}$
Plot 1	61	71	66	66
Plot 2	72	81	75	76
Col. $\overline{X}$	66.5	76	70.5	

III. Both variables appear to have main effects, plot 2 producing more than plot 1 and type of corn also producing different amounts in the order H_1, H_3, and H_2. Plot and type of corn appear to combine additively, the two curves being approximately parallel. Thus, the magnitude of the superiority of H_2 over H_3 and H_3 over H_1 is the same in each plot.

(c) I.

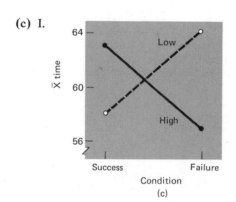

(c)

II. Table of $\overline{X}$'s

	Self-esteem		
	High	Low	Row $\overline{X}$
Success	63	58	60.5
Failure	57	64	60.5
Col. $\overline{X}$	60	61	

III. A crossover interaction between the variables can be observed in the graph. High self-esteem subjects work faster after failure than after success, whereas low self-esteem subjects work faster after success than after failure. These opposite effects cancel out any main effects of success and failure, the row means turning out to be identical. Similarly, self-esteem also has no main effect, the column means being nearly equal.

2. **(a)** In order to emphasize differences in ratings of the two tapes within each of the three groups of majors, the experimental conditions have been plotted along the baseline in the graph below. It would be acceptable, however, to indicate major along the baseline.

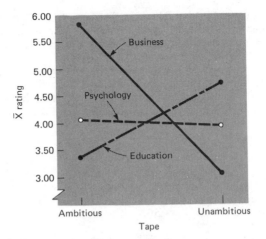

(b) 1. Main effect of tape. $H_0: \mu_A - \mu_U = 0$. Collapsing over raters' type of major, the population mean of the ratings of the tape of the ambitious man is the same as the mean of the ratings of the tape of the unambitious man. 2. Main effect of raters' major. $H_0: \mu_B = \mu_E = \mu_P$. Collapsing over the two tapes, the population means of the ratings of the three types of major are all equal. 3. Interaction between tape and major. H_0: the two variables combine additively. **(c)** $SS_{tot} = (703 + 299 + 387 + 212 + 499 + 358) - [(117 + 67 + 81 + 60 + 95 + 78)^2/120] = 2458 - 2066.7 = 391.3$. $SS_{bg} = (117^2/20) + (67^2/20) + (81^2/20) + (60^2/20) + (95^2/20) + (78^2/20) - 2066.7 = 2172.4 - 2066.7 = 105.7$. $SS_{wg} = (703 - 117^2/20) + (212 - 60^2/20) + (299 - 67^2/20) + (499 - 95^2/20) + (387 - 81^2/20) + (358 - 78^2/20) = 285.6$, or $SS_{wg} = SS_{tot} - SS_{bg} = 391.3 - 105.7 = 285.6$. $SS_{Tape} = (117 + 67 + 81)^2/60 + (60 + 95 + 78)^2/60 - 2066.7 = 8.5$. $SS_{Major} = (117 + 60)^2/40 + (67 + 95)^2/40 + (81 + 78)^2/40 - 2066.7 = 2071.35 - 2066.7 = 4.6$. $SS_{Tape \times Maj} = SS_{bg} - SS_{Tape} - SS_{Major} = 92.6$.

Source	SS	df	MS	F	p*
Tape	8.5	1	8.5	3.40	> .05
Major	4.6	2	2.3	.92	> .05
Interaction	92.6	2	46.3	18.52	< .01
Within groups	285.6	114	2.5		
Total	391.3	119			

---

*For $df = 1$ and 114, $F_{.05} = 4.00$ and for $df = 2$ and 114, $F_{.05} = 3.15$ and $F_{.01} = 4.98$. (Note: when df is not tabled, the next lowest df in the table is used to find the critical F value.)

(d) Neither main effect was significant, but the interaction was highly significant, suggesting that type of tape did affect ratings, but not in the same way in all three majors. Inspection of the graph of $\overline{X}$'s in (a) shows how the significant interaction came about. The business majors rated the ambitious man much more favorably than the nonambitious man, whereas for the education majors, the difference was in the reverse direction. Compared to each of the other two groups, psychology majors gave moderately favorable ratings to both types of tapes. Apparently business majors like men more who are highly ambitious, education majors like men who are less ambitious, and psychology majors like both about equally. (e) For $k = 6$ groups, $df = 114$ and $\alpha = .05$, $hsd = 4.16 \sqrt{2.5/20} = 1.47$. The difference between the means for the two business groups (2.85) exceeded this value and was thus significant. The difference between the means of the two education groups was 1.40 and therefore was close to being significant at the .05 level. The difference for the psychology majors was .15, far short of 1.47, and thus not significant.

CHAPTER 14

1. (a) There is no guessing bias; the probability of guessing head is the same as guessing tail. $H_0: p(H) = p(T) = .50$.

   (b)

	Guess H	Guess T	Total
Observed	118	82	200
Expected	100	100	200
$(O - E)$	+18	−18	
$(O - E)^2$	324	324	
$(O - E)^2/E$	3.24	3.24	

   $\chi^2 = 3.24 + 3.24 = 6.48$. For 1 df, $\chi^2_{.05} = 3.84$. H_0 can be rejected at $\alpha = .05$. It is concluded that there is a bias for guessing "heads."

2. (a)

	Club	Spade	Heart	Diamond	Total
Observed	43	51	54	52	200
Expected	50	50	50	50	200

   $\chi^2 = (-7)^2/50 + (1)^2/50 + (4)^2/50 + (2)^2/50 = 1.40$. For $df = 3$, and $\alpha = .05$, $\chi^2 = 7.82$. Since the obtained χ^2 is less than this value, the hypothesis is retained. (b) It is concluded that in guessing the suit of a card, there is probably no bias.

3. (a) χ^2 test of independence, or test of significance of difference between groups.

   (b)

	Same		Different		Total
	O	E	O	E	
Physicians	14	13	6	7	20
Lawyers	12	13	8	7	20
Total	26		14		40

   $\chi^2 = (1)^2/13 + (-1)^2/13 + (-1)^2/7 + (1)^2/7 = .44$. For 1 df, and $\alpha_{.05}$, $\chi^2 = 3.84$; the hypothesis is accepted. It is concluded that the proportion of those who would make the same choice again does not differ in the two professions.

   (c) $\chi^2 = \dfrac{40 \{(14)(8) - (6)(12)\}^2}{(14 + 6)(12 + 8)(14 + 12)(6 + 8)} = .44$.

   As it should be, the answer is the same as in (b).

4. (a) $\chi^2 = (11 - 7.97)^2/7.97 + (4 - 7.03)^2/7.03 + (13 - 18.06)^2/18.06 + (21 - 15.94)^2/15.94 + (10 - 7.97)^2/7.97 + (5 - 7.03)^2/7.03 = 6.60$. (b) With $df = 2$, $\chi^2_{.05} = 5.99$ and $\chi^2_{.01} = 9.21$. If $\alpha = .05$ is selected, the hypothesis that age is independent of choice would be rejected; it would be concluded that whereas 3-year-olds choose small, medium, and large approximately equally on the test trial following training, 6-year-olds more often choose the medium size stimulus.

5. $df = (5 - 1)(10 - 1) = 36$. In order to evaluate χ^2 based on greater than 30 df, χ^2 must be converted into a z score and its probability determined from the normal curve table. By Formula 14.5, $z = \sqrt{2(55.05)} - \sqrt{2(36) - 1} = 2.06$. At $\alpha = .05$, $z = 1.96$. The obtained z is greater than this value so that the hypothesis is rejected at .05 significance level. It is concluded that the proportion of students choosing various fields is related to geographic region.

CHAPTER 15

1. (a) The assumptions of equal variance and normality appear to be violated. (b) Mdn $= 27.5$. $\chi^2 \doteq 15.2$. Since a χ^2 of 9.21 is required with $df = 2$ and $\alpha = .01$, the null hypothesis of equal enjoyment of competition by the three groups is rejected. Apparently football players enjoy such competition much more than either of the other two groups.

2. $U_1 = 69$, $U_2 = 27$, $U_E = 48$, and $\sigma_U = 12.96$, making $z = 1.62$ with U_1 and $z = -1.62$ with U_2. Since the absolute value of z must exceed 1.96 when $\alpha = .05$, we accept the null hypothesis of no difference in recall for the two groups.

3. $R_1 = 33.5$, $R_2 = 38$, and $R_3 = 48.5$, making H $= 1.18$, which must be compared with a required χ^2 of 5.99 for $df = 2$ and $\alpha = .05$. Therefore H_0 is accepted, and we conclude that neither of the sucrose groups could detect its presence, at least enough to produce the expected increase in speed when sucrose is present as a reward.

4. T $= 3.0$, and $N_{S-R} = 9$. Since Table J shows that T must be less than or equal to 6 to be significant with this value of N_{S-R} and $\alpha = .05$, we reject H_0. We conclude that recognition is facilitated by presenting an associated word before the test word.

5. $\chi^2 = 5.20$, which is not significant with $df = 2$ and $\alpha = .05$, since a χ^2 of 5.99 is required in this situation.

Index